# Modelling Flows in Pipes and Channels

*Guest Editor*
Kamil Urbanowicz
Department of Mechanical
Engineering and
Mechatronics
West Pomeranian University
of Technology Szczecin
Szczecin
Poland

*Editorial Office*
MDPI AG
Grosspeteranlage 5
4052 Basel, Switzerland

This is a reprint of the Special Issue, published open access by the journal *Fluids* (ISSN 2311-5521), freely accessible at: https://www.mdpi.com/journal/fluids/special_issues/L8CJ4QCMCP.

For citation purposes, cite each article independently as indicated on the article page online and as indicated below:

| Lastname, A.A.; Lastname, B.B. Article Title. *Journal Name* **Year**, *Volume Number*, Page Range. |
|---|

**ISBN 978-3-7258-7086-8 (Hbk)**
**ISBN 978-3-7258-7087-5 (PDF)**
**https://doi.org/10.3390/books978-3-7258-7087-5**

Cover image courtesy of Kamil Urbanowicz

# Modelling Flows in Pipes and Channels

Guest Editor

**Kamil Urbanowicz**

MDPI

Basel • Beijing • Wuhan • Barcelona • Belgrade • Novi Sad • Cluj • Manchester

# Contents

**Kamil Urbanowicz**
Advances in Pipe and Channel Flow Modeling
Reprinted from: *Fluids* **2026**, *11*, 37, https://doi.org/10.3390/fluids11020037 . . . . . . . . . . . . 1

**José González, Andrés Meana-Fernández, Iván Vallejo Pérez and Jesús M. Fernández Oro**
Minor Loss Coefficient for Abrupt Section Changes in a Cylindrical Pipe Using a Numerical Approach
Reprinted from: *Fluids* **2024**, *9*, 152, https://doi.org/10.3390/fluids9070152 . . . . . . . . . . . . 9

**M. R. Ajith Kumar, Prashanth R. Hanmaiahgari and Jaan H. Pu**
Pressure and Velocity Profiles over a Weir Using Potential Flow Model
Reprinted from: *Fluids* **2024**, *9*, 182, https://doi.org/10.3390/fluids9080182 . . . . . . . . . . . . 28

**George Sofiadis, Antonios Liakopoulos, Apostolos Palasis and Filippos Sofos**
Turbulent Micropolar Open-Channel Flow
Reprinted from: *Fluids* **2024**, *9*, 202, https://doi.org/10.3390/fluids9090202 . . . . . . . . . . . . 47

**Vicente S. Fuertes-Miquel, Oscar E. Coronado-Hernández and Alfonso Arrieta-Pastrana**
Effects of Expelled Air during Filling Operations with Blocking Columns in Water Pipelines of Undulating Profiles
Reprinted from: *Fluids* **2024**, *9*, 212, https://doi.org/10.3390/fluids9090212 . . . . . . . . . . . . 64

**Alan E. Vardy**
On the Modelling of Asymptotic Wavefronts in Long Ducts with Chambers
Reprinted from: *Fluids* **2024**, *9*, 240, https://doi.org/10.3390/fluids9100240 . . . . . . . . . . . . 89

**Pedro Leite Ferreira and Dídia Isabel Cameira Covas**
Mesh Sensitivity Analysis of Axisymmetric Models for Smooth–Turbulent Transient Flows
Reprinted from: *Fluids* **2024**, *9*, 268, https://doi.org/10.3390/fluids9110268 . . . . . . . . . . . . 107

**Sergey Dmitriev, Alexey Sobornov and Andrey Kurkin**
Features of Motion and Heat Transfer of Swirling Flows in Channels of Complex Geometry
Reprinted from: *Fluids* **2024**, *9*, 293, https://doi.org/10.3390/fluids9120293 . . . . . . . . . . . . 127

**Juan Mairal, Javier Murillo and Pilar Garcia-Navarro**
A Validation of the Junction Riemann Problem in Networks of Channels Under Transitory Flow Conditions
Reprinted from: *Fluids* **2025**, *10*, 53, https://doi.org/10.3390/fluids10030053 . . . . . . . . . . . . 154

**Emilian Popov, Nicholas Mecham and Taylor Grubbs**
High $y^+$ Shear-Stress Turbulence Implementation for High Flux Isotope Reactor Narrow Channel Flows
Reprinted from: *Fluids* **2025**, *10*, 85, https://doi.org/10.3390/fluids10040085 . . . . . . . . . . . . 174

**Nuno M. C. Martins, Dídia I. C. Covas, Bruno Brunone, Silvia Meniconi and Caterina Capponie**
CFD-Based Transient Analysis for the Detection and Characterisation of Extended Partial Blockages in Pipes
Reprinted from: *Fluids* **2025**, *10*, 291, https://doi.org/10.3390/fluids10110291 . . . . . . . . . . . . 199

**Sergio Gandía-Barberá and Sergio Hoyas**
Spatial and Energetic Organization of Coherent Structures in Couette–Poiseuille Turbulent Channels
Reprinted from: *Fluids* **2026**, *11*, 18, https://doi.org/10.3390/fluids11010018 . . . . . . . . . . . . 212

**André Nachbin**
Modelling Water Waves on Graphs
Reprinted from: *Fluids* **2025**, *10*, 140, https://doi.org/10.3390/fluids10060140 . . . . . . . . . . . . **246**

fluids

MDPI

Editorial

# Advances in Pipe and Channel Flow Modeling

**Kamil Urbanowicz**

Faculty of Mechanical Engineering and Mechatronics, West Pomeranian University of Technology in Szczecin, 70-310 Szczecin, Poland; kamil.urbanowicz@zut.edu.pl

## 1. Introduction

The fluid flow within confined conduits (pipes and channels) remains a cornerstone of hydraulic engineering, underpinning the design, analysis and safe operation of countless industrial, environmental and infrastructural systems. The ongoing progress in computational power, measurement techniques and multi-physics understanding continuously refines our ability to model these flows, from predictable steady-state operations to complex transient events. This Special Issue, "Modelling Flows in Pipes and Channels" seeks to capture and contribute to these advancements by presenting cutting-edge research that bridges fundamental insight with practical application.

To illustrate this, we confine our focus to pressurized flows, where the modeling of transient phenomena like water hammer exemplifies the domain's vibrant challenges and innovations. The ongoing research is pushing boundaries on multiple fronts: refining unsteady friction models for more accurate damping prediction [1–3]; developing novel analytical and quasi-two-dimensional solutions for rapid benchmarking and deeper physical insight [4–6] and understanding complex interactions in series-connected pipelines of different materials [7–9]. Furthermore, advanced studies focus on the cavitation effects, employing sophisticated two-phase models and experimental visualization [10–13], while high-fidelity CFD simulations provide unprecedented spatial detail of wave propagation and fluid–structure interaction [14–17]. Critical operational scenarios like rapid filling and emptying, with their inherent air–water interactions, are also being elucidated through combined experimental and 3D computational approaches [18–21]. This has naturally led to applied research on innovative damping and mitigation strategies, from compliant polymeric liners and bypass systems to system vibration analysis [22–25].

This Special Issue serves as a forum for the latest developments across this broad spectrum of problems. We invite contributions that address mathematical model development, innovative numerical schemes, advanced experimental validation and novel applications related to both steady and transient flows in all types of conduits. By collating diverse perspectives, we aim to foster the necessary interdisciplinary dialog to tackle the enduring and emerging challenges in pipe and channel flow modeling, ultimately contributing to more resilient, efficient and sustainable engineering systems.

## 2. Review of Contribution to the Special Issue

The study by González et al. (Contribution 1) introduces a practical and efficient numerical methodology for predicting minor losses in pipes with abrupt section changes. Both sudden contraction and sudden expansion configurations are investigated through a combined theoretical and numerical approach. An initial one-dimensional theoretical model is developed to provide physical insight, while its limitations are clearly identified. The main contribution of the work is a two-dimensional axisymmetric CFD model that is capable of accurately describing pressure losses in both geometries. The numerical results for sudden contractions show strong agreement with established experimental data for both air and water flows. The model demonstrates robust mesh independence, even producing reliable results with relatively coarse discretizations. Comparisons among the Spalart–Allmaras, k–ε and k–ω SST turbulence models reveal negligible differences in the predicted loss coefficients. Additional simulations using a Reynolds stress model improve the accuracy of loss predictions for the intermediate contraction ratios. The evolution of the loss coefficient as a function of the diameter ratio emerges as the most relevant quantitative finding of the study. Although the model tends to overpredict losses in pipe expansions, it remains a valuable and rapid tool for the preliminary design and analysis of geometrical changes in pipe flows.

The paper by Kumar et al. (Contribution 2) introduces a novel semi-inverse potential flow framework that models flow over round-crested weirs by constructing only streamlines, rather than a full flow net. A key innovation is the use of the Serre–Green–Naghdi equation, instead of traditional Bernoulli-type Boussinesq formulations, to generate a more accurate initial free-surface profile. The study advances earlier potential flow models by integrating this initial profile into Boadway's potential flow formulation. The numerical accuracy is improved through the use of a five-point central finite difference scheme and fourth-order spatial discretization of the Laplace equation. The approach enables a precise computation of pressure and velocity fields in transcritical flow conditions. Unlike prior studies, the model explicitly quantifies the influence of streamline curvature on pressure variations along the weir bed. The methodology is demonstrated across multiple weir geometries, including Gaussian, parabolic and semicircular shapes. A dedicated semicircular weir experiment provides direct validation of the predicted pressure and velocity profiles. The close agreement between numerical results and experimental data highlights the robustness of the proposed 2-D ideal flow model. Overall, the study establishes a computationally efficient and physically insightful alternative to the existing potential flow approaches for analyzing complex weir flows.

Sofiadis et al.'s contribution (Contribution 3) presents one of the first applications of the micropolar fluid model to the study of the turbulent open-channel flows that are relevant to environmental applications. The main novelty lies in treating particulate and secondary-phase effects within a fully Eulerian framework, avoiding the complexity of traditional Lagrangian approaches. The micropolar model is coupled with direct numerical simulations (DNS) to rigorously assess its capability in describing turbulent wall-bounded flows. The study demonstrates that the micropolar formulation can accurately reproduce both unladen and particulate flow characteristics at moderate Reynolds numbers. Validation against experimental data and previous DNS results shows excellent agreement, often outperforming existing turbulence models. A key contribution is the systematic investigation of the effect of micropolar viscosity as a proxy for secondary-phase properties. The results clarify previously contradictory findings by revealing mixed velocity behavior between particulate and single-phase flows in the inner and outer regions. The proposed model provides improved physical interpretation of particle–turbulence interactions near the wall. Turbulence intensities and Reynolds shear stresses are captured consistently, highlighting the model's robustness. Overall, the work establishes the micropolar model as a computationally efficient and physically reliable alternative for simulating turbulent particulate open-channel flows.

Fuertes-Miquel et al.'s paper (Contribution 4) introduces a generalized one-dimensional mathematical model to analyze transient events caused by entrapped air pockets during filling operations in pipelines with undulating profiles. The main novelty lies in explicitly incorporating the effects of blocking water columns, which are neglected in previously

https://doi.org/10.3390/fluids11020037

developed filling models. The proposed formulation accounts for the interaction between multiple air pockets and air valves within a single unified framework. Unlike in earlier approaches, the model includes the dynamic behavior of expelled air through air valves during transient events. A polytropic formulation is adopted to realistically describe the air pocket compression and expansion processes. The methodology is applied to a pipeline configuration with two entrapped air pockets and an air valve located at the highest point. Experimental data from a laboratory-scale installation are used to assess the pressure response associated with blocking columns. The results demonstrate that air valves can significantly reduce pressure peaks when air expulsion is unrestricted, but this may lead to new transients upon valve closure. The study further examines the influence of air valve size and discharge characteristics on pressure surges. Overall, this work provides a practical modeling framework that enhances the understanding and prediction of hazardous pressure peaks during pipeline filling operations.

Vardy's paper (Contribution 5) introduces a novel numerical methodology for determining the possible asymptotic shapes of pressure wavefronts in long ducts without simulating their full propagation history. The key innovation lies in predicting final wavefront states directly, rather than computing their evolution from the prescribed initial disturbances. By avoiding long-domain unsteady simulations, the method overcomes the severe computational limitations associated with extremely fine spatial discretization. The approach is specifically tailored to study asymptotic behavior and is therefore complementary to conventional cause-and-effect wave propagation models. A major advantage of the methodology is its ability to achieve very high accuracy using minimal computational resources. The analysis requires simulations that only use short duct segments and short time intervals, even when using grid sizes that are orders of magnitude smaller than the duct diameter. The method is shown to be numerically stable, despite relying on extrapolation for two governing parameters. A detailed assessment demonstrates that the influence of these extrapolations on the results is minimal and acceptable. The framework is applied to ducts with enclosed air chambers: a configuration that was previously difficult to analyze accurately, due to CPU constraints. The study reveals unique one-to-one relationships between asymptotic wavefront pressure amplitude, propagation speed and maximum steepness, highlighting the physical characteristics that do not exist in non-asymptotic wavefronts.

The paper by Ferreira and Covas (Contribution 6) introduces a systematic and comprehensive assessment of radial mesh sensitivity in axisymmetric quasi-2D models for smooth–turbulent transient pipe flows. The main novelty lies in establishing clear guidelines to balance the numerical accuracy and computational efficiency when simulating hydraulic transients. An extensive parametric study is carried out, using 80 different radial meshes and varying the number of cylinders, the geometric common ratio and the axial discretization. The study demonstrates that geometric sequence meshes significantly improve the resolution near the pipe wall, where unsteady shear stress gradients are the most critical. A key contribution is the identification of optimal common ratio values that allow for accurate transient predictions with a substantially reduced computational cost. The comparison between the two best-performing meshes highlights that increasing the common ratio can halve the computation time while maintaining comparable accuracy. The work shows that discrepancies between meshes are confined to a very short time window after the first pressure surge. The transient response is characterized in detail through piezometric head, wall shear stress and velocity histories, confirming physical consistency. The analysis provides new insights into the localization of shear stress gradients within the viscous and buffer layers. Overall, the study offers practical and transferable mesh design recommendations for efficient axisymmetric modeling of turbulent hydraulic transients.

https://doi.org/10.3390/fluids11020037

The study by Dmitriev et al. (Contribution 7) presents a combined experimental and numerical investigation of swirling single-phase coolant flow and heat transfer in channels of complex geometry under nuclear power plant operating conditions. The main novelty lies in the systematic comparison of swirl intensifiers with constant and variable twist pitches in a realistic heat exchanger configuration. Variable pitch swirlers are shown to enhance heat transfer efficiency while limiting hydraulic power penalties. The study introduces a detailed experimental validation of temperature fields and pressure drops for multiple swirling configurations. Advanced turbulence modeling, including an omega-based Reynolds stress model and an SST model with curvature correction, is applied and critically assessed. The work demonstrates that Reynolds stress modeling captures swirling flow structures and boundary layers more accurately than two-equation models. Numerical simulations reveal previously undocumented flow and thermal features arising from changes in swirl pitch along the channel. An effectiveness criterion is introduced to quantify the performance of each swirl intensifier, relative to a smooth pipe. The results highlight the trade-off between computational cost and accuracy among turbulence modeling approaches. Overall, the study provides new insights into the design and modeling of swirl-enhanced heat exchangers for high-performance thermal systems.

Mairal et al.'s study (Contribution 8) presents a one-dimensional methodology for modeling transitory and supercritical flow regimes in channel networks, using the Junction Riemann Problem (JRP). The novelty of the work lies in the inclusion of limiting coefficients in the junction coupling conditions, allowing the method to handle a wide range of flow regimes robustly. The method is validated against reference 2D numerical solutions in challenging geometries, demonstrating its ability to reproduce the main features of wave propagation. Unlike conventional 1D methods, this approach captures the long-term behavior of flow in networks with multiple junctions. The proposed formulation offers a computationally efficient alternative to full 2D simulations while maintaining high accuracy for practical applications. The study emphasizes the transitory flow conditions, showing that shock waves at junctions are properly managed by the JRP method. The methodology provides a framework to include energy conservation in 1D network models, though frictional losses and local reflections are not yet considered. The results highlight the potential of 1D methods for complex channel networks, even under supercritical conditions. Overall, this work demonstrates that 1D modeling, when combined with the JRP, can be a reliable and efficient tool for engineering and environmental studies.

Popov et al.'s paper (Contribution 9) focuses on improving engineering predictions of turbulence in narrow, curved channels, specifically in high-flux isotope reactor (HFIR) geometries. The study enhances a standard Reynolds-averaged Navier–Stokes (RANS) k-ω shear-stress transport (SST) turbulence model by modifying the dissipation equation's loss term and using high wall-distance (high $y^+$) computational grids. These changes enable fast, computationally efficient simulations while maintaining sufficient accuracy for engineering purposes. The methodology was validated against high-resolution numerical data, showing improved prediction of mean flow parameters and turbulence behavior in parallel channel flows. The approach allows for practical computation on coarse meshes with modest computational resources, making it suitable for reactor thermal hydraulics and other industrial channel flow applications. The study demonstrates that proper turbulence modeling is essential for predicting heat transfer and pressure losses in narrow channels, providing a foundation for safe and efficient reactor operation, particularly in the transition to low-enriched fuels.

Martins et al.'s study (Contribution 10) investigates the transient behavior of extended partial blockages (EPBs) in pressurized pipe systems, using 3D CFD simulations based on compressible Navier–Stokes equations in OpenFOAM. EPBs, which represent the longitudinal growth of discrete blockages, significantly alter flow dynamics and pressure responses. The simulations show that EPBs produce distinct pressure signatures compared to discrete blockages: the first reflected pressure wave is similar but inversely proportional to the blockage diameter, while the second peak can exceed conventional predictions due to multiple wave interactions. The length of the EPB strongly affects peak amplitudes and the complexity of the pressure response. These findings demonstrate that transient test-based techniques can accurately detect and characterize EPBs, allowing water utilities to implement targeted maintenance, optimize hydraulic efficiency, reduce energy consumption and prevent blockages from progressing to total obstruction. The study provides a foundation for improved non-invasive diagnostic strategies and proactive water network management.

Gandía-Barberá and Hoyas' study (Contribution 11) investigates coherent structures in turbulent channel flows across a transition from pure Poiseuille to pure Couette flows, at a friction Reynolds number of approximately 250. Using DNS data and a percolation-based method to identify structures, the authors analyze low-speed streaks and intense Reynolds stress structures, focusing on their geometry, energy content and spatial organization. The results show that increasing the mean shear in Couette-like flows enhances the energetic intensity and spanwise organization of the attached streaks and Q structures, while Poiseuille-like flows exhibit a more balanced distribution between the attached and detached structures. Despite changes in shear, the geometric characteristics of the streaks and Q structures, including dimensions, volume ratios and angular inclination, remain largely consistent. The wall normal distribution of Q structures reveals that near-wall regions are dominated by Q4 sweeps, whereas further from the wall, Q2 bursts prevail, with the overall intensity increasing with shear and reaching a maximum in pure Couette flows. The spacing between same-type Q pairs decreases in the streamwise direction with the increasing shear, while Q2–Q4 pairs tend to align spanwise and increase their distance from each other, reflecting the influence of the very large streamwise rolls that are present in Couette-like configurations. Overall, the study demonstrates that the mean shear primarily modifies the energetic intensity and spatial distribution of coherent structures, whereas their geometry and topology remain robust, providing new insight into the interaction between large-scale rolls and near-wall turbulence in mixed Couette–Poiseuille flows and laying the groundwork for future investigations at higher Reynolds numbers and with refined detection techniques.

Nachbin's contribution (Contribution 12) addresses the modeling of water waves on graphs, focusing on the reflection and transmission of waves at graph vertices. This is an area that has received limited attention in the literature. The author compares different mathematical approaches on a two-dimensional "fattened" graph, from hyperbolic conservation laws to weakly nonlinear and weakly dispersive systems, and explores the reduction in these models to one-dimensional graphs. A novel aspect of the work is the systematic inclusion of junction angles in the 1D graph model, which allows for a more general compatibility condition at the vertex when compared with the 2D fattened-graph solutions. The discussion highlights that the existing wave-on-graph models do not account for junction angles or provide detailed comparisons with mesoscale 2D models, which significantly affects the reflection–transmission dynamics. Nachbin proposes that future work steps back from weakly dispersive Boussinesq systems and studies nondispersive hyperbolic systems, using 2D shallow water equations, including more comprehensive comparisons between thick-Y-graph solutions and reduced 1D models with variable angles. Other modeling extensions that are under consideration include channels with non-rectangular or spatially varying cross sections, as well as different types of solutions beyond solitary waves, particularly to understand discrepancies in asymmetric junctions where forking angles affect the accuracy of 1D–2D model agreement. The study thus lays the groundwork

https://doi.org/10.3390/fluids11020037

for the improved modeling of wave propagation on networks, highlighting both the current gaps and some promising directions for future research.

## 3. Conclusions

This Special Issue presents a diverse array of innovative numerical and experimental methodologies that address key challenges in fluid mechanics. Collectively, the contributions advance the prediction of complex flows, from internal pipe systems to open channels and turbulent structures (Table 1). Several studies introduce efficient and accurate computational frameworks to model transient phenomena and complex geometries, while others successfully apply advanced physical models, validating them against experimental data.

**Table 1.** Short overview of published papers.

| # | Author/s | Flow Type | Flow Regime | System Type | Main Focus/Phenomena |
|---|---|---|---|---|---|
| 1 | González et al. | Turbulent | Steady | Pressurized | Minor loss coefficients for abrupt pipe section changes (CFD-RANS) |
| 2 | Kumar et al. | Potential/Ideal | Steady | Open Channel | Pressure/velocity profiles over a weir, using potential flow model |
| 3 | Sofiadis et al. | Turbulent (Micropolar) | Steady | Open Channel | Turbulent open-channel flow with suspended sediments (micropolar theory) |
| 4 | Fuertes-Miquel et al. | Transient Mixed-Phase | Transient | Pressurized | Effects of expelled air during pipeline filling in undulating profiles |
| 5 | Vardy | Wave Propagation | Transient | Pressurized/Ducts | Modeling asymptotic pressure wavefronts in long ducts with chambers |
| 6 | Ferreira and Covas | Transient Turbulent | Transient | Pressurized | Mesh sensitivity analysis for axisymmetric modeling of smooth–turbulent transients |
| 7 | Dmitriev et al. | Turbulent | Steady | Complex Channel | Swirling flows and heat transfer in channels of complex geometry (PWR-related) |
| 8 | Mairal et al. | Shallow Water Flow | Transient | Open Channel (Network) | Validation of Junction Riemann Problem solver for transitory flows in channel networks |
| 9 | Popov et al. | Turbulent | Steady (RANS) | Pressurized (Narrow) | High $y^+$ turbulence modeling for narrow channel flows in nuclear reactors (HFIR) |
| 10 | Martins et al. | Transient Turbulent | Transient | Pressurized | CFD-based detection and characterization of extended partial blockages in pipes |
| 11 | Gandía-Barberá and Hoyas | Turbulent | Steady (DNS) | Channel Flow | Spatial/energetic organization of coherent structures in turbulent channel flow |
| 12 | Nachbin | Wave Propagation | Steady/Quasi | Open Channel (Graph) | Mathematical modeling of water wave propagation on network graphs |

A strong focus is placed on enhancing the safety and design of engineering systems. The research provides critical insights into transient pressure mechanisms, the characterization of flow obstructions and the performance of flow-control devices. Furthermore, the volume offers practical improvements to turbulence modeling for industrial applications and explores fundamental coherent structures in shear flows. By also examining the foundational methodologies for complex networked systems, this Special Issue suggests essential future research directions. This collection bridges fundamental research and engineering application, delivering both practical tools and a deeper physical understanding across a wide spectrum of fluid flow phenomena.

https://doi.org/10.3390/fluids11020037

**Funding:** This research received no external funding.

**Conflicts of Interest:** The author declares no conflicts of interest.

**List of Contributions**

1. González, J.; Meana-Fernández, A.; Pérez, I.V.; Oro, J.M.F. Minor Loss Coefficient for Abrupt Section Changes in a Cylindrical Pipe Using a Numerical Approach. *Fluids* **2024**, *9*, 152. https://doi.org/10.3390/fluids9070152.
2. Kumar, M.R.A.; Hanmaiahgari, P.R.; Pu, J.H. Pressure and Velocity Profiles over a Weir Using Potential Flow Model. *Fluids* **2024**, *9*, 182. https://doi.org/10.3390/fluids9080182.
3. Sofiadis, G.; Liakopoulos, A.; Palasis, A.; Sofos, F. Turbulent Micropolar Open-Channel Flow. *Fluids* **2024**, *9*, 202. https://doi.org/10.3390/fluids9090202.
4. Fuertes-Miquel, V.S.; Coronado-Hernández, O.E.; Arrieta-Pastrana, A. Effects of Expelled Air during Filling Operations with Blocking Columns in Water Pipelines of Undulating Profiles. *Fluids* **2024**, *9*, 212. https://doi.org/10.3390/fluids9090212.
5. Vardy, A.E. On the Modelling of Asymptotic Wavefronts in Long Ducts with Chambers. *Fluids* **2024**, *9*, 240. https://doi.org/10.3390/fluids9100240.
6. Ferreira, P.L.; Covas, D.I.C. Mesh Sensitivity Analysis of Axisymmetric Models for Smooth–Turbulent Transient Flows. *Fluids* **2024**, *9*, 268. https://doi.org/10.3390/fluids9110268.
7. Dmitriev, S.; Sobornov, A.; Kurkin, A. Features of Motion and Heat Transfer of Swirling Flows in Channels of Complex Geometry. *Fluids* **2024**, *9*, 293. https://doi.org/10.3390/fluids9120293.
8. Mairal, J.; Murillo, J.; Garcia-Navarro, P. A Validation of the Junction Riemann Problem in Networks of Channels Under Transitory Flow Conditions. *Fluids* **2025**, *10*, 53. https://doi.org/10.3390/fluids10030053.
9. Popov, E.; Mecham, N.; Grubbs, T. High y+ Shear-Stress Turbulence Implementation for High Flux Isotope Reactor Narrow Channel Flows. *Fluids* **2025**, *10*, 85. https://doi.org/10.3390/fluids10040085.
10. Martins, N.M.C.; Covas, D.I.C.; Brunone, B.; Meniconi, S.; Capponi, C. CFD-Based Transient Analysis for the Detection and Characterisation of Extended Partial Blockages in Pipes. *Fluids* **2025**, *10*, 291. https://doi.org/10.3390/fluids10110291.
11. Gandía-Barberá, S.; Hoyas, S. Spatial and Energetic Organization of Coherent Structures in Couette–Poiseuille Turbulent Channels. *Fluids* **2026**, *11*, 18. https://doi.org/10.3390/fluids11010018.
12. Nachbin, A. Modelling Water Waves on Graphs. *Fluids* **2025**, *10*, 140. https://doi.org/10.3390/fluids10060140.

## References

1. Jiang, D.; Zeng, C.; Lu, Q.; Guo, Q. Water hammer in pipelines based on different friction models. *Sci. Rep.* **2024**, *14*, 953. [CrossRef]
2. Mohsin, Z.; Sulaiman, Z.; Khilqa, S. Comparative analysis of steady and unsteady friction models in numerical simulation of water hammer phenomenon. *Eng. Appl. Sci. Res.* **2025**, *52*, 207–220. Available online: https://ph01.tci-thaijo.org/index.php/easr/article/view/259832 (accessed on 10 January 2026).
3. Urbanowicz, K.; Bergant, A.; Stosiak, M.; Deptuła, A.; Karpenko, M.; Kubrak, M.; Kodura, A. Water Hammer Simulation Using Simplified Convolution-Based Unsteady Friction Model. *Water* **2022**, *14*, 3151. [CrossRef]
4. Urbanowicz, K.; Di Nucci, C.; Sharma, B.K.; K., A.; Kawiak, M.; Krajewski, S. Revisiting Yamaguchi-Ichikawa water hammer model. *Chin. J. Phys.* **2025**, *94*, 487–503. [CrossRef]
5. Brühl, M. Analytical Solution for Laminar Water Hammer With Frequency-Dependent Friction. *J. Fluids Eng.* **2022**, *144*, 111302. [CrossRef]
6. Bordbar, M. Analytical Solution of Quasi-Two-Dimensional Method for Unsteady Transient Pressurized Pipe Flows in the Time Domain. *J. Hydraul. Eng.* **2025**, *151*, 04025035. [CrossRef]
7. Kubrak, M.; Kodura, A.; Malesinska, A.; Urbanowicz, K. Water Hammer in Steel–Plastic Pipes Connected in Series. *Water* **2022**, *14*, 3107. [CrossRef]
8. Lewandowski, M.; Adamkowski, A. Experimental Verification of 1D-Simulation Method of Water Hammer Induced in Two Series-Connected Pipes of Different Diameters: Determination of the Pressure Wave Speed. *Appl. Sci.* **2024**, *14*, 7173. [CrossRef]

https://doi.org/10.3390/fluids11020037

9. Kraiem, D.; Triki, A. Tackling of hydraulic cavitation in pressurized pipe flow using high- or low-density polyethylene penstock and short-section. *AQUA Water Infrastruct. Ecosyst. Soc.* **2023**, *72*, 2126–2137. [CrossRef]
10. Sun, Q.; Wang, F.; Wu, Y.B.; Xu, Y.; Hao, Y. Energy analysis of transient flow with cavitation by considering the effect of water temperature in viscoelastic pipes. *J. Hydroinform.* **2023**, *25*, 2034–2052. [CrossRef]
11. Jansson, M.; Andersson, M.; Karlsson, M. High-Speed Imaging of Water Hammer Cavitation in Oil–Hydraulic Pipe Flow. *Fluids* **2022**, *7*, 102. [CrossRef]
12. Urbanowicz, K.; Bergant, A.; Kodura, A.; Kubrak, M.; Malesińska, A.; Bury, P.; Stosiak, M. Modeling Transient Pipe Flow in Plastic Pipes with Modified Discrete Bubble Cavitation Model. *Energies* **2021**, *14*, 6756. [CrossRef]
13. He, J.; Li, C.; Guo, Y. Modeling Transient Vaporous Cavitating Flow in Pipelines by a Two-Phase Homogeneous Flow Model. *Processes* **2025**, *13*, 3510. [CrossRef]
14. Neyestanaki, M.K.; Dunca, G.; Jonsson, P.; Cervantes, M.J. A Comparison of Different Methods for Modelling Water Hammer Valve Closure with CFD. *Water* **2023**, *15*, 1510. [CrossRef]
15. Cao, Y.; Zhou, L.; Ou, C.; Fang, H.; Liu, D. 3D CFD simulation and analysis of transient flow in a water pipeline. *AQUA Water Infrastruct. Ecosyst. Soc.* **2022**, *71*, 751–767. [CrossRef]
16. Zhang, Y.; Duan, H.F.; Keramat, A. CFD-aided study on transient wave-blockage interaction in a pressurized fluid pipeline. *Eng. Appl. Comput. Fluid Mech.* **2022**, *16*, 1957–1973. [CrossRef]
17. Kumar, A.M.R.; Pu, J.H.; Hanmaiahgari, P.R.; Lambert, M.F. Insights into CFD Modelling of Water Hammer. *Water* **2023**, *15*, 3988. [CrossRef]
18. Kumar, A.M.R.; Hanmaiahgari, P.R.; Lambert, M.F. CFD modelling of transient flow during rapid filling of horizontal pipeline with air. *Ocean Eng.* **2026**, *349*, 124174. [CrossRef]
19. Chen, X.; Hou, Q.; Laanearu, J.; Tijsseling, A.S. Experimental data on filling and emptying of a large-scale pipeline. *Sci. Data* **2024**, *11*, 603. [CrossRef] [PubMed]
20. Ferreira, J.P.; Ferràs, D.; Covas, D.I.C.; Kapelan, Z. Air entrapment modelling in water supply networks during pipe filling events. *Urban Water J.* **2024**, *21*, 685–697. [CrossRef]
21. Paternina, D.A.; Coronado-Hernández, O.E.; Aguirre-Mendoza, A.M.; Espinoza-Román, H.G.; Fuertes-Miquel, V.S. Three-Dimensional Simulation of Transient Flows during the Emptying of Pipes with Entrapped Air. *J. Hydraul. Eng.* **2023**, *149*, 04023007. [CrossRef]
22. Kriaa, K.; Elgamal, M.; Farouk, M. Experimental study on reducing water hammer effects in uPVC pipes using rubber bypass tubes. *Ain Shams Eng. J.* **2024**, *15*, 102562. [CrossRef]
23. Vardy, A.E. On Sources of Damping in Water-Hammer. *Water* **2023**, *15*, 385. [CrossRef]
24. Kubrak, M. Water hammer mitigation by internal rubber hose. *Arch. Civ. Eng.* **2024**, *70*, 275–288. [CrossRef]
25. Henclik, S.; Janicki, W. Water hammer runs in elastically supported pipeline and the impact of system vibrations on pressure amplitudes reduction. *Int. J. Eng. Sci.* **2026**, *218*, 104400. [CrossRef]




https://doi.org/10.3390/fluids11020037

fluids

Article

# Minor Loss Coefficient for Abrupt Section Changes in a Cylindrical Pipe Using a Numerical Approach

José González [1,*], Andrés Meana-Fernández [2], Iván Vallejo Pérez [3] and Jesús M. Fernández Oro [1]

1 Área de Mecánica de Fluidos (Energía), Universidad de Oviedo, Campus de Viesques, 33204 Gijón, Spain; jesusfo@uniovi.es
2 Área de Máquinas y Motores Térmicos (Energía), Universidad de Oviedo, Campus de Viesques, 33204 Gijón, Spain; andresmf@uniovi.es
3 Volkswagen-Navarra, 31170 Pamplona, Spain; ivanvallejoperez14@gmail.com
* Correspondence: aviados@uniovi.es

**Abstract:** Abrupt section changes are a classic problem in the study of flow in cylindrical ducts or pipes. For its analysis, there are a wide set of exiting data from previous studies, among which some authors stand out and will be mentioned. Those previous works have been used to obtain reliable results for the resolution of section changes along a pipe, either due to cross area increases or reductions on a 1D basis. It is also known that a numerical 2D axisymmetric simulation (CFD) could find a consistent result compared to experimental data in almost all fluid flow fields. The main novelty of the present study is the development of a simple numerical approach used to solve the minor loss calculation. Firstly, a theoretical analysis is developed, and then the results of the numerical simulations carried out on the behavior that affects the water and air flow rate in an abrupt section change, for both contraction and expansion problems, are presented. In both cases, the results are analyzed with different meshes (discretizations) and turbulence models. Finally, the obtained numerical results are compared with those in the technical literature. Also, a theoretical approach is shown in order to show a whole frame of the discussion. The core results are the loss coefficient evolution as a function of the section change both for the sudden contraction and the expansion of a pipe flow. As the results follow the existing experimental values, it is concluded that the developed model provides a feasible and quick design tool to analyze possible geometrical changes without the need for further experiments.

**Keywords:** pipe flow; numerical flow simulations; CFD analysis; turbulent models; abrupt section change; area ratio; pipe flow patterns

## 1. Introduction

Due to different elements used in pipe flows to correctly deliver or handle a given fluid, not only the pipe's linear (or major) losses but also the minor ones have been extensively studied and constitute a classic problem in Fluid Mechanics. Among the minor loss calculations, the sudden change in cross-section at a given pipe constitutes a good example of an extensively studied problem. Nevertheless, not many numerical approaches have tackled it and, to the authors knowledge, some innovative approaches in the loss coefficient calculations are still to be developed. Obviously, using a fully 3D solution would be the best option, but our interest was to develop a quicker and more engineering-like valid solution.

The section changes in a cylindrical pipe can be branched out in different cases depending on the how such section change is defined [1]. Within those classifications, the following can be considered, among others: abrupt contraction, rounded contraction, conical contraction, and beveled (flanged) contraction. In this article, the analysis will be focused on the abrupt contraction, whose model is shown in Figure 1.

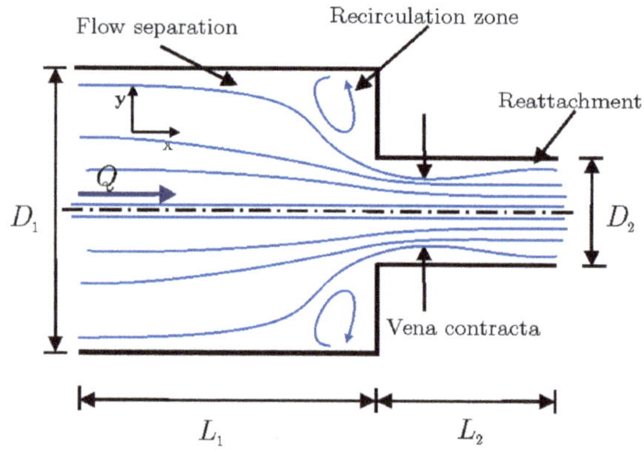


**Figure 1.** Streamline sketch through an abrupt contraction in the pipe (sketch adapted from [2]).

The main fluid feature behind the solution for the abrupt contraction flow, is the fact that this section change promotes the sum of two effects: a first contraction of the flow that produces an acceleration, followed by an expansion that produces a deceleration [1–4]. The point where the change from acceleration to deceleration occurs is the area where the flow reaches its smallest section, known as the contracted vein or "vena contracta", see [1].

In this phenomenon, shown in Figure 1, the total head loss is the sum of the losses due to the contraction between the last point of adherence (Section 1) and the "vena contracta" (Section 2), plus the losses of the expansion from the area of contracted vein (Section 2) and the exit of the flow or reattachment area (Section 3).

The sudden expansion process can be defined, parallelly to the contraction geometry, on the basis of the section exchange rate [1]. The following geometrical options are available in this classification: abrupt expansion, direct expansion with conical diffuser, multi-stage expansion with conical diffuser, expansion with curved wall diffuser, and pipe reducer, among others. In the present article, the analysis will be based on the geometry of an abrupt expansion, whose model can be observed in Figure 2.

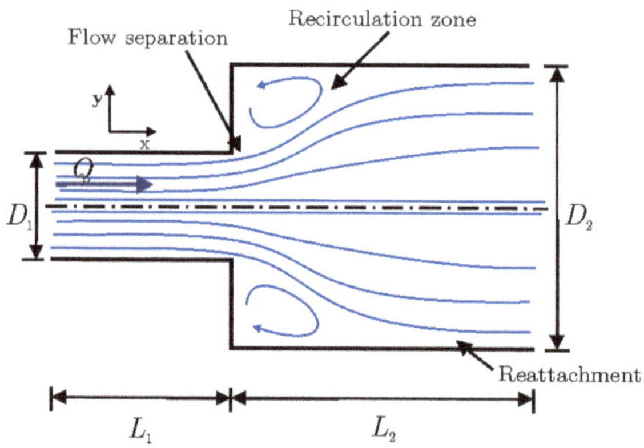


**Figure 2.** Streamline sketch through an abrupt pipe expansion (sketch adapted from [2]).

The main flow feature for the abrupt expansion is the separation of particles from the imposed geometry change at the separation point which is located just at the corner defined by the expansion zone (abrupt section change, as shown in Figure 2). That separation induces the eddy generation in a recirculation region, also called the free mixing zone. Downstream of separating point and the so-defined recirculation zone, the flow returns to the imposed geometrical condition at the reattachment point (right-hand side in Figure 2). The critical issue there is to correctly describe the reattachment point.

As depicted before, flows through section changes have been widely studied by numerous authors in previous works. In such works, difficulties came from the accurate flow description when the case of abrupt section changes were considered. The correct location of the separation and reattachment positions are always the key issue in many studies. A first part with initial approaches (from the early XX century) is developed here, and more recent values are then discussed in the final part of the article. In order to show an evolution of this subject over the years, the contributions of some of the most important authors in this field are reviewed chronologically.

An interesting review in the study of the flows along a cylindrical duct, calculating the so-called minor losses, as those generated due to contractions and expansions, can be found in [3].

Previously, the BSc thesis in [4] develops an experiment about abrupt section changes in the water flow through a pipe. From those results, it is concluded that the losses due to expansion are approximately three times those produced during contraction. The study relies on previous studies by Gibson (in 1910 and 1912) and Weisbach (in 1855), as stated and reordered in [3]. In addition, [4] determines two exponential formulas for the estimation of the load losses due to the sudden section change, following:

$$\begin{cases} \text{Expansion}: & h_{loss} = 0.00932\ R^{0.413} V^{1.885} \\ \text{Contraction}: & h_{loss} = 0.00187\ R^{0.88} V^{1.84} \end{cases} \quad (1)$$

In [5], an equation is provided for the calculation of energy losses due to abrupt expansion. This relationship has the particularity of depending on the speed of flow in both sections:

$$h_{\text{loss}} = \frac{(V_2 - V_1)^{1.919}}{2g} \quad (2)$$

Later on, [6] provided experimental results about load losses and the influence of the Reynolds number on them. To analyze the influence of the Reynolds number, changes in the speed profiles were examined. In the textbook [7], it is determined that the coefficient due to abrupt expansion is determined by the following equation:

$$K_{\text{cont}} = C_{\text{a}}\left(\frac{A_1}{A_2} - 1\right)^2 \quad (3)$$

where $C_{\text{a}}$ is an experimental factor (dimensionless) that is determined by an experimental graph [7].

On the other hand, [8] started to use the non-dimensional equation to determine the loss coefficient, that is:

$$h_{\text{loss}} = \frac{V_2^2}{2g} \quad (4)$$

The main goal in [2] was to define a coefficient to determine the losses during the contraction. Thus, the following relation is obtained from data obtained through experimentation, using water and air as fluids, and their extrapolation is:

$$h_{\text{exp}} = 0.57806 + 0.39543\beta - 4.53854\beta^2 + 14.24265\beta^3 - 19.22214\beta^4 + 8.54038\beta^5 \quad (5)$$

Then, in 1969, Idelcik published his well-known work, "Memento des pertes de charge" (in English: "Handbook of hydraulic resistance" [9]), in which one chapter deals with the flow in pipes and channels. In the chapter, Idelcik proposes a value for the coefficient of head losses in various types of contraction, including the case of an abrupt contraction.

This same equation is later proposed by Reza (1985) and takes the following formulation ($Re > 10^4$):

$$h_{\text{loss}} = \frac{\gamma V_{D2}^2}{2g}\left(\frac{1}{2}\left(1 - \frac{A_2}{A_1}\right)\right) \tag{6}$$

In [10], a work is developed from the results and data provided by [6], and based on his conclusions, two different approaches to the two considered problems are proposed, both for expansion and contraction, including the consideration of the "vena contracta".

As stated in [3], Weisbach (in 1972) provided a summary of the previous work on this subject, publishing a book containing the results of all the previous authors. He also contributed new experiments carried out by his team, which were adjusted from [1].

In [11], the authors studied the differences in the definition of the loss coefficient from a one-dimensional analysis of incompressible flow, highlighting some difficulties with the geometry of the contraction and flow for real pipes and fluids. Those authors have published many articles on the topic. In particular, in [12], they used the equation by Bullen et al. (back in 1983) to obtain, in an experimental and theoretical way, coefficients using different area ratios and an Re range that oscillates between $4 \times 10^4$ and $2 \times 10^5$. The definition selected for the pressure loss coefficient in the contraction is the following:

$$K_{\text{cont}} = \frac{2gA_2^2 h_{\text{loss}}}{Q^2} - \left[1 - \left(\frac{A_2}{A_1}\right)^2\right] \tag{7}$$

The study concludes that the values of pressure loss coefficients from the static wall pressure distribution are directly proportional to the Re (in a study range of $4 \times 10^4 < Re < 2 \times 10^5$). It also determines that the pressure loss is affected by how abrupt the contraction is. It also ensures that the results are valid with an accuracy of $\pm 5\%$ to $\pm 10\%$ at high Reynolds numbers. Using low Reynolds numbers, this hypothesis should be used with caution.

In more recent years, numerous studies and articles have been presented with the aim of making new contributions in this field, as well as compiling information from previous work, some of which are briefly mentioned in what follows.

As explained in [13], other contributions used the equation from the work of [5] to calculate the energy losses caused by abrupt expansion. In [14] and in [15], it is established that the loss coefficient due to abrupt expansion might be determined by means of a graph, based on the area ratio between the two conduits. For instance, [16] compiles the previous values and provide as new data a value of the pressure loss coefficient for the inlet from a pipe to a large tank, where the following condition is imposed: $D_2 >> D_1$ (so $D_1/D_2 \approx 0$). In [17], on the basis of previous documents, the following equation is proposed for the loss coefficient in an abrupt contraction, namely:

$$K_{\text{cont}}\left(\frac{V_2^2}{2g}\right) \approx 0.42\left[1 - \left(\frac{D_2}{D_1}\right)^2\right]\left(\frac{V_1^2}{2g}\right) \tag{8}$$

Also, in [1], some previously developed studies were considered to finally obtain the "shrinkage loss coefficient" based on a document dated from "The Crane Company" (1957). In [1], there is a study that adjusts the data obtained in [3]. From that study, a large number of equations and tables were developed to obtain the total losses in different types of contraction and expansion (varying according to the model of section change). The following equation is finally proposed for calculating the losses in an abrupt contraction:

$$h_{\text{loss}} = K_{\text{acc}}\left(1 - \beta^5\right)\lambda^2 + (\lambda - 1)^2 \tag{9}$$

where $K_{\text{acc}}$ represents the coefficient of losses due to the acceleration of flow and is taken as a value of 0.0696. In addition, $\beta = D_2/D_1$, and the flow speed ratio $\lambda$ is found as:

$$\lambda = 1 + 0.622\left(1 - 0.215\beta^2 - 0.785\beta^5\right) \tag{10}$$

In [1], a value is also provided for pressure losses in an abrupt expansion based on the Borda–Carnot Equation [18]:

$$h_{\text{loss}} = \frac{(V_1 - V_2)^2}{2g} \tag{11}$$

Some of the previously mentioned results are plotted in Figure 3, in the same kind of graph and variables as were chosen in [12], namely the loss coefficient as a function of the area ratio. Although a complete list would be almost impossible to compile, and more than the ones mentioned have been considered, it could be worth mentioning here the results in [1,2,6,9]. The results of [2] come from the plot of Equation (5). Even though the equation is a continuous curve, only four discrete values have been plotted for clarity (not to be confused with the values from [1]).

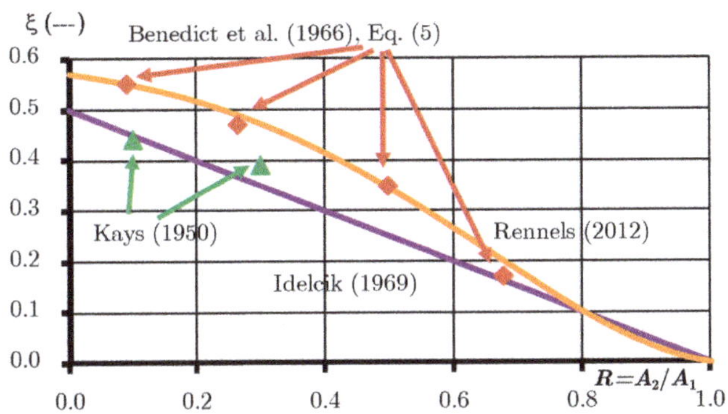


**Figure 3.** Comparison of the experimental loss coefficients in an abrupt contraction [1,2,6,9].

Recent numerical approaches for similar problems have been proposed by different authors: for a numerical strategy, see [19], or for experimental data in convergent–divergent flows, see the work by [20], which considers a light slope instead of a sudden section change. A basic approach intended for CFD teaching was also proposed by [21], and a solution similar to the one presented here but, once again, for a progressive section change, not for an abrupt one and in the frame of a porous media kind of application, was published by [22].

From the previously mentioned bibliographical study, a theoretical 1D model is developed in the following text (Part 2) to then show a 2D numerical axisymmetric model (Part 3) and both analyze its results and compare it with the exiting data in the bibliography (Part 4). The 2D numerical model developed in the present article offers a practical and efficient method to analyze pressure losses for contraction or expansions in pipe flow. Since the numerical results follow the existing experimental data, it can be used as a quick loss predictor for geometrical changes in conduits without the need for further experimentation. A model verification with existing bibliographic values is shown as a conclusion.

## 2. Theoretical Model

The two different geometries are now considered on a theoretical 1D basis in an effort to have a clear insight into the involved variables. What becomes straightforward when reading the different available bibliography is that the abrupt expansion is a much easier problem than the sudden contraction. The main reason for this is the actual complexity that the contraction adds to the problem in terms of the description of the "vena contracta" [23]. Therefore, even though both have been considered in theoretical description, only the equations for the sudden contraction will be shown and described.

Both problems will be handled using the energy equation with the losses of the element and the control volume momentum conservation equation. A classical solution that goes back to the Borda–Carnot Equation for a given element in an incompressible flow ($V$ being the flow average velocity in the element), namely:

$$h_{\text{loss}} = \frac{(V - V_2)^2}{2g} \tag{12}$$

The calculation for the losses starts with the control volume definition. In this part of the present study, a control volume is chosen to handle the geometry on the basis that the friction losses are negligible, and that then, the only effects to be considered are the boundary conditions (at flow inlet and outlet) and the pressure in the surface after the sudden contraction (see Figure 4).

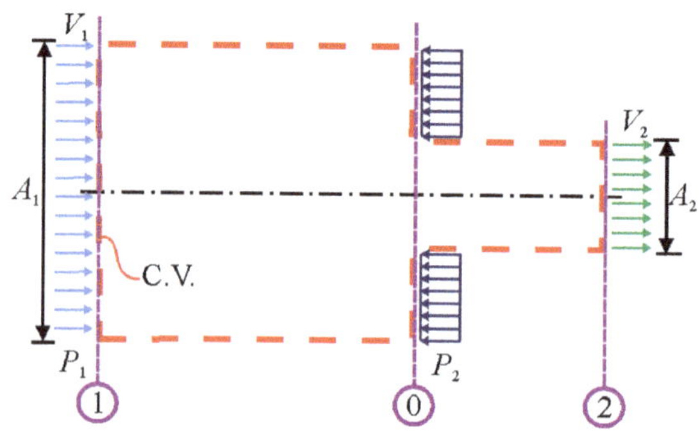


**Figure 4.** Control volume for the sudden contraction geometry (1D model).

The first approach is a basic calculation, considering that $V_2 > V_1$, and the other boundary conditions are shown in Figure 4. And then, the forces equilibrium is derived as:

$$P_1A_1 - P_2A_2 - P_0(A_1 - A_2) = \rho Q(V_2 - V_1) \tag{13}$$

If the linear losses after the expansion are to be neglected, ($P_0 \approx P_2$), and then:

$$P_1A_1 - P_2A_2 - P_2(A_1 - A_2) = (P_1 - P_2)A_1 = \rho Q(V_2 - V_1) = \rho A_1V_1(V_2 - V_1) \tag{14}$$

And using the energy equation from (1) to (2) (Figure 4), the following can be obtained:

$$h_{\text{loss}} = \left(\frac{f_1L_1}{D_1}\right)\frac{1}{2}\frac{V_1^2}{g} + \left(\frac{f_2L_2}{D_2} + \xi_{\text{cont}}\right)\frac{1}{2}\frac{V_2^2}{g} \tag{15}$$

Now, the value of (15) is substituted in (14) considering the boundary conditions. Those include the inherent potential energy or height variations equal to zero. The linear (or major) losses that are non-relevant for the given geometry are also considered. Then, the losses in the expansion are found as:

$$\frac{V_1(V_2 - V_1)}{2g} = \frac{V_2^2 - V_1^2}{2g} + \xi_{\text{cont}}\frac{1}{2}\frac{V_2^2}{g} \tag{16}$$

And the final expression for the coefficient is then obtained as:

$$\xi_{\text{cont}}\left(\frac{1}{2}\frac{V_2^2}{g}\right) = \frac{V_2^2}{2g}\left[1 - \frac{A_2}{A_1}\right]^2, \qquad \xi_{\text{cont}} = \left[1 - \frac{A_2}{A_1}\right]^2 \tag{17}$$

Obviously, this solution is quite symmetrical to the case of the expansion or enlargement, but it does not look very realistic as the flow overcomes many more different features, such as the stagnation of the flow and subsequent "vena contracta", which seems again to be in the root of such behavior. The question lays then on the value chosen for the velocity on the initial section (see Figure 4). Accordingly, a table can be built by choosing different plausible values. The results are shown in Table 1, which include as a final row the value of (17), as developed in the present model, and will furtherly be referred to as "simple 1D theory", in the final figures of the article.

**Table 1.** Different losses according to the bibliography and 1D developed model (Equation (17)).

| **Value of $V_0$, (m/s)** | **Losses, $h_{Loss}$ (m)** | Cont (---) | |
|---|---|---|---|
| $V_0 = \frac{V_1+V_2}{2}$ | $\frac{1}{2}\frac{V_2^2}{g}\left[\frac{1}{2}-\frac{1}{2}\left(\frac{A_2}{A_1}\right)\right]^2$ | $\left[\frac{1}{2}-\frac{1}{2}\left(\frac{A_2}{A_1}\right)\right]^2$ | Equation (6) by [9] |
| $V_0 = V_1 + \frac{V_2}{2}$ | $\frac{1}{2}\frac{V_2^2}{g}\left[\frac{1}{2}-\left(\frac{A_2}{A_1}\right)\right]^2$ | $\left[\frac{1}{2}-\left(\frac{A_2}{A_1}\right)\right]^2$ | |
| $V_0 = \frac{V_1(A_1-A_2)+V_2A_2}{A_1}$ | $\frac{1}{2}\frac{V_2^2}{g}\left[1-\left(\frac{A_2}{A_1}\right)\right]^4$ | $\left[1-\left(\frac{A_2}{A_1}\right)\right]^4$ | Equation (3) by [7] |
| $V_0 = V_2\frac{A_2}{A_1}$ | $\frac{1}{2}\frac{V_2^2}{g}\left[1-\left(\frac{A_2}{A_1}\right)\right]^2$ | $\left[1-\left(\frac{A_2}{A_1}\right)\right]^2$ | 1D model Equation (17) |

## 3. Numerical Model

### *3.1. Geometry*

The simulations and the computational solutions have been developed from the basic geometries shown in Figure 5, which is a simple sketch (not scaled). As the geometrical values are a critical issue for the final results, and considering a 2D axisymmetric model, the cylindrical pipe with the sudden cross-section change (expansion/contraction) is then numerically treated, and details are given in what follows. For the contraction geometry (right), the velocity profiles will be shown in x = 1.0 m, x = 1.7 m and x = 3.5 m (end of the domain), while the data reduction for the loss coefficient is done in x = 1.7 m and x =1.8 m. For the expansion geometry (left), the velocity profiles will be analyzed in x = 1.0 m, x = 1.7 m and x = 3.5 m (end of the domain), while the data reduction for the loss coefficient is done in x = 1.55 m and x = 1.95 m.

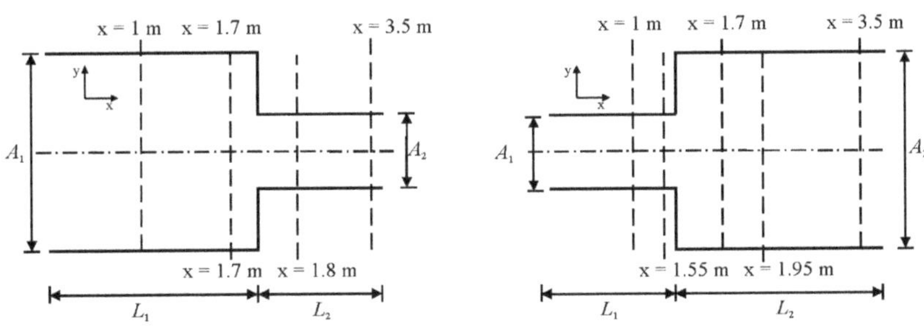


**Figure 5.** Data recording positions in the two geometries: contraction (**left**), expansion (**right**).

The geometries have been generated using the software application Gambit 2.4.6 (2007), and the contraction (Figure 6, left) consists of two pipes: a first one with diameter ($D_1$ = 0.025 m) and a second one, whose diameter ($D_2$) varies depending on the case of the simulation, as shown in Table 2. A global arrangement and mesh detail is shown in Figure 6. The total length of the computing domain is 3.5 m, ensuring the axial length necessary to achieve a fully developed flow at the outlet, and placing the section change close to the midpoint of the total length of the duct. More details will be given when boundary conditions are discussed later on.

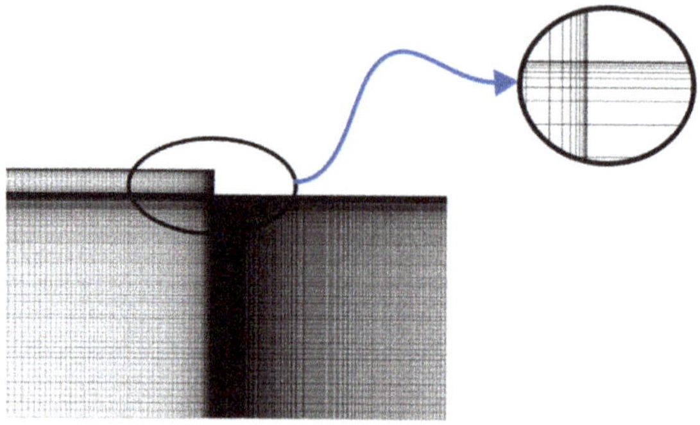

**Figure 6.** Global mesh and detail for a given case (contraction), in the present model.

**Table 2.** Summary of the generated meshes.

| Studied Cases | Mesh Type | Number of Cells |
|---|---|---|
| Geometry with $D_2 = 0.9D_1$ | Single | 322,050 |
| Geometry with $D_2 = 0.7D_1$ | Single | 410,250 |
| Geometry with $D_2 = 0.5D_1$ | Coarse<br>Standard<br>Refined | 245,850<br>420,679<br>694,028 |
| Geometry with $D_2 = 0.3D_1$ | Single | 395,000 |
| Geometry with $D_2 = 0.1D_1$ | Single | 356,250 |
| No step geometry | Coarse<br>Standard<br>Refined | 70,000<br>115,000<br>210,000 |

For the cases of no section change or only a linear loss geometry ($D_2 = D_1$) and contraction pipes with an area ratio of 0.5 ($D_2 = 0.5D_1$), three different meshes have been generated, increasing the number of cells in each one and using these cases to ensure that the results obtained are independent of the mesh.

The program used for the simulations was Ansys-Fluent 6.3.26 (2005). The 2D axisymmetric and stationary model has been selected to analyze the flow in the contraction and expansion study. The size of the meshes in the area near the wall is smaller, in order to adapt it to the thin layer with higher gradients expected there, as explained in Figure 6, mainly at the critical part of the domain, where the section change is produced. The cell size of wall-adjacent meshes in all solid boundaries has been fixed to preserve y+ values below unity.

### *3.2. Mesh Independence Study*

In order to study the independence of the mesh, pressure gradients and pressure losses will be analyzed throughout the analysis domain of the pipe, both for the phenomenon of contraction and for that of expansion, using different meshes in the case of $0.5D_1$ (coarse, medium, and fine meshes) and always considering the turbulence model of κ-ε.

The study of the independence of the mesh has been carried out through the three meshes for the case of $0.5D_1$, simulated from the turbulence model of κ-ε. The results of both the pressure drop and subsequent pressure loss have been analyzed using the data obtained from the simulations with the different meshes, and it will be proved that the results are independent of the mesh used for the simulations.

As can be seen in Table 3, the values of pressure and head differences between the inlet and outlet are similar in all the meshes used for each phenomenon. Therefore, it can be concluded that the simulations are correct and that the results obtained with the refined model are representative of the solution to the studied problem.

**Table 3.** Mesh independence tests for the case $D_2 = 0.5D_1$.

| Geometry | Mesh Type | Pressure Losses (Pa) |
|---|---|---|
| Sudden contraction | Coarse | 36,395.11 |
| | Standard | 36,422.75 |
| | Refined | 36,286.85 |
| Sudden expansion | Coarse | 20,304.47 |
| | Standard | 20,295.86 |
| | Refined | 20,501.45 |

### 3.3. *Turbulence Model*

The physical characteristics that allow the identification of a turbulent flow have been defined by many scholars (see, for instance, [24]) and are the following: irregularity, high diffusivity, high Reynolds number values, high dissipation, and continuity. The turbulent model is the term often used in numerical flow simulation to handle a proper approach to the turbulent shear stress tensor, that is, with the tensor as:

$$\overline{\tau_{i,j}} = -\rho\overline{u_i' u_j'} \tag{18}$$

Such a tensor is often referred to in the bibliography as the Reynolds stress tensor, and it shows up along the averaging process of the Navier–Stokes Equations. That averaging process is quite convenient to solve the incompressible turbulent flow by splitting a given unsteady variable into an average term plus an oscillating term.

On the present article, the three turbulent models that are compared are the Spalart–Allmaras model (S-A) and two different two-equation models (κ-ε and the κ-ω). Therefore, for that two-equation models, the presented simulations will lie in the RANS (integral zone in the graph in Figure 7). The two equation models are used independently, not as in the Menter's SST model [25].

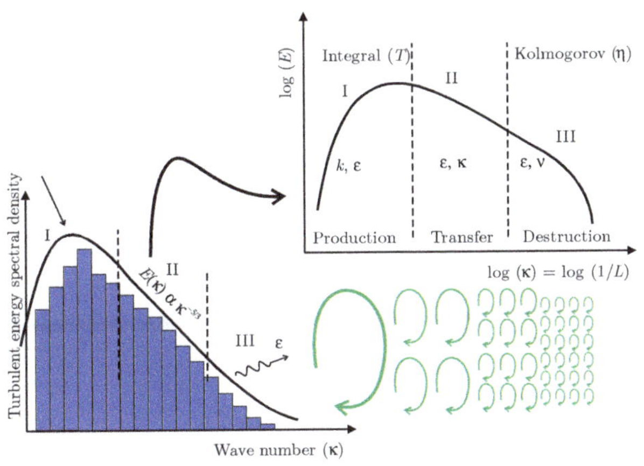


**Figure 7.** Different levels in the approach of the turbulence effects (adapted from [25]).

As a final complement and to check the influence of the possible two-equation closure conditions [26], a calculation with a Reynolds stress model (RSM) is performed. Usually, the RSM is better indicated for 3D geometries, but in the case of 2D axisymmetric ones, it seems that accurate results can also be obtained. Moreover, this RSM looks to be best suited to predict recirculation and reattachment. Therefore, a comparison of the results for the studied geometry will be shown.

The RSM calculations were implemented in the Ansys Fluent 5-Equation model, [27]. It uses a linear pressure-strain approach, with wall boundary conditions from the κ-Equation and wall reflection effects, as developed by [28]. The near wall treatment was chosen to be the enhanced one.

### 3.4. Fluid Properties

Regarding the conditions of the fluid, data have been taken from the program for liquid water (water-liquid), which are, among others, the following values, summarized in Table 4.

**Table 4.** Fluid properties implemented in the numerical model for water flow.

| Property | Symbol | Value (unit) |
|---|---|---|
| Density | $\rho$ | 998.2 $kg/m^3$ |
| Pressure coefficient | $C_p$ | 4182 (J/(kg K)) |
| Viscosity | $\mu$ | $1.0003 \times 10^{-3}$ kg/(m s) |
| Molecular weight | M | 18.052 (g/mol) |

### 3.5. Boundary Conditions and Geometry

The general convergence criterion used when carrying out numerical flow simulation (CFD) analysis is variable throughout the literature in this area, as previous works use different methods to either verify or validate their results. In this work, numerical computations are considered to have converged when the residues of the different variables are less than six orders of magnitude in all cases (sum of residuals in the used domain under $10^{-6}$), reaching down to eleven orders of magnitude in some variables and simulations ($10^{-11}$). To carry out the simulations, it is necessary to establish the boundary conditions. The following ones were considered:

1. Domain outlet: Atmospheric pressure conditions are stablished.
2. Domain inlet: It is necessary to discern the cases of contraction and expansion. The details of the conditions are explained in the strands (3) and (4), which follow.
3. Inlet velocity for the sudden contraction: before the simulations results are considered valid, the speed profile at the exit of the finest straight pipe mesh is obtained. This speed profile is imposed as the input speed of the contraction simulation. In such a way, the velocity profiles at the inlet are fully developed ones.
4. Inlet velocity for the sudden expansion: before the recorded/final simulations, the speed profile at the exit of the finest pipe mesh in the contraction is obtained. This speed profile is imposed as the input speed of the expansion simulation. Therefore, the velocity profile at the inlet are again fully developed ones.
5. Wall condition, for the pipe surface, in the different domain limits that correspond to the solid wall condition.
6. Axial symmetry of the whole domain, as global condition for the flow behavior.

The numerical model firstly managed to obtain a series of main parameters to be fixed, namely the number of cells and global lay-out of the mesh to be ready for the calculations. After that, following the model's detailed setup as already explained in the previous sections, a double-step protocol was started up: firstly, to stablish the best option for the turbulence model, and secondly, to run the model for the different geometrical arrangements. For the last one, two different kinds of results will be shown both for the contraction and the expansion geometries, namely the global results and the study of flow fields, basically the velocity and pressure fields. The studied geometries are considered in Table 5.

**Table 5.** Considered geometries in the present study.

| | | | | | | | |
|---|---|---|---|---|---|---|---|
| Contraction (water) | $D_2/D_1$ | 0.1 | 0.3 | 0.5 | 0.7 | 0.9 | 1.0 |
| Contraction (air) | $D_2/D_1$ | | 0.32 | 0.55 | 0.71 | 0.84 | 0.95 |
| Expansion | $D_1/D_2$ | 0.1 | 0.3 | 0.5 | 0.7 | 0.9 | 1.0 |

The total length of the model is fixed to 3.5 m, leaving the quotient of the length over diameter at least 60 times higher from the contraction/expansion to the beginning/end of the domain (according to Figure 5, for the expansion geometries, the value was $\frac{L_2}{D_1} = 68$, and for the contraction geometries, the values becomes $\frac{L_1}{D_1} = 64$). In such a way, the effect of the boundary conditions it is considered to be far enough from the core region, where the change in the cross-section is produced.

## 4. Numerical Model Results

The analysis of the results begins by carefully studying the speed and pressure profiles for each of the turbulence models. After a global frame is stablished, the local velocity profiles are obtained and conclusions on the turbulent model are drawn. Then, a deep look at the velocity fields and streamlines is proposed. Finally, pressure losses are obtained, and core results of the article lead to final conclusions. All results are shown for the finest mesh.

### *4.1. Pressure and Velocity Fields*

From the numerical results, a wide range of data became available, and the post-processing is advanced towards a comprehensive data generalization. As a starting point, the pressure and velocity fields are observed, and a global frame is stablished. For instance, Figure 8 shows the pressure field and velocity values for a contraction with $D_2 = 0.5D_1$. The first ones are plotted in the lower part of the figure, within a range of [0, 1.13 $\times$ $10^5$ Pa]. The expected pressure drop is numerically obtained.

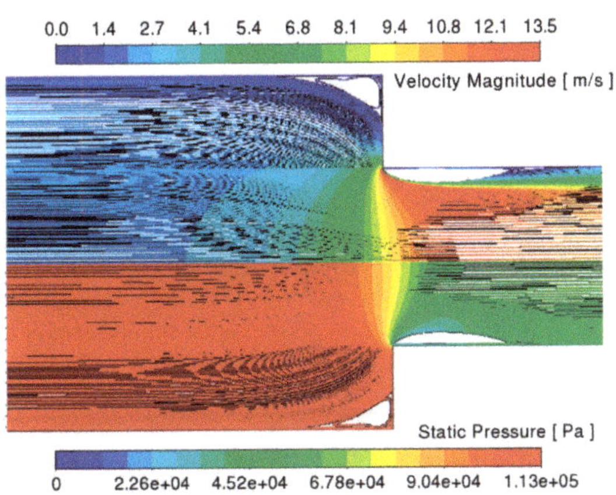


**Figure 8.** Pressure (**lower** part) and velocity (**upper** part) distributions for $D_2 = 0.5D_1$.

Also in Figure 8, in the upper part of the figure, the velocity values on top of the streamlines are shown for a particular contraction case. The velocity ranges for this particular case were (according to initial numerical trials) within the limits [0, 13.5 m/s]. Those ranges fitted for most of the studied cases and a proper analysis was carried out to determine the validity of the results considering the global expected trends for the velocity and pressure distributions. Based on a deeper analysis of the obtained results, velocity profiles at three different sections will be shown in what follows. The previous split on the different available numerical values will be kept, showing first the results for the contraction and then for the pipe expansion. A non-dimensional representation is chosen in the local velocity analysis to better analyze the obtained results. To account for such idea, non-dimensional velocity is plotted vs. a normalized position from the pipe wall.

Regarding the contraction, the comparison between the different velocity profiles at some points of the pipe for the models used in the simulation is shown in Figure 9, retrieving the expected turbulent-like profiles for the case $D_2 = 0.5D_1$. A fairly good agreement with the theoretical profiles was observed, and in the detailed comparison, almost equal values for the three different models was obtained. Almost identical behavior was observed for the x = 1 m location (before contraction), while more relative differences were found for the x = 1.7 m and x = 3.5 m (domain exit). For those two locations, the κ-ω model is the one with higher losses, with higher velocity gradient in the near-wall region. However, the three models provide similar values in all locations.

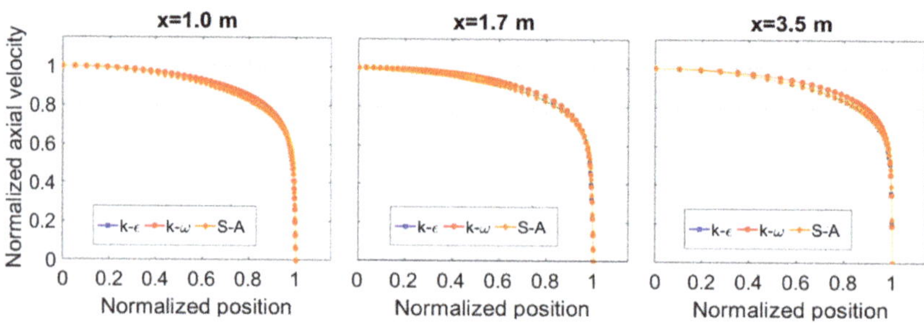


**Figure 9.** Sudden contraction ($D_2 = 0.5D_1$) velocity profiles for the three turbulence models at different locations, namely x = 1 m, x = 1.7 m, and x = 3.5 m.

From the values in Figure 9 and similar ones for the other studied contractions, the κ-ε model was found to be a good approach and was chosen for the final core results comparison. On the other hand, as the differences encountered were not so big, the final model selection was not so critical. It was then used for the subsequent data reduction, together with the finest grid, as explained in the previous part.

A plot with the velocity fields is then proposed in Figure 10. The field is made non-dimensional using the inlet velocity, and then five different contractions are studied ($D/D_1$). Similarly to what was proposed in Figure 8, the lower part of each figure is used to plot this non-dimensional field, while the upper part of each figure is used to plot the streamlines.

As expected, the velocity increase is higher for higher contractions, reaching its maximum at $D_2 = 0.5D_1$. Both patterns, streamlines and velocities, are also in agreement with the trends typically expected in the defined problem. That is, from the literature review and the flow behavior in the different cases, it can be concluded that (at least) the obtained results would be considered as feasible.

In Figure 10, for the cases with higher cross-section change, there is an upstream effect with high velocities even for x positions before the section change is produced. For the lower cross-section changes (cases with $D_2 \leq 0.5D_1$), the geometry effect on the flow velocities resembles a practically symmetric potential source with a monopole vortex (cases with $D_2 \geq 0.5D_1$). On the other hand, the low-speed corner, just before the contraction, becomes lower with decreasing contraction degree, being minimum for the value at $D_2 = 0.9D_1$. The minor losses of the element are then a straightforward effect of the shown velocity changes and, as a matter of fact, the higher the decrease in the cross-section is, the higher the velocity gradients retrieved are, and therefore the higher the losses will be.

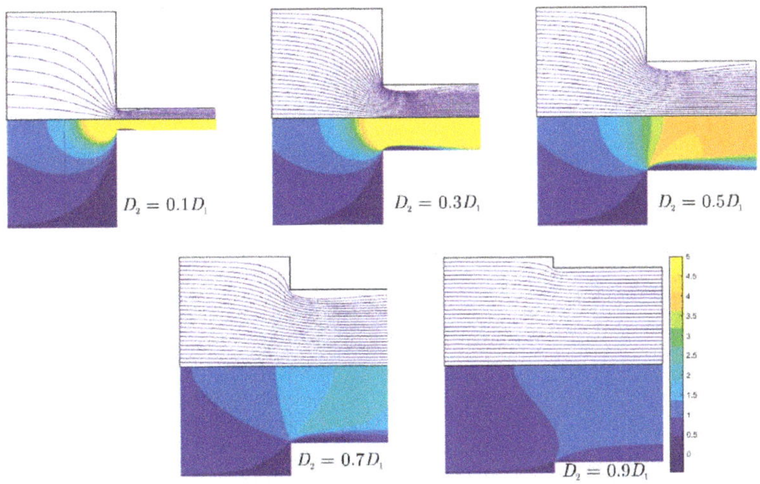


**Figure 10.** Streamlines and velocity values for the contraction (κ-ε model).

In what refers to expansion, a comparison between the different velocity profiles at three points of the pipe for the used turbulence modeling in the simulation are provided; the results, again on a non-dimensional basis, are shown in Figure 11. Parallelly to the studied contraction, the analysis of the results of the expansion process begins by reviewing the speed and pressure profiles for each of the turbulence models.

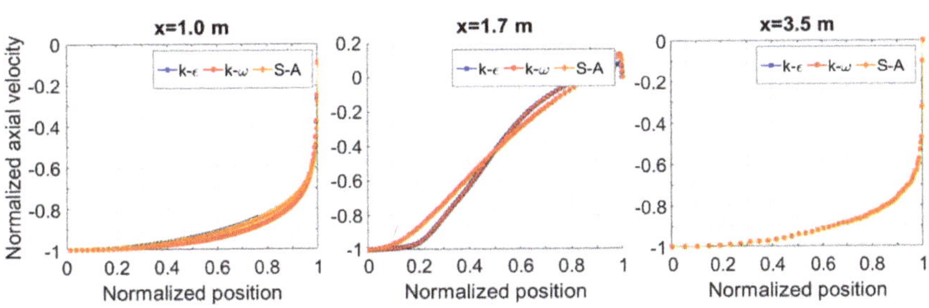


**Figure 11.** Sudden contraction ($D_2$ = 0.5$D_1$) velocity profiles for the three turbulence models at different locations, namely x = 1 m, x = 1.7 m and x = 3.5 m.

In Figure 11, the κ-ε model provides the lowest friction losses in the near-wall region for the $D_2$ = 1.5$D_1$ at x = 1 m. Then, at x = 1.7 (just after the expansion), both κ-ε and κ-ω provide almost the same velocity profile, which includes some negative values as the recirculation bubble is placed in that section. In section x = 1.7, only the κ-ω and the Spalart–Allmaras (S-A) models are shown as the κ-ε model provides very similar values to the κ-ω one. Finally, for the x = 3.5, all three models practically collapse into only one curve. After a full study of the different cases for the two possible changes in section (namely Figures 9 and 11 and equivalent ones for other diameter ratios), the κ-ε model was considered for further analysis.

A non-dimensional plot of the velocities is presented in Figure 12, using as a reference the outlet velocity for the different expansions. Again, as in Figure 10, the upper part of the plots is used to plot the streamlines and the lower part shows the non-dimensional velocity distributions, namely the values $V/V_{\text{exit}}$.

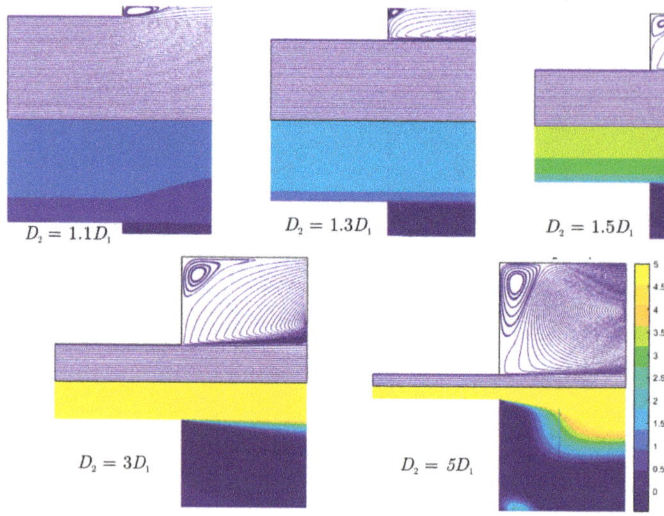


**Figure 12.** Streamlines and velocity values for the expansion (κ-ε model).

Almost all cases (except the one with lower section change, that is, the one with $D_2 = 1.1D_1$) show a big recirculation bubble, this being the main part of a pair of contra-rotating vortices. Also, the flow field is divided into two main zones: the core or jet region and the low-speed corner, with the recirculation bubble. For the lower cross-section changes (cases with $D_2 \leq 1.5D_1$), there is low interaction between the two zones, while for cases with $D_2 \geq 1.5D_1$, the interaction is higher. Again, as expected, the higher $D_2/D_1$, the higher the losses.

The re-attachment position, not shown in Figure 12, follows the expected trend [26]. It is not shown for the sake of uniformity in the representation, to set the focus on the expansion zone.

### 4.2. Pressure Losses for the Abrupt Section Change

From the already discussed derivations, the best parameter to compare the different values would be the minor loss coefficient, defined as follows both for the pipe contraction and expansion:

$$\xi = \frac{h_{\text{loss}}}{\left(\frac{1}{2}\frac{V_2^2}{g}\right)} \tag{19}$$

And then, from the developed numerical models for the contraction and the expansion geometries, a possible comparison of the numerically obtained results with the existing bibliography becomes available. The only considered issue is that both conditions differ in the value of the higher diameter: in the contraction, the reference value is the initial one and, for the expansion, the higher one is the final diameter. Bearing that in mind, and using Equation (19), two different figures are plotted: Figure 13 for the pipe contraction and Figure 14 for the pipe expansion. The minor loss coefficient is plotted versus the change in diameter. Such a representation is considered to be the most relevant output from the performed study and resembles the values previously considered in the bibliographical review (see Figure 3, where the variable used was the cross-section ratio, namely the parameter $R = \beta^2$).

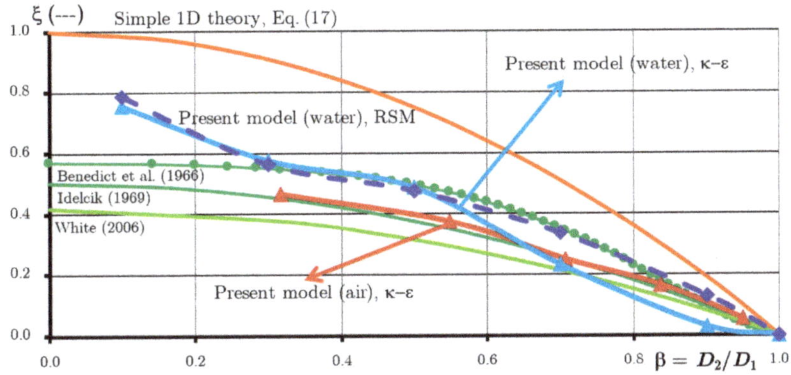


**Figure 13.** Comparison of the minor loss coefficient vs. β = $D_2/D_1$, for a sudden contraction [2,9,17].

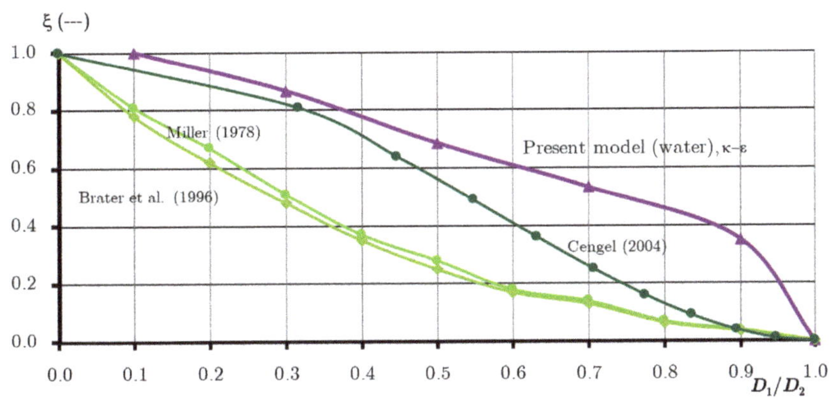


**Figure 14.** Comparison of the minor loss coefficient vs. $D_1/D_2$, for a sudden expansion [8,29,30].

Figure 13 plots, on one hand, some of the existing published results from [2,9] and [17]. Then, the values developed in Equation (17) with the simple 1d theory are plotted. From the numerical calculation, the deeper analyzed values were the ones using the κ-ε modeling, and two different curves are shown, i.e., one for water as fluid and another one for air, in order to check possible density influence on the results. Finally, a curve using the RSM is also plotted. Those last three curves are named as the "Present model", with the different particular conditions.

Although several other values are available in the literature for such problems, three have been selected for comparison, namely the values found in [2] and the derived and proposed values by [9] and the more recent [17]. The results from the developed model are plotted considering water and air as fluids. The water model seems to follow the expected trend, but only catches the actual values in a range for the diameter ratio [0.3, 0.7]; for lower values, it overpredicts the losses and, what is more critical, for higher values, it underpredicts the losses. On the other hand, the air model follows quite closely the values in [9] and reaches quite good results at high diameter ratios (close to 1). Globally speaking, both models show a good trend in the loss prediction, and it seems that the water model better finds the experimental values for the range [0.3, 0.6], while it finds a drop in the losses that is not so coherent for diameter ratios higher than 0.7.

The simple theoretical equation derived in previous sections (Equation (17)) is also plotted, and although it follows the global trend, the error is too high to be considered as a design tool. In other words, up to a given global view, it somehow follows the expected trend, but it does not catch the actual values at all, providing a loss coefficient almost double in comparison with any of the other previously plotted and commented values (loss overprediction, particularly at low values of β).

For high contractions (β < 0.5), the results from RSMs replicate the behavior observed previously with the κ-ε models. In addition, at contractions below β < 0.3, there is a significant overprediction of all the numerical models tested with respect to the experimental observations. On the other hand, when the contraction is moderate to weak (β > 0.5), RSMs estimate a higher value of the loss coefficient, very similar to the results of Benedict et al. [2]. Moreover, turbulence closure with κ-ε models clearly underpredict the loss coefficient if β > 0.5. In those situations, with such slight diameter reductions, it is presumable that Reynolds stress models (RSMs) can capture mild anisotropic effects better than two-equation models, as it happens in the presented results.

Similarly, in Figure 14, the comparison of the loss coefficient for different bibliographical values and the present model is performed. The coordinate system is equivalent to Figure 13, and the minor loss coefficient is plotted versus the quotient $D_1/D_2$.

In Figure 14, the values provided by [8,29,30] are plotted together with the prediction by [30]. In fact, the last one is the closest to the present model's results. Nevertheless, it seems that the model overpredicts the losses for the whole range of the diameter ratio. Only for the lower values, in the range [0.1, 0.4], can it be considered that there is a kind of agreement between the model values and the ones obtained in the equation by [30].

It seems that the location of the reattachment point does affect the reliability of the numerical model results. It is likely not even the flow solution, which the authors consider to be up-to-date as regards the present technology, but the way the data reduction is performed and the right calculation of such a point that may include some numerical additional uncertainties.

Globally speaking, the comparisons in Figures 13 and 14 show how the 2D model can capture the physics behind the flow behaviors in a sudden pipe contraction, while the same model for a sudden expansion finds more difficulties and fails to accurately describe the losses in the whole $D_1/D_2$ range, clearly overpredicting the losses.

## 5. Conclusions

The numerical modeling of a sudden or abrupt change in a pipe section presents a widespread application for pipe flow. However, a quick but accurate numerical procedure for the calculation of such losses for different practical situations is not so common in the bibliography. In the present article, minor losses are considered for the sudden change in a section of a pipe, both for the contraction and for the expansion geometries. An initial 1D model is developed and its limits are clearly shown.

The developed numerical 2D axisymmetric model yields a very accurate description of both practical situations, which can be very useful to understand the physical mechanisms governing the losses in such classical Fluid Mechanics problems. Those minor losses referred to have been widely studied on an experimental and theoretical basis, but not so much using a numerical approach. Within the proposed methodology, a numerical calculation of the losses for the two possible pipe section changes has been shown.

The numerical model has proven its robustness in what refers to mesh independency on the final results. For the obtained sensitivity of the developed model, even a lower number of cell meshes could be enough to reach the same values. A wide study has been carried out on such mesh independency, with several tests to prove the conclusion.

A wide set of both velocity and pressure maps have been obtained for the two practical situations, revealing the influence of the turbulence model being used. Three different models have been tested, namely the Spalart–Allmaras, κ-ε, and κ-ω (SST) models, which were used alongside the calculations. Although slight changes in the near-wall velocity profiles were found, when calculating the head losses and loss coefficient, almost equal values were retrieved for the three turbulent models, with less than 0.5% variations. A final check with an RSM was also conducted for the contraction flow condition, and the results show the consistency of the developed models, with an improvement in the prediction for the RSM in the range $\beta = [0.5, 1]$, for which the two-equation models underpredict the losses.

The most relevant finding in this study is the evolution of the loss coefficient versus the diameter quotient for the different geometries. Regarding the pipe contraction geometry, a double simulation was performed using air and water as fluids. The results for both show quite good agreement with the existing data. A third approach in terms of a theoretical development was introduced, although that final value does not fit the experimental and numerical results.

For the pipe expansion, the question regarding the reattachment point has been a cornerstone throughout the different studies and becomes a critical issue in any possible approach to the problem. The comparison in this problem was not so promising as that for the previous one. As can be observed, quite different values and trends were obtained and reported in the existing values. However, the presented 2D model somehow overpredicts the losses.

The global picture of the head losses for a pipe section change has been numerically studied, and particular values for loss coefficient predictions are available. A comparison with the existing bibliographical experimental values has also been tackled. The actual comparison shows good agreement for the contraction and is fairly different to experimental values for the expansion.

**Author Contributions:** Conceptualization, J.G. and J.M.F.O.; methodology, J.M.F.O.; software, A.M.-F. and I.V.P.; validation, I.V.P. and A.M.-F.; formal analysis, J.M.F.O.; resources, J.G.; data curation, I.V.P. and A.M.-F.; writing—original draft preparation, J.G. and A.M.-F.; writing—review and editing, J.M.F.O. and I.V.P.; visualization, I.V.P.; supervision, J.G.; project administration, J.G. and J.M.F.O.; funding acquisition, J.M.F.O. All authors have read and agreed to the published version of the manuscript.

**Funding:** This research received no external funding.

**Data Availability Statement:** Data will be shared under requirement.

**Acknowledgments:** A grateful acknowledgement should be given to the financial support from Project ENE2017-89965-P provided by the Spanish Ministry of Economy, Industry and Competitiveness. It has also important contribution the economic support of the "Agencia Estatal de Investigación" (AEI) of the Spanish Ministry of Science and Innovation, in the context of the State Program to Promote Scientific-Technical Research and its Transfer, through the Project "Optimization through flow control techniques of a vertical axis wind turbine for urban environments" (ref. TED2021-131307B-I00), included in the NextGenerationEU funds of the European Community. Some additional contributions of the University Institute for Industrial Technology of Asturias (IUTA), through projects SV-19-GIJON-1-15, SV-20-GIJON-1-17 and SV-22-GIJON-1-02, have allowed the development of this study and are also appreciated.

**Conflicts of Interest:** The authors declare no conflicts of interest.

## Nomenclature

| | |
|---|---|
| $A_1$ | Inlet section contraction/outlet expansion, ($m^2$). |
| $A_2$ | Outlet section contraction/inlet expansion, ($m^2$). |
| $C_a$ | Experimental factor, (---), from [7]. |
| $C_p$ | Pressure coefficient, (---). |
| C.V. | Control volume, as often used in Fluid Mechanics. |
| $D_1$ | Inlet diameter, (m). |
| $D_2$ | Outlet diameter, (m). |
| $f$ | Friction coefficient, (---). |
| $h_{exp}$ | Experimental head loss, (m). |
| $h_{loss}$ | Head loss when changing section, (m). |
| $K_{cont}$ | Head loss coefficient, (---). |

| | |
|---|---|
| $K_{acc}$ | Head loss coefficient due to flow acceleration, (---). |
| $L_1$ | Length inlet contraction/outlet expansion, (m). |
| $L_2$ | Length output contraction/inlet expansion, (m). |
| $M$ | Molecular weight, (g/mol). |
| $P$ | Pressure, (Pa). |
| $P_0$ | Pressure in the contraction section, (Pa). |
| $P_1, P_2$ | Pressure at the inlet/outlet sections, (Pa). |
| $Q$ | Flow rate, ($m^3$/s). |
| $R$ | Cross-section quotient, $R = A_2/A_1 = \beta^2$, (---). |
| Re | Reynolds number, (---). |
| RSM | Reynolds stress model. |
| ′ | Fluctuating velocity field, (m/s). |
| S-A | Spalart–Allmaras model. |
| $V_0, V_{exit}$ | Inlet velocity inlet (contraction), exit velocity (expansion), both in (m/s). |
| $V_1$ | Inlet velocity, contraction/outlet velocity, expansion (m/s). |
| $V_2$ | Outlet velocity, contraction/inlet velocity, expansion (m/s). |
| $\mu$ | Fluid viscosity, (kg/(m s)). |
| x, y | Coordinates, axial and radial one, (m). |
| y+ | Wall distance, (---). |
| t | Time, (s)-. |
| g | Gravity acceleration constant, ($m/s^2$). |
| $\beta$ | Diameter ratio, $D_2/D_1$, (---). |
| $\varepsilon$ | Specific dissipation rate, (1/s). |
| $\xi$ | Pressure loss coefficient, (---). |
| $\xi_{cont}$ | Pressure loss coefficient for a contraction, (---). |
| $\kappa$ | Turbulent kinetic energy, (J/kg = $m^2/s^2$). |
| $\lambda$ | Flow speed ratio, Equation (10), (---). |
| $\rho$ | Fluid density, ($kg/m^3$). |
| $\tau_{i,j}$ | Stress tensor elements, (Pa). |
| $\omega$ | Specific viscous dissipation rate or specific vorticity fluctuations, ($s/m^2$). |

## References


1. Rennels, D.C.; Hudson, H.M. *Pipe Flow: A Practical and Comprehensive Guide*; John Wiley & Sons: New York, NY, USA, 2012; Chapter 10: Contractions.
2. Benedict, R.P.; Carlucci, N.A.; Swetz, S.D. Flow losses in abrupt enlargements and contractions. *Trans. ASME J. Eng. For. Power* **1966**, *88*, 73–81. [CrossRef]
3. Weisbach, J. *Mechanics of Engineering*; (Trans. Coxe, E.B.); van Nostrand Book Co.: New York, NY, USA, 1992.
4. Polkowski, H. Effect of Sudden Expansion or Contraction in a Pipe upon the Flow of Water. Bachelor's Thesis, University of Illinois, Champaign, IL, USA, 1912.
5. Archer, W.H. Loss of head due to enlargements in pipes. *Trans. ASCE* **1913**, *76*, 999–1026.
6. Kays, W.M. Loss coefficient for abrupt changes in flow cross section cross section with low Reynolds number flow in single and multiple tube systems. *Trans. ASME* **1950**, *72*, 1067–1074. [CrossRef]
7. Vennard, J.K.; Street, R.L. *Elementary Fluid Mechanics*, 4th ed.; John Wiley & Sons: New York, NY, USA, 1961.
8. Brater, E.F.; King, H.W.; Lindell, J.E.; Wei, C.Y. *Handbook of Hydraulics*; McGraw-Hill: New York, NY, USA, 1996.
9. Idelcik, I.E. *Memento des Pertes de Charge*; Meury, M., Translator; Eyrolles: Paris, France, 1969. (In French)
10. Martin, J.J. Expansion and contraction losses in fluid flow: An example of engineering analysis. *Chem. Eng. Educ.* **1972**, *8*, 138–140.
11. Bullen, P.R.; Cheeseman, D.J.; Hussain, L.A. A study of turbulent flow in pipe contractions. *Proc. IMEchE Part E J. Process Mech. Eng.* **1996**, *210*, 171–180. [CrossRef]
12. Bullen, P.R.; Cheeseman, D.J.; Hussain, L.A.; Ruffell, A.E. The determination of pipe contraction pressure loss coefficients for incompressible turbulent flow. *Int. J. Heat Fluid Flow* **1987**, *8*, 111–118. [CrossRef]
13. Schmidt, J.; Friedel, L. Two-phase pressure drop across sudden contractions in duct areas. *Int. J. Multiph. Flow* **1997**, *23*, 283–299. [CrossRef]
14. Fox, R.W.; Pritchard, P.J.; McDonald, A.T. *Introduction to Fluid Mechanics*; John Wiley & Sons: Hoboken, NJ, USA, 2006.
15. Munson, B.R.; Young, D.F.; Okiishi, T.H. *Fundamentals of Fluid Mechanics*; John Wiley & Sons: New York, NJ, USA, 1998.
16. Barrero, A.; Pérez-Saborid, M. *Fundamentos y Aplicaciones de la Mecánica de Fluidos*; Mc. Graw-Hill Iberoamericana: Madrid, Spain, 2005. (In Spanish)
17. White, F.M. *Fluid Mechanics*; Mc Graw-Hill: New York, NJ, USA, 2006.

18. Wikipedia. Article on the Borda-Carnot Equation. Available online: https://en.wikipedia.org/wiki/Borda%E2%80%93Carnot_equation (accessed on 20 December 2020).
19. Darihaki, F.; Zhang, J.; Shirazi, S.A. Application of Scale-Resolving Simulations and Hybrid Models for Contraction-Expansion Pipe Flows. In Proceedings of the ASME 2021 Fluids Engineering Division Summer Meeting (FEDSM2021), Virtual, 10–12 August 2021; Volume 1, p. V001T02A026.
20. Ye, J.; Cheng, Y.; Xie, J.; Huang, X.; Zhang, Y.; Hu, S.; Salem, S.; Wu, J. Effects of divergent angel on the flow behaviors in low-speed wind accelerating ducts. *Renew. Energy* **2020**, *152*, 1292–1301. [CrossRef]
21. Kaaushik, V.V.R.; Ghosh, S.; Das, G.; Das, P.K. CFD modelling of water flow through sudden contraction and expansion in a horizontal pipe. *Chem. Eng. Educ.* **2011**, *45*, 30–36.
22. Rashid, F.L.; Alhabeeb, B.A.; Alhwayzee, M.; Mohammed, A.A. Study the sudden expansion and contraction in the pipeline on the distribution of pressure at the presence of a porous media by using CFD simulation. *J. Eng. Sci. Technol.* **2021**, *16*, 3960–3973.
23. Wikipedia. Article on the "Vena Contracta" (Basic Idea on the Phenomena and Implications). Available online: https://en.wikipedia.org/wiki/Vena_contracta (accessed on 3 March 2024).
24. Leschziner, M.A. *Statistical Turbulence Modelling for the Computation of Physically Complex Flows*; Lecture Series; von Karman Institute: Sint-Genesius-Rode, Belgium, 2001.
25. Davidson, L. Fluid Mechanics, Turbulent Flow and Turbulence Modelling. Master's Thesis, Chalmers University, Göteborg, Sweden, 2017; pp. 1–363.
26. Chieng, C.C.; Launder, B.E. On the calculation of turbulent transport downstream from an abrupt pipe expansion. *Num. Heat Transf.* **1980**, *3*, 189–207.
27. Menter, F.R.; Lechner, R.; Matyushenco, A. Best Practice: RANS Turbulence Modeling in Ansys CFD (Version 1.0). ANSYS-Fluent Tutorial. Available online: https://www.afs.enea.it/project/neptunius/docs/fluent/html/th/node78.htm (accessed on 3 March 2024).
28. Gibson, M.M.; Launder, B.E. Ground effects on Pressure fluctuations in the atmospheric boundary layer. *J. Fluid Mech.* **1978**, *86*, 491–511. [CrossRef]
29. Miller, D.S. *Internal Flow Systems*; BHRA Information Services: Bedford, UK, 1978.
30. Cengel, Y.A.; Cimbala, J.M. *Fluid Mechanics: Fundamentals and Applications*; Mc. Graw-Hill: New York, NJ, USA, 2004.

*fluids*

MDPI

*Article*

# Pressure and Velocity Profiles over a Weir Using Potential Flow Model

**M. R. Ajith Kumar [1], Prashanth R. Hanmaiahgari [1,*] and Jaan H. Pu [2,*]**

[1] Department of Civil Engineering, Indian Institute of Technology Kharagpur, Kharagpur 721302, India; ajithk1998@kgpian.iitkgp.ac.in

[2] Faculty of Engineering and Digital Technologies, University of Bradford, Bradford BD7 1DP, UK

* Correspondence: hpr@civil.iitkgp.ac.in (P.R.H.); j.h.pu1@bradford.ac.uk (J.H.P.)

**Abstract:** A potential flow model of the semi-inverse type is proposed to simulate flow over round crested weirs. This technique involves the construction of only streamlines over the weir instead of constructing the entire flow net. A Serre–Green–Naghdi (SGN) equation is employed to determine the initial free-surface profile, which is solved using a combined finite volume-finite difference scheme. The potential flow equations were numerically solved using a five-point central finite difference scheme. The model was applied to define the pressure and velocity fields in channel controls involving transcritical flow, such as the Gaussian weir, parabolic weir, and semicircular weir. The impact of streamline curvature on pressure and velocity distributions was investigated in the study. The curvature of the streamline strongly influenced the rise and drop of the bed pressures along the test section. A semicircular weir experiment was also conducted to validate the pressure and velocity profiles obtained using the proposed 2-D fluid flow model. The computed pressure and flow profiles from the solution of the potential flow equation agree perfectly with the present experiment and similar experiments available in the literature. In conclusion, the SGN equation provides an excellent initial profile to solve a 2-D ideal fluid flow numerically.

**Keywords:** semi-inverse mapping; transcritical flow; potential flow; Laplace equation; round crested weir

## 1. Introduction

Weirs are the best hydraulic structures widely used for water discharge measurements due to their ability to measure a wide range of flow. The shape of the crest in the flow direction classifies weirs, which are then further subdivided based on the cross-sections of the crest. Among these, short-crested weirs differ significantly from the broad-crested weirs and the sharp-crested weirs characterized by sharply bent streamlines and non-hydrostatic pressure conditions across stream nappe over the crest surface [1]. The centrifugal force induced by the stream curvature in transcritical flow over the weir forms a non-hydrostatic pressure and non-uniform velocity [2,3]. Transcritical flow is an undular flow motion observed when the flow depth is close to the critical depth in an open channel; a small change in specific energy can induce a large change in flow depth. Incompressible, non-viscous, and irrotational flow equations are used to describe the transition of flow from subcritical to supercritical throughout a critical point [4]. The potential flow theory, a simple mathematical model to obtain the detailed flow features, assumes the flow is inviscid and irrotational, as compared with the advanced turbulence model [5]. The potential flow model involves determining the potential and stream functions at every location in the flow domain, thereby generating pressure and velocity fields [2,6,7]. Furthermore, Rouse [6] and, more recently, Pu et al. [8], Pu [9], and Zheng et al. [10] used various incompressible, non-viscous, irrotational flow and turbulence equations to describe the 2D flow behaviour of open channel flow.

The availability of experimental data in transcritical flow to describe the flow profile, velocity, and pressure distributions is limited. Messe [11] demonstrated that transition from subcritical to supercritical regimes is possible with a continuous surface profile and finite surface slope while maintaining the hydrostatic-pressure distribution. Andersen [12] investigated the flow curvature's influence on the non-hydrostatic pressure profile at the location where a change in slope occurs. The relaxation approach was utilized by Southwell and Vaisey [13] to solve the partial differential equation for the stream function $\psi$ on the physical plane $(x, y)$. But solving the Laplace equation on physical plane $(x, y)$ is very difficult. Markland [14] solved the Laplace equation for dependent variables $x$ and $y$ in this complex potential plane $(\phi, \psi)$ using a finite difference technique and a relaxation procedure for spillways and free overfall problems. However, discretizing the equation over curved beds and selecting an effective relaxation pattern are time-consuming processes [15]. Chan et al. [16] utilized the finite element approach to model flow under gates. Cheng et al. [17] used the boundary integral approach to model flow across spillways. A different approach was suggested by Thom and Apelt [15], transforming the relationship between dependent and independent variables where a rectangular strip on the complex potential plane was used to map the real plane. Chow and Han [18] used an approach where the hodograph of free overfall was plotted on a rectangular strip, solving the Laplace equation for stream function with hodograph variables as an independent variables using relaxation techniques.

The Laplace equation for $\psi$ on the physical plane $(x, y)$ was modified by Boadway [19] into a relation that expresses $y$ as a function of $\psi$ and $x$. Montes [20] used this semi-inverse transformation of the Laplace equation derived by Boadway to simulate flow over curved beds. A review of existing potential flow solutions focused on flow profile and bottom-pressure simulations, and they were not employed to examine other internal flow properties [21,22]. Castro-Orgaz [22] used the Boadway model for analyzing transcritical flow over a weir where specific energy is the minimum at the crest. Due to the existence of an uneven free surface, the potential flow solution for open-channel flow is complicated [22,23]. Castro-Orgaz [22] also considered the Boussinesq-type energy equation to estimate the initial flow profile.

The flow curvature formed over a round-crested control structure results in non-uniform velocity and non-hydrostatic pressure [2]. The divergence of pressure from hydrostatic conditions is not applicable to the Saint-Venant equation in these structures [24]. For shallow flows with a horizontal channel bottom or with minimum inclination $\theta$, Saint-Venant equations provide reasonable approximations of solutions [25]. The Saint-Venant equations, as modified by Boussinesq [26], included the curvature effect on particle velocities. Dressler [25] derived the generalized nonlinear shallow-flow equations in a bottom-fitted curvilinear coordinate system. Dressler-derived equations simplified the Saint-Venant approach for channel reaches by having a constant bottom slope [27]. On the other hand, Naghdi and Vongsarnpigoon [28] numerically solved the Boussinesq-type energy equation, where the model was applied to a rigid hump case. Matthew [29] derived a higher-order equation for potential flow in open channels in Cartesian coordinates using Picard's iterative procedure. It was found that in the Bernoulli solver, the presumed upstream flow depth of the hydrostatic condition is insensible, causing the downstream flow depth and total head calculations to be error-prone [30]. The advantage of the Serre–Green–Naghdi solver over the Bernoulli solver is that the computed flow profile has less impact through incorrect estimation of upstream flow depth [24]. The Serre–Green–Naghdi (SGN) equations are a set of nonlinear partial differential equations used to model shallow water flow, mostly used in marine hydraulics. There were very few previous applications of a steady or unsteady version of the SGN equation found for modelling flow over round-crested weirs. The SGN equation was first solved for flow over curved beds using a finite volume-finite difference scheme using the MUSCL-Hancock scheme by Castro-Orgaz and Cantero-Chinchilla [31]. The objective of the present study is to utilize the initial profile solved using a similar approach to complete Boadway's potential flow model. The present study differs from the previous potential flow model by Montes [20,32] and Castro-Orgaz [22] by preferring the Serre–Green–Naghdi equation over the Bernoulli-type Boussinesq equation to determine the initial free surface profile. An experiment is conducted to determine velocity and pressure profiles at selected sections of the semicircular weir in detail. Also, a 2-D ideal fluid flow solution is used to validate velocity and pressure profiles by the present experimental work to show the curvature of the streamline effect on pressure profiles at selected sections of the semicircular weir in detail.

## 2. Governing Equation

### *2.1. The x-ψ Transformation*

Southwell and Vaisey [13] considered the Laplace equation for $\psi$ on the $x$-$y$ plane:

$$\nabla^2\psi = \frac{\partial^2\psi}{\partial x^2} + \frac{\partial^2\psi}{\partial y^2} = 0 \tag{1}$$

where $\nabla^2$ is a second-order differential operator commonly known as the Laplacian operator. As mentioned earlier, an irregular computational mesh was required to solve the equation; therefore, the process of solution in the physical plane near curved boundary streamlines was laborious [15,22]. Boadway [19] transformed Equation (1) into a relationship expressing $y$ as a function of $\psi$ and $x$. The advantage of this kind of transformation is that the initial free surface profile may be specified considerably more easily as a function of $x$ than that of $\phi$. BC in Figure 1 represents the free surface, and AB and CD are two vertical sections where the potential flow is assumed to be parallel to the boundary with uniform velocity and hydrostatic-pressure distributions. This geometry is represented by the $x$-$\psi$ plane rectangular strip ABCD. The semi-inverse transformation of the Laplace equation $y = y(x, \psi)$ is expressed as [19]:

$$\nabla^2 y = \frac{\partial^2 y}{\partial x^2}\left(\frac{\partial y}{\partial \psi}\right)^2 + \frac{\partial^2 y}{\partial \psi^2}\left[1 + \left(\frac{\partial y}{\partial x}\right)^2\right] - 2\frac{\partial^2 y}{\partial x \partial \psi}\frac{\partial y}{\partial x}\frac{\partial y}{\partial \psi} = 0 \tag{2}$$

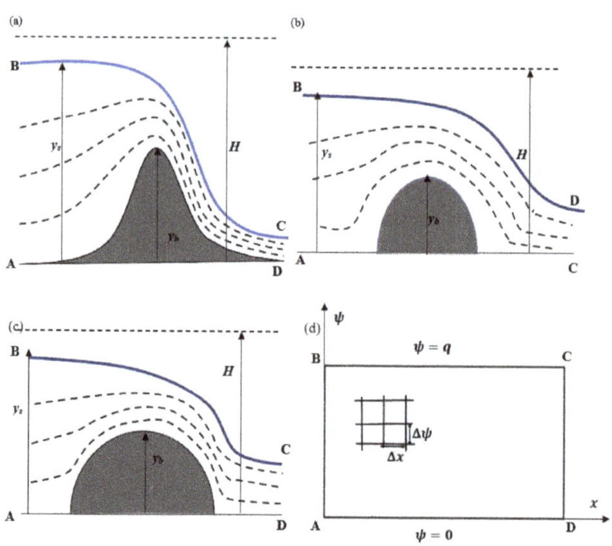


**Figure 1.** Potential flow definition sketch in the $x$-$y$ plane for (**a**) Gaussian weir; (**b**) parabolic weir; (**c**) semicircular weir. (**d**) Potential flow in $x$-$\psi$ plane subdivided in finite difference grid.

Boadway's mapping allows modellers to simplify the complete flow net to only determine the streamline pattern; as a result, the potential function can be ignored [22].

The potential flow's velocity and pressure distributions are generated once the equation for each streamline has been solved. The vertical and horizontal velocity components in terms of $x$, $y$ and $\psi$ are

$$v = u\frac{\partial y}{\partial x} \tag{3}$$

$$u = \frac{\partial \psi}{\partial y} \tag{4}$$

## 2.2. Boundary Conditions

### 2.2.1. Upstream and Downstream Boundaries

The boundary conditions are taken at where streamlines are aligned with the channel bottom. They are given as follows: (1) The vertical ordinate $y(x, 0)$ at all points in the lower boundary, where $\psi$ =0 is prescribed by $y_b$; (2) the vertical ordinates $y(x_0, q)$ and $y(x_n, q)$ at upstream and downstream sections of the free surface, respectively. Let $j$ represent the mesh's node index in the $\psi$-direction, as shown in Figure 2. The variation of $y$ at point $j$ in the upstream and downstream sections is given by Equation (5)

$$y_j = y_b + \Delta y\left(\frac{\psi_j}{\psi_n}\right) \tag{5}$$

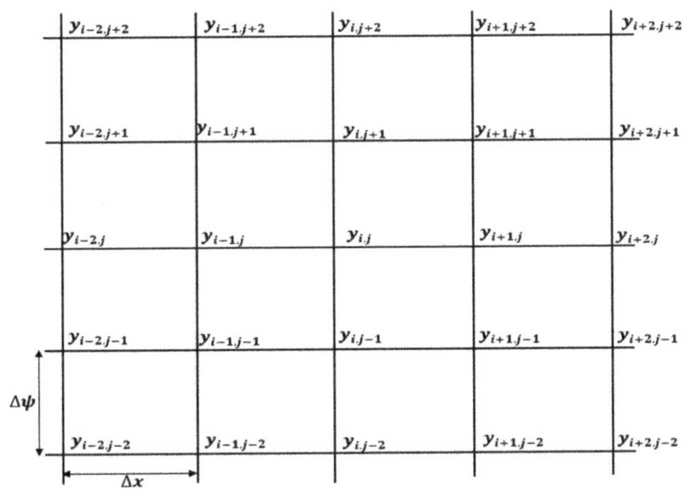


**Figure 2.** Notations for indicating the position of streamlines of adjacent nodes in the array.

### 2.2.2. Free Surface Boundary

The first step in a potential flow model is determining the initial flow profile which is a crucial part of the model. Previous potential flow solvers utilized experimental data or initial profiles obtained from other models to estimate the initial flow profile [22]. It is an essential step because it determines the solution's accuracy. The selection of a poor initial flow profile could lead to the failure of the solution. Montes [32] used the energy equation to find the vertical ordinate $y$ of the free surface.

$$y = H - y_b - \frac{u_s^2}{2g}\left(1 + \left(\frac{dy_b}{dx}\right)^2\right) \tag{6}$$

where $y_b$ = bed elevation; $\psi_n = q$; $H$ = total head; $y$ = water depth; $g$ = acceleration due to gravity; and $u_s$ = horizontal surface velocity. Castro-Orgaz [22] recommended an initial free-surface $y_s$ and energy head $H$ for the selected discharge per unit width $q$ to start the potential flow solution. He used a second-order equation for potential flow developed by Matthew [29], similar to that of the Boussinesq-type Bernoulli equation, to obtain the initial free-surface profile.

$$H = y_b + y + \frac{q^2}{2gy^2}\left(1 + \frac{2y(d^2y/dx^2) - (dy/dx)^2}{3} + y\left(d^2y_b/dx^2\right) + (dy_b/dx)^2\right) \quad (7)$$

where $x$ is the longitudinal space coordinate; $d/dx$ indicates ordinary differentiation with respect to $x$. The above equation considers the hydrostatic condition $H = y_1 + \frac{q^2}{2gy_1^2}$ as one of the boundary conditions, which assumes $\frac{dy}{dx} = \frac{d^2y}{dx^2} = 0$ at $x = x_1$. However, if the upstream water depth $y_1$ is inaccurate, then $d^2y/dx^2 \neq 0$, and nonlinear and periodic waves will occur upstream, causing the calculation of $H$ to be incorrect, too [24]. As proven, the error involved with the incorrect estimation of upstream flow depth is lesser in the SGN solver [24]. This paper adopts a steady SGN solver to estimate the initial flow profile over curved beds to start the solution of Boadway's semi-inverse transformation of the Laplace equation.

The SGN equations are widely employed to calculate shallow and moderate ocean waves [24,33,34]. The free surface analysis in the undular jumps is one of the applications of the steady SGN equations in open-channel hydraulics by Hosoda and Tada [35] and in depth-averaged flow models to estimate dune form and flow resistance by Onda and Hosoda [36]. Castro-Orgaz and Cantero-Chinchilla [31] were the first to apply the steady and unsteady versions of the SGN equation to compute the flow characteristics over curved beds. The steady SGN for 1D water waves travelling across any topography [31,36]:

$$\frac{d}{dx}\left[\frac{y^2}{2} + \frac{q^2}{gy}\left(1 + \frac{y\left(\frac{d^2y}{dx^2}\right) - \left(\frac{dy}{dx}\right)^2}{3} + \frac{y\left(\frac{d^2y_b}{dx^2}\right)}{2} - \frac{\left(\frac{dy}{dx}\right)\left(\frac{dy_b}{dx}\right)}{2}\right)\right] = -\left[y + \frac{q^2}{2gy^2}\left(2y(d^2y_b/dx^2) + y(d^2y/dx^2) - (dy/dx)^2 - 2(dy_b/dx)(dy/dx)\right)\right]\frac{dy_b}{dx}. \quad (8)$$

Equation (7) is the first integral of Equation (8), where total head $H$ is the constant of integration of Equation (8). This relation is verified by differentiating Equation (8) once with respect to $x$(the derivation is given in Appendix A). Both differential equation features are different during numerical simulations; however, they produce the same final converged solutions after the correct estimation of upstream flow depth [24]. Equation (9) is obtained after neglecting the second-order products in derivatives of $y$ and $y_b$, as is shown in Appendix B.

$$\frac{q^2}{g}\left(\frac{1}{3}\left(d^3y/dx^3\right) + \frac{1}{2}\left(d^2y_b/dx^2\right)\right) + \left(y - \frac{q^2}{gy^2}\right)(dy/dx) = -y(dy_b/dx). \quad (9)$$

Equation (9) is a third-order ordinary differential equation which requires three boundary conditions. Komai [37] set the experimental values of flow depth as boundary conditions in upstream and downstream sections and $dy/dx = 0$ at the upstream section and solved the system of implicit non-linear equations by the Newton–Raphson iteration technique. A similar procedure is adopted in this work to solve the non-linear implicit equation to obtain an initial flow profile over a curved weir.

## 3. Solution Techniques

### 3.1. Solution of $x$-$\psi$ Transformation of Laplace Equation

The solution for the semi-inverse transformation of the Laplace equation for the $y$-coordinate in terms of $\psi$ and $x$ is obtained once all the values of $y$ are prescribed along the contour A-B-C-D-A [Figure 1d] in the computational domain of the $x$-$\psi$ plane. Let $i$ represent the node index of the mesh in the longitudinal direction with grid size $\Delta x$, and $j$ represent the node index of the mesh in the vertical direction with grid size $\Delta\psi$ as shown in Figure 2. The five-point finite difference approach [38] was used to estimate the terms for the computational node in the region of the internal boundary section. The partial derivative terms of Equation (2) are replaced with fourth-order accurate central finite difference approximations in order to reduce truncation error (i.e., derivation of Equation (12) is given in Appendix C). They produce higher accuracy by including more considered points when compared to the three-point central finite difference scheme.

$$\frac{\partial y}{\partial x} = \frac{-y_{i+2,j} + 8y_{i+1,j} - 8y_{i-1,j} + y_{i-2,j}}{12\Delta x} \tag{10}$$

$$\frac{\partial y}{\partial \psi} = \frac{-y_{i,j+2} + 8y_{i,j+1} - 8y_{i,j-1} + y_{i,j-2}}{12\Delta \psi} \tag{11}$$

$$\frac{\partial^2 y}{\partial x \partial \psi} = \frac{1}{144\Delta x \Delta \psi}\begin{pmatrix} -y_{i-2,j+2} + 8y_{i-2,j+1} - 8y_{i-2,j-1} + y_{i-2,j-2} + \\ 8y_{i-1,j+2} - 64y_{i-1,j+1} + 64y_{i-1,j-1} - 8y_{i-1,j-2} \\ +y_{i+2,j+2} - 8y_{i+2,j+1} + 8y_{i+2,j-1} - y_{i+2,j-2} - \\ -8y_{i+1,j+2} + 64y_{i+1,j+1} - 64y_{i+1,j-1} + 8y_{i+1,j-2} \end{pmatrix} \tag{12}$$

$$\frac{\partial^2 y}{\partial x^2} = \frac{-y_{i+2,j} + 16y_{i+1,j} - 30y_{i,j} + 16y_{i-1,j} - y_{i-2,j}}{12(\Delta x)^2} \tag{13}$$

$$\frac{\partial^2 y}{\partial \psi^2} = \frac{-y_{i,j+2} + 16y_{i,j+1} - 30y_{i,j} + 16y_{i,j-1} - y_{i,j-2}}{12(\Delta \psi)^2} \tag{14}$$

$$A_1 = -y_{i+2,j} + 8y_{i+1,j} - 8y_{i-1,j} + y_{i-2,j} \tag{15}$$

$$B_1 = -y_{i+2,j} + 16y_{i+1,j} + 16y_{i-1,j} - y_{i-2,j} \tag{16}$$

$$C_1 = -y_{i,j+2} + 8y_{i,j+1} - 8y_{i,j-1} + y_{i,j-2} \tag{17}$$

$$D_1 = -y_{i,j+2} + 16y_{i,j+1} + 16y_{i,j-1} - y_{i,j-2} \tag{18}$$

$$E_1 = -y_{i+2,j+2} + 8y_{i+2,j+1} - 8y_{i+2,j-1} + y_{i+2,j-2} \tag{19}$$

$$F_1 = -8y_{i+1,j+2} + 64y_{i+1,j+1} - 64y_{i+1,j-1} - 8y_{i+1,j-2} \tag{20}$$

$$G_1 = +y_{i-2,j+2} - 8y_{i-2,j+1} + 8y_{i-2,j-1} - y_{i-2,j-2} \tag{21}$$

$$H_1 = 8y_{i-1,j+2} - 64y_{i-1,j+1} + 64y_{i-1,j-1} - 8y_{i-1,j-2} \tag{22}$$

Substituting Equations (10)–(22) into Equation (2), the system of linear equations results for $y_{i,j}$ at the finite difference mesh's nodes are written as

$$y_{i,j} = \frac{6B_1\left(C_1^2\right) + 6D_1\left(144(\Delta x)^2 + A_1^2\right) - (E_1 + F_1 + G_1 + H_1)(A_1)(C_1)}{180\left(C_1^2 + A_1^2 + 144(\Delta x)^2\right)} \tag{23}$$

The convergent solution of Equation (23) is obtained by an iterative method, as the $y_{i,j}$ value is dependent on the four points surrounding the mesh. To avoid fictitious nodes beyond the computational domain, the terms for the computational node in the area of the extreme boundary section were estimated using the three-point finite difference method. Replacing the partial derivatives of Equation (2) by three-point central finite difference approximations such as [20,22,32]:

$$\frac{\partial y}{\partial x} = \frac{y_{i+1,j} - y_{i-1,j}}{2\Delta x} \tag{24}$$

$$\frac{\partial y}{\partial \psi} = \frac{y_{i,j+1} - y_{i,j-1}}{2\Delta \psi} \tag{25}$$

$$\frac{\partial^2 y}{\partial x^2} = \frac{y_{i+1,j} - 2y_{i,j} + y_{i-1,j}}{(\Delta x)^2} \quad (26)$$

$$\frac{\partial^2 y}{\partial \psi^2} = \frac{y_{i,j+1} - 2y_{i,j} + y_{i,j-1}}{(\Delta \psi)^2} \quad (27)$$

$$\frac{\partial^2 y}{\partial x \partial \psi} = \frac{y_{i+1,j+1} - y_{i-1,j+1} - y_{i+1,j-1} + y_{i-1,j-1}}{4\Delta x \Delta \psi} \quad (28)$$

$$A' = y_{i+1,j} - y_{i-1,j} \quad (29)$$

$$B' = y_{i+1,j} + y_{i-1,j} \quad (30)$$

$$C' = y_{i,j+1} - y_{i,j-1} \quad (31)$$

$$D' = y_{i,j+1} + y_{i,j-1} \quad (32)$$

$$E' = y_{i+1,j+1} - y_{i-1,j+1} - y_{i+1,j-1} + y_{i-1,j-1} \quad (33)$$

Substitution of Equations (24)–(33) into Equation (2), the system of linear equations results for $y_{i,j}$ at the nodes of the FD mesh.

$$y_{i,j} = \frac{2B'\left((C')^2\right) + 2D'\left(4(\Delta x)^2 + (A')^2\right) - E'(A')(C')}{4\left((C')^2 + (A')^2 + 4(\Delta x)^2\right)} \quad (34)$$

### 3.2. Velocity and Pressure Distributions

The hydraulic characteristics of flow at transitions with moderate to fast accelerating flow inhibiting the formation of the boundary layer are accurately predicted using potential-flow techniques [32]. After obtaining the values of $y_{i,j}$ at all grid points, the horizontal and vertical velocity fields of the interior nodes are computed from the fourth-order central discretization of Cartesian velocity components:

$$u_{i,j} = \frac{12\Delta\psi}{-y_{i,j+2} + 8y_{i,j+1} - 8y_{i,j-1} + y_{i,j-2}} \quad (35)$$

$$v_{i,j} = u_{i,j}\frac{-y_{i+2,j} + 8y_{i+1,j} - 8y_{i-1,j} + y_{i-2,j}}{12\Delta x} \quad (36)$$

Let $N$ represent the highest value of $j$ at the free surface. The horizontal velocity at the free surface and bottom are computed from the fourth-order discretization with the five-point one-sided stencil scheme:

$$u_s = \frac{12\Delta\psi}{25y_{i,N} - 48y_{i,N-1} + 36y_{i,N-2} - 16y_{i,N-3} + 3y_{i,N-4}} \quad (37)$$

$$u_b = \frac{12\Delta\psi}{-25y_{i,0} + 48y_{i,1} - 36y_{i,2} + 16y_{i,3} - 3y_{i,4}} \quad (38)$$

Let $M$ represent the highest value of $i$ at the upstream boundary. The vertical velocity at the upstream and downstream boundary sections is computed from the fourth-order discretization with a five-point one-sided stencil scheme:

$$v_d = u_{M,j}\frac{25y_{M,j} - 48y_{M-1,j} + 36y_{M-2,j} - 16y_{M-3,j} + 3y_{M-4,j}}{12\Delta x} \quad (39)$$

$$v_u = u_{0,j}\frac{-25y_{0,j} + 48y_{1,j} - 36y_{2,j} + 16y_{3,j} - 3y_{4,j}}{12\Delta x} \quad (40)$$

Once the surface profile and velocity fields are computed, the pressure $p$ is given by using the condition of energy conservation:

$$\frac{p_{i,j}}{\gamma} = H - y_{i,j} - \frac{u_{i,j}^2 + v_{i,j}^2}{2g}. \tag{41}$$

Due to the nonlinearity in the equations for boundary conditions, correcting the free surface location is likely an essential aspect of implementing this sort of numerical method [20]. When the local energy value $H(x)$ differs from the upstream energy level $H$, i.e., when $p_s/\gamma$ at the surface is not equal to zero, the flow profile must be recalculated. Hence, it must be modified iteratively until $p_s/\gamma \approx 0$ at the free surface with some tolerance [22].

The location of each free surface node was calculated using the Newton–Raphson algorithm [17]. However, this method becomes unstable at certain channel sections, making corrections time-consuming [32]. To address this issue, a perturbation procedure [22] is adopted to estimate the free-surface correction $\Delta y_s$. The details of the perturbation procedure are found in the work of Castro-Orgaz [22].

## 4. Results and Discussion

### 4.1. Flow over Gaussian Weir

The experimental measurements of the flow profile and bottom pressure distributions in flow over a symmetrical hump $y_b = 0.2 \exp[-0.5(x/0.24)^2]$ (m) by Sivakumaran et al. [27] are shown in Figure 3, where the potential flow model results are validated with the experimental observations. The results accord well with the experimental measurements. The discharge per unit width is 0.112 $m^2/s$. The positions of the upstream and the downstream boundary sections are $x$(m) = $\pm 1$. In this paper, flow is modelled using 30 streamlines. The calculated pattern of streamlines is illustrated in Figure 3a, which is the solution of the semi-inverse transformation of the Laplace equation. The computed flow profile from the steady SGN equation, computed free surface velocity distribution and computed bottom pressure distributions from potential flow solutions are displayed in Figure 3b and validated with the experimental results of Sivakumaran et al. [27]. In Figure 3b, $z$ denotes the vertical coordinate, and $y_c$ denotes the critical depth. The estimated bottom pressure distribution coincides with the free surface profile close to the upstream and downstream sections, which indicates the hydrostatic pressure distribution in these regions. At the region just upstream of the weir, the streamlines have convex curvature as the flow accelerates towards the crest, leading to a decrease in bed pressure. At the crest of the weir, the radius of curvature is small, causing more pressure drop in the region. At the downstream region from $x/y_c = 3$ to $x/y_c = 6$, the streamline follows a concave path, and the radius of curvature is larger, causing the pressure rise in the region, as shown in Figure 3b. The flow surface is converging marginally at the upstream of the weir; hence, flow velocity remains constant till the crest of the weir.

The computed horizontal velocity distributions plotted at eight representative sections of the Gaussian weir are shown in Figure 4a. In the figure, $\eta$ denotes the vertical coordinate above the bed, and $U_c$ is the critical velocity. The calculated values of $u$ velocity agree well with the velocity profiles obtained from the Boussinesq approximation of Castro-Orgaz and Hager [39]. In the sections $x = -0.4$ m to $x = -0.1$ m, the velocity profiles are non-uniform. The $u$ velocity profile changes rapidly within the transcritical flow over the weir crest. The $u$ velocity is minimum at the surface and increases towards the bed at the crest of the weir. The shape of the velocity profile remains unchanged till $x = 0.2$ m. At the section $x = 0.3$ m, the velocity distribution is almost uniform in the $y$-direction. The computed vertical velocity distributions plotted at eight representative sections of the Gaussian weir are shown in Figure 4b. The calculated $v$ velocities are in good agreement with the velocity components originating from the Picard iteration [39]. The velocity profile is non-uniform with a value close to zero at the surface and increases to a positive value at the bed in the section $x = -0.4$ m to $x = -0.1$ m. The $v$ velocity is non-uniform at the weir crest with a value close to zero in the bed and decreases to a negative value on the surface. In the section $x = 0.1$ m to $x = 0.3$ m, the v velocity profile is nearly uniform with a negative value. The computed pressure distributions plotted at nine representative sections of the Gaussian weir are shown in Figure 4c. The effect of streamline curvature on pressure distribution using the Gaussian weir is observed in this study. At sections $x = -0.4$ m to $x = -0.2$ m, the pressure distribution is hydrostatic where the water surface coincides with the bottom pressure distributions. However, at section $x = -0.1$ m, the pressure profile is below hydrostatic. At the crest section, the pressure profile is below hydrostatic because centrifugal acceleration induced by convex streamline curvature acts upward, opposite to gravity. In the sections $x = 0.1$ m and $x = 0.2$ m, the pressure profile is still below hydrostatic because of the presence of convex streamline curvature. At the section $x = 0.3$ m, the pressure profile is just above hydrostatic. At section $x = 0.5$ m, the pressure profile is above hydrostatic because centrifugal effects induced by concave streamline curvature act downward, in the direction of gravity.

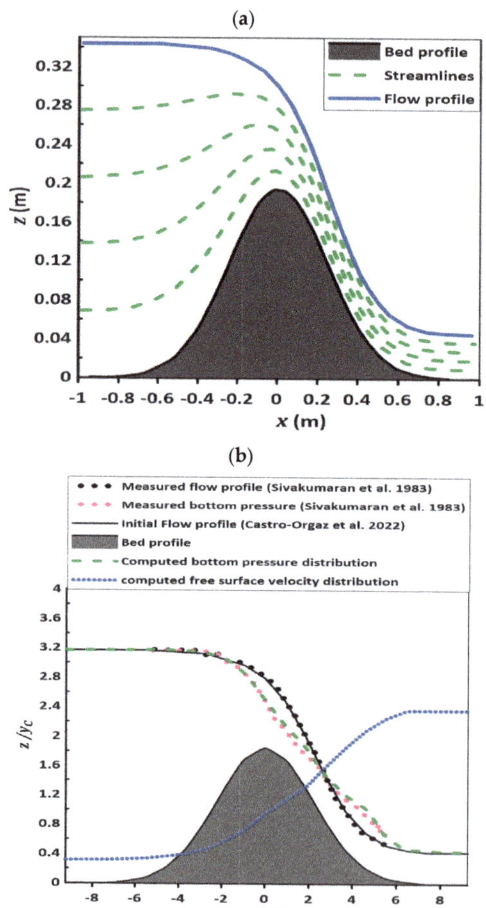


**Figure 3.** Flow over Gaussian weir of Sivakumaran et al. [27]. (**a**) Computed streamlines. (**b**) Computed flow profile [24], surface velocity and bottom-pressure distributions, validated with experimental measurements.

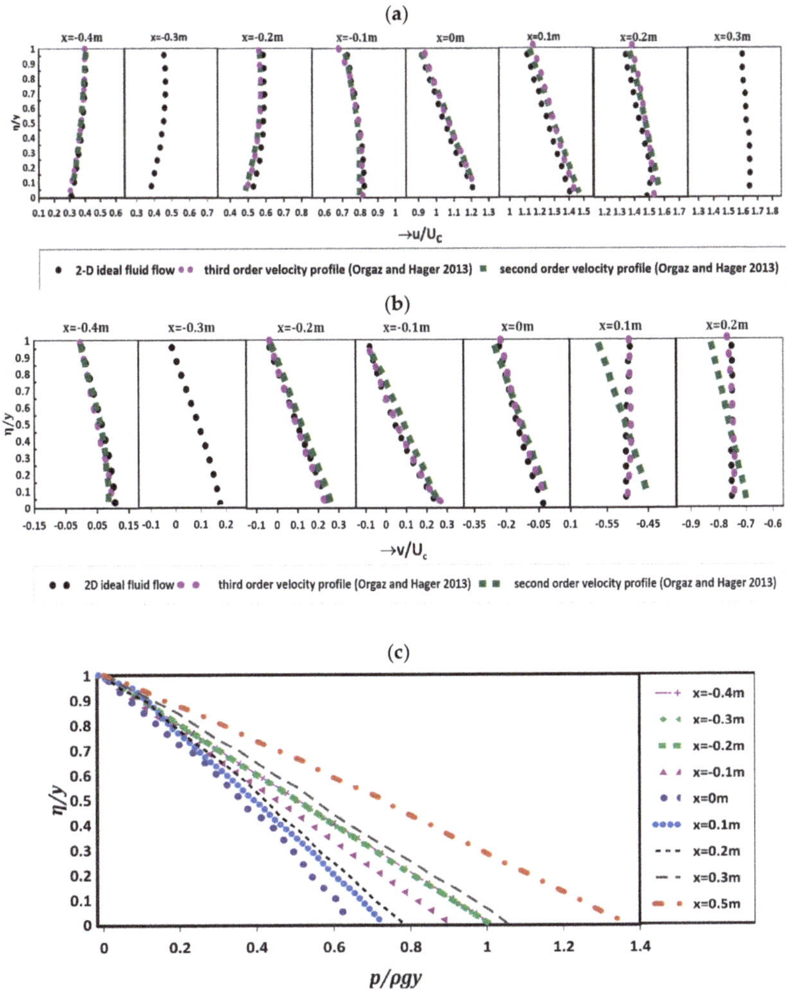


**Figure 4.** Flow features along the Gaussian weir domain computed with 2D fluid flow solution and validated with the results of Castro-Orgaz and Hager [39]: (**a**) horizontal velocity distributions; (**b**) vertical velocity distributions; (**c**) pressure distributions.

## *4.2. Flow over Parabolic Weir*

The experimental measurements of the flow profile and bottom pressure distributions in flow over a parabolic weir $y_b = 0.3216 - 0.54525x^2$ (m) by Blau [40] are shown in Figure 5; also, the findings of the potential flow model are validated with the measured observations, as shown in the figure. The results accord well with the experimental measurements. The discharge per unit width is 0.51 $m^2/s$. The position of the upstream and the downstream boundary sections are $x$ (m) = $\pm 1.5$. The calculated pattern of streamlines is presented in Figure 5a, which is the solution of the semi-inverse transformation of the Laplace equation. The estimated flow profile from the SGN equation, computed free surface velocity distribution and computed bottom pressure distributions from the potential flow model are displayed in Figure 5b and validated with the experimental results of Blau [40]. The bottom pressure distribution matches the free surface profile near the sections both upstream and downstream. This indicates that the hydrostatic pressure distribution is present in these areas. At the section just upstream of the weir, the streamlines exhibit a convex curvature as the flow accelerates towards the crest, leading to a decrease in the bed pressure. At the weir's crest, where the radius of curvature is small, the pressure drop is more significant. At the downstream section, the streamline follows a concave path, and the radius of curvature is larger, causing the pressure rise in that region as shown in Figure 5b.

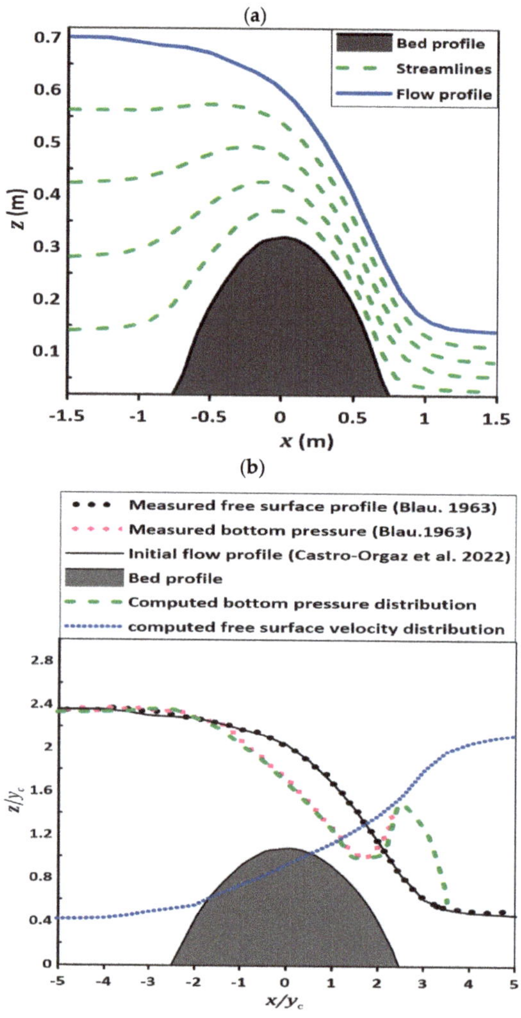


**Figure 5.** Flow over parabolic weir of Blau [40]. (**a**) Computed streamlines; (**b**) computed flow profile [24], surface velocity and bottom-pressure distributions, validated with experimental measurements.

The computed horizontal velocity distributions plotted at eight representative sections of the parabolic weir are shown in Figure 6a. In the sections $x = -0.75$ m and $x = -0.45$ m, the velocity distribution is non-uniform, as shown in the figure. At the section $x = -0.3$ m, the $u$ velocity profile tends to be constant. The streamwise velocity distribution changes rapidly within the transcritical flow over the weir's crest. At the weir crest, the $u$ velocity is minimal at the surface and increases towards the bed. The shape of the velocity profile remains unchanged for $x = 0.15$ m and $x = 0.45$ m. At the section $x = 0.6$ m, the velocity profile attains a maximum value at some distance above the bed. At the section $x = 0.75$ m, the velocity profile varies with the maximum value at the surface and decreases towards the bed. The calculated vertical velocity distributions plotted at eight representative sections of the parabolic weir are shown in Figure 6b. At section $x = -0.75$ m to $x = -0.3$ m, the velocity profile is non-uniform with a value close to zero at the surface and increases to a positive value at the bed, as shown in the figure. The $v$ velocity is non-uniform from the weir crest to $x = 0.15$ m with a value close to zero at the bed and decreases to a negative value on the surface. From the sections $x = 0.45$ m to $x = 0.75$ m, the $v$ velocity profile is nearly uniform with a negative value. The computed pressure distributions plotted at eight representative sections of the parabolic weir are shown in Figure 6c. The curvature of the streamlines' influence on the vertical pressure profile is demonstrated. In the reach between $x = -0.75$ m and $x = -0.45$ m, the pressure profile is hydrostatic, where the water surface coincides with the bottom pressure distributions. At section $x = -0.3$ m, the pressure distribution is just below hydrostatic. At the crest, the pressure profile is below hydrostatic because centrifugal acceleration induced by convex streamline curvature acts upward, in the opposite direction of gravity. In the sections $x = 0.15$ m and $x = 0.6$ m, the pressure profile is still below hydrostatic because of convex streamline curvature. At section $x = 0.75$ m, the pressure profile is above hydrostatic because centrifugal effects induced by concave streamline curvature act downward, in the direction of gravity.

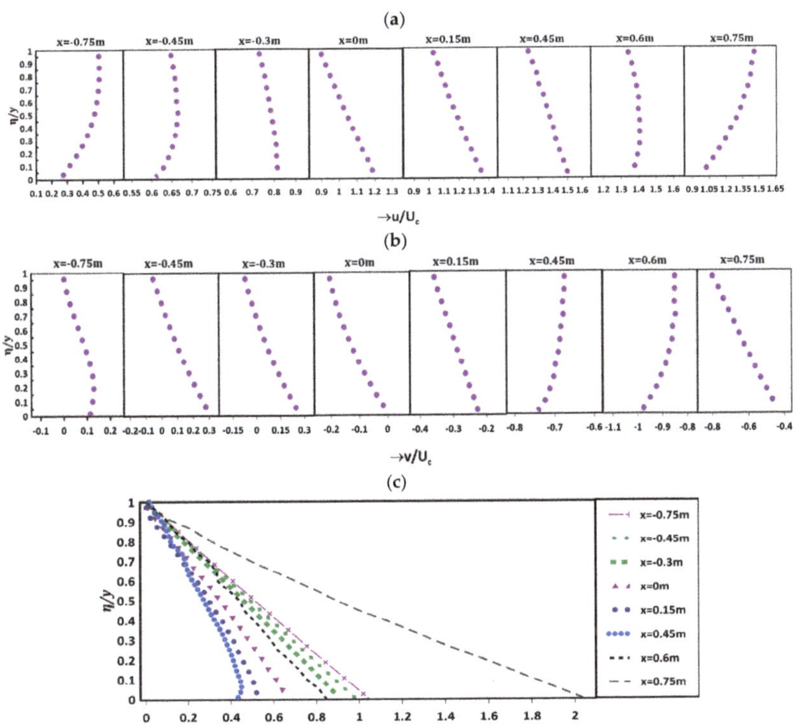


**Figure 6.** Flow features along the parabolic weir domain computed with 2D fluid flow solution: (**a**) horizontal velocity distributions; (**b**) vertical velocity distributions; (**c**) pressure distributions.

## *4.3. Flow over Semi-Circular Weir*

The experimental measurements of the flow profiles and velocity distributions over a semicircular weir were taken in a tilting flume, set at an angle of 0.115 degrees as shown in Figure 7. The flume was 4.87 m long, 0.08 m wide, and 0.25 m high with bed and walls made of transparent Perspex. The base of the weir was attached to the flume, and the position was marked. Finally, water flow was given over the semicircular weir. The discharge per unit width is 0.022875 $m^2/s$. The positions of the upstream and the downstream boundary sections are $x$ (m) = $\pm 0.1$. The weir is located at a distance of 1.63 m from the downstream end of the flume. During the experiment, the water surface depths at respective sections are taken with the help of a point gauge. Velocity measurements are taken with the help of pitot tube arrangements. Flow is modelled using 30 streamlines. The calculated streamline flow pattern is presented in Figure 8a, which is the solution of the semi-inverse transformation of the Laplace equation. The estimated surface velocity, measured flow profile, and computed bottom pressure profiles from potential flow solutions are presented in Figure 8b. As shown in Figure 8b the pressure distribution coincides with the free surface flow, indicating the hydrostatic condition in the upstream section. The pressure drop at the weir crest is significantly greater because of strong streamline curvature. A large pressure rise is observed in the downstream section where the flow follows the concave streamline.

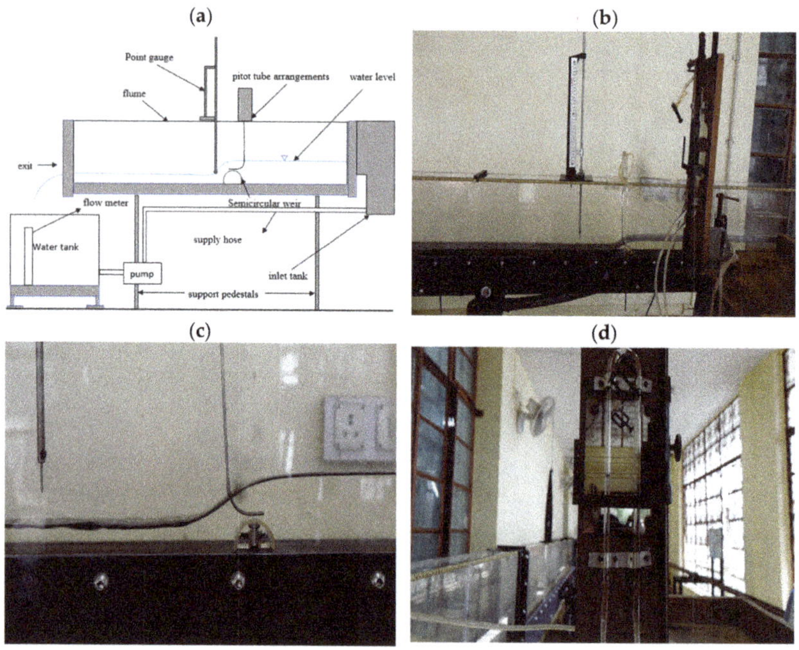


**Figure 7.** (**a**) Schematic layout of the experimental setup. (**b**) Hydraulic flume with point gauge and pitot tube arrangements. (**c**) The semicircular weir in the experiment. (**d**) Pitot tube arrangements for velocity readings.

The computed horizontal velocity distributions plotted at four representative sections of the semicircular weir and compared with the experiments are depicted in Figure 9a. The calculated values agree well with the experimental measurements. At section $x = -0.1$ m, the velocity distribution is non-uniform, as shown in Figure 9a. The streamwise velocity profile changes rapidly within the transcritical flow over the weir crest. At the weir crest, the $u$ velocity is minimum at the surface and increases towards the bed. At section $x = 0.02$ m, the velocity profile changes rapidly, attaining the maximum value around mid-depth. At section $x = 0.06$ m, the estimated velocity profile tends to be uniform; however, the velocity decreases towards the bed in the experimental measurement due to frictional resistance offered to flow by the channel boundary. The computed vertical velocity distributions plotted at four representative sections of the semicircular weir are shown in Figure 9b. At section $x = -0.1$ m, the vertical velocity profile is uniform with a value close to zero, as shown in Figure 9b. The $v$ velocity is non-uniform at the crest of the weir with a value close to zero in the bed and decreases to a negative value on the surface. At the section $x = 0.02$ m, the $v$ velocity profile is non-uniform with a negative value throughout the depth. At section $x = 0.06$ m, the v velocity is non-uniform with a value close to zero at the bed and decreases to a negative value on the surface. The computed pressure distributions plotted at six representative sections of the parabolic weir are shown in Figure 9c. The influence of streamline curvature on pressure distribution is demonstrated in Figure 9c. At section $x = -0.1$ m to $x = -0.02$ m, the pressure profile is hydrostatic, where the free surface coincides with the bottom pressure distributions as illustrated in Figure 9c. The pressure profile is below hydrostatic at section $x = -0.01$ m. At the crest, the pressure profile is below hydrostatic because centrifugal acceleration induced by convex streamline curvature acts upward, in the direction of gravity. At section $x = 0.02$ m, the pressure profile is above hydrostatic because centrifugal effects caused by concave streamline curvature act downwards, in the direction of gravity.

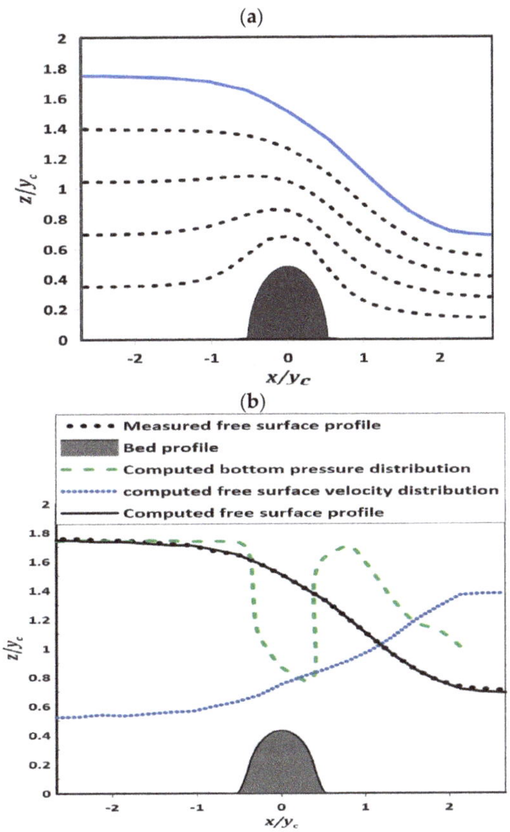


**Figure 8.** Flow over semi-circular weir: (**a**) Computed streamlines; (**b**) Computed flow surface velocity and bottom-pressure distributions, validated with experimental measurements.

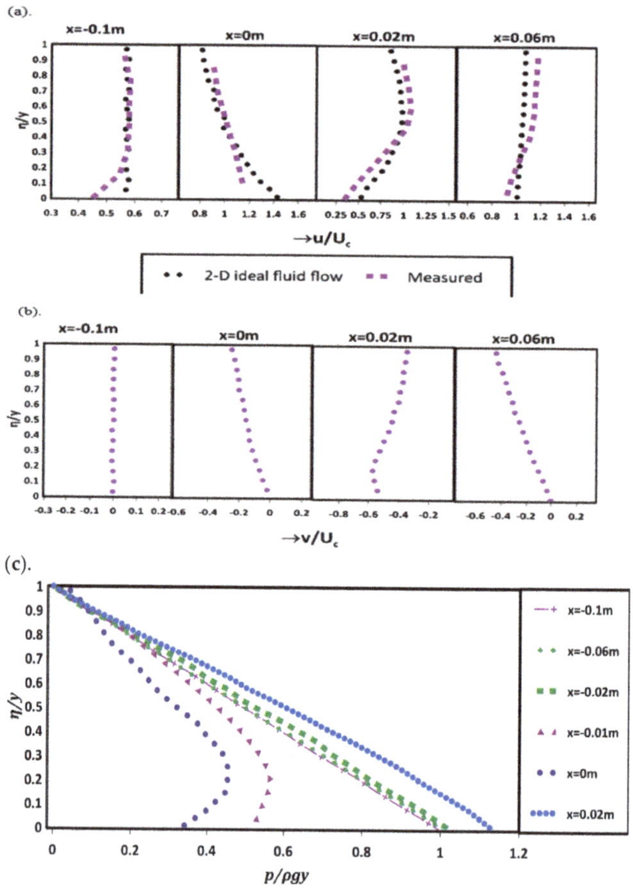


**Figure 9.** Flow features along the semi-circular weir domain computed with 2D fluid flow solution and compared with experimental measurements: (**a**) horizontal velocity distributions; (**b**) vertical velocity distributions; (**c**) pressure distributions.

## 5. Summary and Conclusions

The present study differs from the previous potential flow model by Montes [20,32] and Castro-Orgaz [22] in the following ways: (1) The present study used the Serre–Green–Naghdi equation to determine the initial water surface profile. (2) The five-point central finite difference scheme was used to discretize the partial derivative terms of the Laplace equation proposed by Boadway [19] in the interior nodes to reduce truncation errors. (3) A semicircular weir experiment was conducted to validate the pressure and velocity profiles obtained using the proposed 2-D fluid flow model.

A potential flow model of $x$-$\psi$ transformation was applied to calculate the patterns of streamlines, velocity, and pressure distributions for the selected test cases, such as Gaussian weir, parabolic weir, and semicircular weir. A steady version of the Serre–Green–Naghdi equation was used to determine the initial free-surface profile. The partial derivative terms of Laplace equations and Cartesian velocity coordinates were replaced with fourth-order accurate finite difference approximations to precisely estimate the streamline positions, thereby calculating the pressure and velocity distributions accurately. The computed flow profiles and bottom pressure distributions of Gaussian and Parabolic weirs match well with the previous experimental observations. The 2-D ideal fluid flow solution was also used to represent velocity and pressure distributions along the cross section at selected sections of all three weirs in detail. The effect of the streamline curvature on pressure distributions was observed for all three weirs discussed in the study. The rise or drop in the bed pressure depends on the nature of free surface curvature. An experiment on flow over a semicircular weir accurately predicted the shift in the velocity within the transcritical flow over the weir and the influence of the streamline curvature on pressure distributions. The computed velocity value for the semi-circular weir close to the bed slightly differed from the experimental velocity profile. Hence, the decrease in velocity near the bed was due to the boundary resistance offered to flow at the bottom.

**Author Contributions:** M.R.A.K.: writing—original draft preparation, writing—review and editing, and data curation; P.R.H.: writing—original draft preparation, writing—review and editing, funding acquisition, project administration, data curation, and supervision; J.H.P.: writing—review and editing, and data curation. All authors contributed to the work. All authors have read and agreed to the published version of the manuscript.

**Funding:** This research did not receive any specific grant from funding agencies in the public, commercial, or not-for-profit sectors.

**Data Availability Statement:** Data of this manuscript will be made available on reasonable request to corresponding authors.

**Acknowledgments:** For the purposes of open access, the authors have applied a Creative Commons Attribution (CC BY) license to any author-accepted manuscript version arising from this submission.

**Conflicts of Interest:** The authors declare no conflicts of interest.

## Appendix A

The steady SGN for 1D water waves travelling across any topography [31,36]:

$$\frac{d}{dx}\left[\frac{y^2}{2}+\frac{q^2}{gy}\left(1+\frac{y(d^2y/dx^2)-(dy/dx)^2}{3}+\frac{y(d^2y_b/dx^2)}{2}-\frac{(dy/dx)(dy_b/dx)}{2}\right)\right] = -\left[y+\frac{q^2}{2gy^2}\left(2y(d^2y_b/dx^2)+y(d^2y/dx^2)-(dy/dx)^2-2(dy_b/dx)(dy/dx)\right)\right]\frac{dy_b}{dx}$$

Applying the derivative to the left side terms,

$$y\left(\frac{dy}{dx}\right)-\frac{q^2}{gy^2}\left(\frac{dy}{dx}\right)+\frac{q^2}{3gy^2}\left[\left(y^2\frac{d^3y}{dx^3}\right)-\left(2y\frac{dy}{dx}\frac{d^2y}{dx^2}\right)+\left(\frac{dy}{dx}\right)^3\right]+\frac{q^2}{2g}\left[\left(\frac{d^3y_b}{dx^3}\right)\right]-\frac{q^2}{2gy^2}\left[\left(y\frac{d^2y}{dx^2}\right)-\left(\frac{dy}{dx}\right)^2\right]\frac{dy_b}{dx} -\frac{q^2}{2gy}\left[\left(\frac{dy}{dx}\frac{d^2y_b}{dx^2}\right)\right] = -y\left(\frac{dy_b}{dx}\right)-\frac{q^2}{2gy^2}\left[\left(y\frac{d^2y}{dx^2}\right)-\left(\frac{dy}{dx}\right)^2\right]\frac{dy_b}{dx}-\frac{q^2}{gy^2}\left[\left(y\frac{d^2y_b}{dx^2}\right)-\left(\frac{dy}{dx}\right)\frac{dy_b}{dx}\right]\frac{dy_b}{dx}$$

Simplifying the equation to

$$y\left(\frac{dy}{dx}\right)+y\left(\frac{dy_b}{dx}\right)-\frac{q^2}{gy^2}\left(\frac{dy}{dx}\right)+\frac{q^2}{3gy^2}\left[\left(y^2\frac{d^3y}{dx^3}\right)-\left(2y\frac{dy}{dx}\frac{d^2y}{dx^2}\right)+\left(\frac{dy}{dx}\right)^3\right]+\frac{q^2}{2gy}\left[y\left(\frac{d^3y_b}{dx^3}\right)-\left(\frac{dy}{dx}\frac{d^2y_b}{dx^2}\right)\right] +\frac{q^2}{gy^2}\left[\left(y\frac{d^2y_b}{dx^2}\right)-\left(\frac{dy}{dx}\right)\frac{dy_b}{dx}\right]\frac{dy_b}{dx}=0 \quad \text{(A1)}$$

The Bernoulli-type Boussinesq equation [28] to estimate initial free surface flow is given as

$$H = y_b + y + \frac{q^2}{2gy^2}\left(1+\frac{2y(d^2y/dx^2)-(dy/dx)^2}{3}+y\left(d^2y_b/dx^2\right)+(dy_b/dx)^2\right) \quad \text{(A2)}$$

Equation (A2) is the first integral of Equation (A1), where $H$ is the constant of integration of Equation (A1). This is verified by differentiating Equation (A2) once with respect to $x$

$$y\left(\frac{dy}{dx}\right)+y\left(\frac{dy_b}{dx}\right)-\frac{q^2}{gy^2}\left(\frac{dy}{dx}\right)+\frac{q^2}{6gy^3}\left[\left(2y^3\frac{d^3y}{dx^3}\right)-4\left(y^2\frac{dy}{dx}\frac{d^2y}{dx^2}\right)+2y\left(\frac{dy}{dx}\right)^3\right]+\frac{q^2}{2g}\left[\left(\frac{d^3y_b}{dx^3}\right)\right]$$
$$-\frac{q^2}{2gy}\left[\left(\frac{dy}{dx}\frac{d^2y_b}{dx^2}\right)\right]+\frac{q^2}{2gy^2}\left[\left(y\frac{d^2y_b}{dx^2}\right)-\left(\frac{dy}{dx}\right)\frac{dy_b}{dx}\right]2\frac{dy_b}{dx}=0$$

It is written as

$$\left(\frac{dy}{dx}\right)+y\left(\frac{dy_b}{dx}\right)-\frac{q^2}{gy^2}\left(\frac{dy}{dx}\right)+\frac{q^2}{3gy^2}\left[\left(y^2\frac{d^3y}{dx^3}\right)-\left(2y\frac{dy}{dx}\frac{d^2y}{dx^2}\right)+\left(\frac{dy}{dx}\right)^3\right]+\frac{q^2}{2gy}\left[y\left(\frac{d^3y_b}{dx^3}\right)-\left(\frac{dy}{dx}\frac{d^2y_b}{dx^2}\right)\right]$$
$$+\frac{q^2}{gy^2}\left[\left(y\frac{d^2y_b}{dx^2}\right)-\left(\frac{dy}{dx}\right)\frac{dy_b}{dx}\right]\frac{dy_b}{dx}=0 \quad \text{(A3)}$$

Hence Equation (A3) is similar to Equation (A1). Hence, Equation (A2) is the first integral of Equation (A1).

## Appendix B

The steady SGNE for 1D water waves travelling across any topography [31,36]:

$$\frac{d}{dx}\left[\frac{y^2}{2}+\frac{q^2}{gy}\left(1+\frac{y(d^2y/dx^2)-(dy/dx)^2}{3}+\frac{y(d^2y_b/dx^2)}{2}-\frac{(dy/dx)(dy_b/dx)}{2}\right)\right]$$
$$=-\left[y+\frac{q^2}{2gy^2}\left(2y(d^2y_b/dx^2)+y(d^2y/dx^2)-(dy/dx)^2-2(dy_b/dx)(dy/dx)\right)\right]\frac{dy_b}{dx}$$

To simplify the equation, neglecting the second-order products of derivatives of $y$ and $y_b$,

$$\frac{d}{dx}\left[\frac{y^2}{2}+\frac{q^2}{gy}\left(1+\frac{y(d^2y/dx^2)}{3}+\frac{y(d^2y_b/dx^2)}{2}\right)\right]=-(y)\frac{dy_b}{dx}$$

Differentiating the left side term with respect to x,

$$y(dy/dx)-\frac{q^2}{gy^2}(dy/dx)+\frac{q^2}{3g}\left(d^3y/dx^3\right)+\frac{q^2}{2g}\left(d^2y_b/dx^2\right)=y(dy_b/dx)$$

Then, one obtains the relation

$$\frac{q^2}{g}\left(\frac{1}{3}\left(d^3y/dx^3\right)+\frac{1}{2}\left(d^2y_b/dx^2\right)\right)+\left(y-\frac{q^2}{gy^2}\right)(dy/dx)=-y(dy_b/dx)$$

## Appendix C

We aim to approximate the mixed second derivative $\frac{\partial^2 y}{\partial x \partial y}$ at the point $(x_i, \psi_j)$ using central differences with five points in each direction.

### *Appendix C.1. Central Difference Approximation for First Derivative*

For the first derivative in the $x$ direction, centred at $(x_i, \psi_j)$, using central differences with five points, we have

$$\frac{\partial y}{\partial x}=\frac{-y_{i+2,j}+8y_{i+1,j}-8y_{i-1,j}+y_{i-2,j}}{12\Delta x}$$

Similarly, for the first derivative in the $\psi$ direction, centered at $(x_i, \psi_j)$, we have

$$\frac{\partial y}{\partial \psi}=\frac{-y_{i,j+2}+8y_{i,j+1}-8y_{i,j-1}+y_{i,j-2}}{12\Delta\psi}$$

*Appendix C.2. Mixed Partial Derivative Approximation*

We now apply these central difference formulas to compute the mixed partial derivative. To approximate $\frac{\partial^2 y}{\partial x \partial \psi}$, we differentiate the central difference formula for $\frac{\partial y}{\partial x}$ with respect to $\psi$:

$$\frac{\partial}{\partial \psi}\left(\frac{\partial y}{\partial x}\right) = \frac{1}{12\psi}\left(\frac{\begin{array}{c} -\left(-y_{i+2,j+2} + 8y_{i+1,j+2} - 8y_{i-1,j+2} + y_{i-2,j+2}\right) + \\ 8\left(-y_{i+2,j+1} + 8y_{i+1,j+1} - 8y_{i-1,j+1} + y_{i-2,j+1}\right) - \\ -8\left(-y_{i+2,j-1} + 8y_{i+1,j-1} - 8y_{i-1,j-1} + y_{i-2,j-1}\right) + \\ -y_{i+2,j-2} + 8y_{i+1,j-2} - 8y_{i-1,j-2} + y_{i-2,j-2} \end{array}}{12\Delta x}\right)$$

Simplifying, we obtain

$$\frac{\partial^2 y}{\partial x \partial \psi} = \frac{1}{144\Delta x \Delta \psi}\begin{pmatrix} -y_{i-2,j+2} + 8y_{i-2,j+1} - 8y_{i-2,j-1} + y_{i-2,j-2} + \\ 8y_{i-1,j+2} - 64y_{i-1,j+1} + 64y_{i-1,j-1} - 8y_{i-1,j-2} \\ +y_{i+2,j+2} - 8y_{i+2,j+1} + 8y_{i+2,j-1} - y_{i+2,j-2} - \\ -8y_{i+1,j+2} + 64y_{i+1,j+1} - 64y_{i+1,j-1} + 8y_{i+1,j-2} \end{pmatrix}$$

## References


1. Vo, N.D. *Characteristics of Curvilinear Flow Past Circular-Crested Weir*; Concordia University: Montreal, QB, Canada, 1992.
2. Vallentine, H.R. *Applied Hydrodynamics*; Butterworths Scientific Publications: London, UK, 1969.
3. Chanson, H. Minimum Specific Energy and Critical Flow Conditions in Open Channels. *J. Irrig. Drain. Eng.* **2006**, *132*, 498–502. [CrossRef]
4. Castro-Orgaz, O.; Chanson, H. Depth-Averaged Specific Energy in Open-Channel Flow and Analytical Solution for Critical Irrotational Flow over Weirs. *J. Irrig. Drain. Eng.* **2013**, *140*, 04013006. [CrossRef]
5. Tadayon, R.; Ramamurthy, A.S. Turbulence Modeling of Flows over Circular Spillways. *J. Irrig. Drain. Eng.* **2009**, *135*, 493–498. [CrossRef]
6. Rouse, H. *Fluid Mechanics for Hydraulic Engineers*; McGraw-Hill: New York, NY, USA, 1938.
7. Pu, J.H.; Huang, Y.; Shao, S.; Hussain, K. Three-Gorges Dam Fine Sediment Pollutant Transport: Turbulence SPH Model Simulation of Multi-Fluid Flows. *J. Appl. Fluid Mech.* **2016**, *9*, 1–10. [CrossRef]
8. Pu, J.H.; Cheng, N.S.; Tan, S.K.; Shao, S. Source Term Treatment of SWEs Using Surface Gradient Upwind Method. *J. Hydraul. Res.* **2012**, *50*, 145–153. [CrossRef]
9. Pu, J.H. Turbulence Modelling of Shallow Water Flows Using Kolmogorov Approach. *Comput. Fluids* **2015**, *115*, 66–74. [CrossRef]
10. Zheng, X.G.; Pu, J.H.; Chen, R.D.; Liu, X.N.; Shao, S.D. A Novel Explicit-Implicit Coupled Solution Method of SWE for Long-Term River Meandering Processinduced by Dambreak. *J. Appl. Fluid Mech.* **2016**, *9*, 2647–2660. [CrossRef]
11. Messe, P. Ressaut et Ligne D'eau Dans Les Cours à Pente Variable [Hydraulic Jump and Free Surface Profile in Channels of Variable Slope]. *Rev. Gén. Hydr.* **1938**, *4*, 7–11.
12. Andersen, V.M. Transition from Subcritical to Supercritical Flow. *J. Hydraul. Res.* **1975**, *13*, 227–238. [CrossRef]
13. Southwell, R.V.; Vaisey, G. Relaxation Methods Applied to Engineering Problems XII: Fluid Motions Characterized by Free Streamlines. *Philos. Trans. R. Soc. Lond.* **1946**, *240*, 6. [CrossRef]
14. Markland, E. Calculation of Flow at a Free Overfall by Relaxation Methods. *Proc. Inst. Civ. Eng.* **1965**, *31*, 71–78. [CrossRef]
15. Thom, A.; Apelt, C. *Field Computations in Engineering and Physics*; Van Nostrand: London, UK, 1961.
16. Chan, S.T.; Larock, B.E.; Herrmann, L.R. Herrmann Free Surface Ideal Fluid Flows by Finite Element Methods. *J. Hydraul. Div.* **1973**, *99*, 959–974. [CrossRef]
17. Cheng, A.H.; Liu, P.L.; Liggett, J.A. Liggett Boundary Calculations of Sluice and Spillway Flows. *J. Hydraul. Div.* **1981**, *107*, 1163–1178. [CrossRef]
18. Chow, W.L.; Han, T. Inviscid Solution for the Problem of Free Overfall. *ASME J. Appl. Mech.* **1979**, *46*, 1–5. [CrossRef]
19. Boadway, J.D. Transformation of Elliptical Partial Differential Equations for Solving Two-Dimensional Boundary Value Problems in Fluid Flow Overfall. *Int. J. Numer. Methods Eng.* **1976**, *10*, 527–533. [CrossRef]
20. Montes, J.S. A Potential Flow Solution for the Free. *Proc. Inst. Civ. Eng.* **1992**, *96*, 259–266. [CrossRef]
21. Cassidy, J.J. Irrotational Flow Over Spillways of Finite Height. *J. Eng. Mech. Div.* **1965**, *91*, 155–173. [CrossRef]

22. Castro-Orgaz, O. Potential Flow Solution for Open-Channel Flows and Weir-Crest Overflow. *J. Irrig. Drain. Eng.* **2013**, *139*, 551–559. [CrossRef]
23. Chanson, H.; Montes, J. Overflow Characteristics of Circular Weirs: Effect of Inflow Conditions. *J. Irrig. Drain. Eng.* **1998**, *124*, 152–162. [CrossRef]
24. Castro-Orgaz, O.; Hager, W.H.; Cantero-Chinchilla, F.N. Shallow Flows over Curved Beds: Application of the Serre–Green–Naghdi Theory to Weir Flow. *J. Hydraul. Eng.* **2022**, *148*, 04021053. [CrossRef]
25. Dressler, R.F. New Nonlinear Shallow-Flow Equations with Curvature. *J. Hydraul. Res.* **1978**, *16*, 205–222. [CrossRef]
26. Boussinesq, J. *Essai Sur La Theorie Des Eaux Courantes*; Memoires Presentes Par Divers Savants a l'Academie Des Sciences; Imprimerie Nationale: Paris, France, 1877; pp. 1–680. Available online: https://gallica.bnf.fr/ark:/12148/bpt6k56673076 (accessed on 11 August 2024).
27. Sivakumaran, N.S.; Tingsanchali, T.; Hosking, R.J. Steady Shallow Flow over Curved Beds. *J. Fluid Mech.* **1983**, *127*, 469–487. [CrossRef]
28. Naghdi, P.M.; Vongsarnpigoon, L. The Downstream Flow beyond an Obstacle. *J. Fluid Mech* **1986**, *162*, 223–236. [CrossRef]
29. Matthew, G.D. Higher Order, One-Dimensional Equations of Potential Flow in Open Channels. *Proc.—Inst. Civ. Eng. Part 2 Res. Theory* **1991**, *91*, 187–201. [CrossRef]
30. Zhu, D.Z.; Lawrence, G.A. Non-Hydrostatic Effects in Layered Shallow Water Flows. *J. Fluid Mech.* **1998**, *355*, 1–16. [CrossRef]
31. Castro-Orgaz, O.; Cantero-Chinchilla, F.N. *Non-Linear Shallow Water Flow Modelling over Topography with Depth-Averaged Potential Equations*; Springer: Amsterdam, The Netherlands, 2020; Volume 20, ISBN 0123456789.
32. Montes, J.S. Potential-Flow Solution to 2D Transition from Mild to Steep Slope. *J. Hydraul. Eng.* **1994**, *120*, 601–621. [CrossRef]
33. Wei, G.; Kirby, J.T.; Grilli, S.T.; Subramanya, R. A Fully Nonlinear Boussinesq Model for Surface Waves. Part 1. Highly Nonlinear Unsteady Waves. *J. Fluid Mech.* **1995**, *294*, 71–92. [CrossRef]
34. Brocchini, M. A Reasoned Overview on Boussinesq-Type Models: The Interplay between Physics, Mathematics and Numerics. *Proc. R. Soc. A Math. Phys. Eng. Sci.* **2013**, *469*, 20130496. [CrossRef] [PubMed]
35. Hosoda, T. Akhide Tada Free Surface Profile Analysis of Open Channel Flows by Means of 1-D Basic Equations with Effect of Vertical Acceleration. *Proc. Hydraul. Eng.* **1994**, *38*, 457–462. [CrossRef]
36. Onda, S.; Hosoda, T. Numerical Simulation on Development Process of Dunes and Flow Resistance. *Proc. Hydraul. Eng.* **2004**, *48*, 973–978. [CrossRef]
37. Komai, Y. Basic Characteristics on the Response of Water Surface Profiles of Open Channel Flows over an Obstacle. Bachelor's Thesis, Undergraduate School of Kyoto University, Kyoto, Japan, 2006.
38. Abramowitz, M.; Stegun, I.A. *Handbook of Mathematical Functions with Formulas, Graphs, and Mathematical Tables*, 10th ed.; Wiley: New York, NY, USA, 1972.
39. Castro-Orgaz, O.; Hager, W.H. Velocity Profile Approximations for Two-Dimensional Potential Open Channel Flow. *J. Hydraul. Res.* **2013**, *51*, 645–655. [CrossRef]
40. Blau, E. *Der Abfluss Und Die Hydraulische Energieverteilung Über Einer Parabelförmigen Wehrschwelle (Distributions of Discharge and Energy over a Parabolic-Shaped Weir)*; Mitteilungen der Forschungsanstalt für Schiffahrt, Wasser- und Grundbau: Berlin, Germany, 1963.

*fluids*

MDPI

*Article*

# Turbulent Micropolar Open-Channel Flow

George Sofiadis [1,*], Antonios Liakopoulos [1], Apostolos Palasis [1,2] and Filippos Sofos [2,*]

[1] Hydromechanics and Environmental Engineering Laboratory, Department of Civil Engineering, University of Thessaly, Pedion Areos, 38334 Volos, Greece; aliakop@uth.gr (A.L.); appalasis@uth.gr (A.P.)
[2] Condensed Matter Physics Laboratory, Department of Physics, University of Thessaly, 35100 Lamia, Greece
* Correspondence: sofiadis@uth.gr (G.S.); fsofos@uth.gr (F.S.)

**Abstract:** The present paper focuses on the investigation of the turbulent characteristics of an open-channel flow by employing the micropolar model. The underlying model has already been proven to correctly describe the secondary phase of turbulent wall-bounded flows. The open-channel case comprises an ideal candidate to further test the micropolar model as many environmental flows carry a secondary phase, the behavior of which is of great interest for applications such as sedimentation transport and debris flow. Direct Numerical Simulations (DNSs) have been carried out on an open channel for $Re_b = 11{,}200$ based on mean crossectional velocity, channel height, and the fluid kinematic viscosity. The simulated results are compared against previous experimental as well as Langrangian DNS data of similar flows, with excellent agreement. The micropolar model is capable of describing the same problem but in an Eulerian frame, thus significantly simplifying the computational cost and complexity.

**Keywords:** open-channel flows; Direct Numerical Simulations (DNSs); micropolar fluids; sediment transport; non-Newtonian flows

## 1. Introduction

Open-channel particulate flows is a demanding field of research which, similarly to many natural processes, requires significant effort to investigate numerically. The complexity of the problem lies in the effort to establish a multiparametric model to accurately describe as many of its physical counterparts as possible. This general type of flow applies to several cases of environmental and industrial flows. Typical examples of open-channel particulate flows include river flows, which, in turn, entail sediment transport or industrial processes and the transportation of granular materials. As can be easily understood from these examples, particle deposition as well as the general behavior of the secondary phase plays a crucial role in the physical development of the flow.

Particle interactions with the flow, especially in the vicinity of the wall, can be very difficult to model. In many cases, the behavior of small scale turbulent structures may be altered, leading to phenomena such as turbulence enhancement or attenuation. Particle–turbulence interactions in an open-channel flow have been the main subject of numerous experimental and numerical studies in recent decades, with the results still being controversial. Early experimental studies have pointed out the importance of the interactions occurring between particles and turbulence structures in the near-wall region [1–3].

In the original experiments of Rashidi et al. [1], two different particle sizes were considered, where different effects were observed for each case. Particles with a larger diameter led to increased turbulence intensities and Reynolds stresses, while smaller size particles led to opposite phenomena. Apart from the particle size, another critical parameter is the concentration of the dispersed phase. Kulick et al. [4] examined the effect of different particle loadings in the two-phase flow and how this parameter could alter the turbulent characteristics. They reached the conclusion that, as particle mass loading increases, turbulence attenuation is observed, a phenomenon that is more pronounced in the transverse flow direction.

From the examples given above, it can be easily understood that the dispersed phase adds complexity to the problem, even in straight open-channel flows. Apart from those mentioned above, other particle characteristics that may influence the flow properties are the degrees of freedom and the spin gradient or inertia properties, many of which are quite cumbersome to measure experimentally. Therefore, many studies have tried to approach the problem numerically, either through Direct Numerical Simulations or by utilizing a turbulence model [5–9]. Once again, these studies have revealed interesting phenomena regarding the interaction of particles with the carrier flow concerns. Dritselis and Vlachos [6] adopted a point-force model in order to examine the effect of small heavy particles on the flow, without finding any evidence of qualitative velocity changes due to particle presence. On the other hand, they pointed out that near-wall velocity-coherent structures seem to be directly affected by the rotational motion of the dispersed particles. Another important parameter that affects the flow phenomena is the clustering and dispersion of the secondary phase, especially close to the bottom-wall area [10,11].

Finally, in the comparative analysis of Yu et al. [12], various studies with controversial results were examined. The controversy of the studies that the authors examined concerned the velocity lag between the laden and unladen flow, as well as the effect of the dispersed phase on the von Kármán velocity distribution constant. In their analysis, they concluded that the addition of particles in the flow should lead to mixed behavior as far as the velocity distribution is concerned; hence, the von Kármán is constant. In the inner region, the velocity values of the particulate flow surpass the respective unladen ones, while in the outer region the exact opposite behavior should be observed.

In the majority of the above mentioned numerical studies, Lagrangian methods have been utilized in order to model the two-phase flow. Although these methods have been well-validated throughout the years, they still require considerable computational resources due to the additional number of equations that need to be solved. Moreover, the increased problem complexity leads to many additional model parameters, which in many cases are difficult to solve. Thus, an Eulerian approach to modeling the particulate phase in the special case of an open-channel flow would have obvious advantages, in terms of speed and accuracy, as well as reduced parameter complexity, provided that it is successfully verified. Eulerian approaches come with some limitations as well, when compared to Lagrangian ones. Although the Eulerian approach is able to take into account the physics of the secondary phase and describe its impact on the general flow physics, it cannot accurately describe particle motion.

A well-documented Eulerian model that is capable of modeling particulate flows is the micropolar model, which has been originally mathematically introduced by Eringen [13] and successfully applied to turbulent wall-bounded (Poiseuille) flows [14,15]. The micropolar model takes into account characteristics of the dispersed phase such as spin gradient, inertia, and particle viscosity but in an Eulerian frame of reference. Since several studies still present conflicting results regarding even the first-order properties of the flow [12], such as the velocity lag between the laden and unladen flow, alternative approaches, like the micropolar model, would be beneficial. The micropolar model comprises an invaluable computational asset for the present problem, capable of diving deep into both the macroscopic and particle properties, without complex modeling requirements.

The idea of the micropolar model being applied to environmental flows has already been revised, e.g., refs. [16,17]. Bender et al. [16] have introduced a novel Smoothed Particle Hydrodynamics (SPH) micropolar model in order to perform numerical simulations of turbulent environmental flows. The results of their study are in good agreement with the literature, without the simulations presenting computational overhead. In addition, Bender et al. [16] have incorporated the micropolar model in combination with the Discrete Element Method (DEM) to perform simulations on the OpenFOAM CFD platform.

Their numerical simulations of two-phase geological flows, with particles of different sizes, were successful, with the results once again being encouraging.

In this study, the micropolar model is exploited for the first time to study a typical case of environmental flows, the turbulent open-channel flow, in combination with fully resolved Direct Numerical Simulations (DNSs). The results of the unladen case have been firstly validated against respective experimental as well as computational ones, taken from well-documented previous studies. In most cases, the present micropolar model achieves better predictions of the flow characteristics compared to existing models in the literature. Furthermore, a case of high micropolar viscosity value, which accounts for a particulate flow, has been examined and compared once again with previous data. The results of the micropolar case seem very encouraging, while the model is able to capture the usual flow characteristics of a particulate flow, giving, in some cases, better physical explanation of the underlying phenomena.

The present paper is organized as follows. The governing equations of the micropolar model, as well as the specifications of the geometry and numerical modeling, are given in Section 2. The results and discussion, which include the validation and comparison against experimental and computational data, are presented in Section 3. Useful conclusions and comments are finally presented in Section 4.

## 2. Materials and Methods

### 2.1. Governing Equations

The numerical modeling of the present study has been kept in alignment with previous DNS studies in micropolar channel flows [14,15]. In this sense, the governing equations included the continuity and micropolar linear and angular momentum equations, as originally proposed by Eringen [13].

$$\nabla \times \mathbf{u} = 0 \tag{1}$$

$$\frac{\partial \mathbf{u}}{\partial t} + (\mathbf{u} \times \nabla)\mathbf{u} = -\nabla P + \frac{1}{Re}\nabla^2 \mathbf{u} + \frac{m}{Re}\nabla \times \boldsymbol{\omega} \tag{2}$$

$$\frac{JN}{m}\left(\frac{\partial \boldsymbol{\omega}}{\partial t} + (\mathbf{u} \times \nabla)\boldsymbol{\omega}\right) = \frac{1}{Re}\nabla^2 \boldsymbol{\omega} + \frac{N}{Re}\nabla \times \mathbf{u} - 2\frac{N}{Re}\boldsymbol{\omega} \tag{3}$$

In Equations (1)–(3), the usual variable of $t$ has been selected to represent time, while two different velocity representations are present. The angular flow velocity, which is also present in the original equation description of Eringen [13], is represented here by $\boldsymbol{\omega}$. Moreover, the classic linear velocity vector and pressure are represented by $\mathbf{u}$ and $P$, respectively. For a detailed derivation of Equations (1)–(3), the interested reader is referred to the Appendix of Sofiadis and Sarris [14].

In order to arrive at the non-dimensional state of the governing set of Equations (1)–(3), a characteristic length ($\delta$) has been chosen. The total height of the channel is set to be equal to $h = 2\delta$, while the characteristic velocity and density are denoted by $U_0$ and $\rho_0$, respectively, where the zero subscript indicates average quantities.

For the open-channel case that is presented in this study, the Reynolds number depends on the channel total height and flow total viscosity and is defined as $Re_h = \frac{\rho U_0 h}{\mu+\kappa}$, while this can be rewritten in terms of ($2h$) as $Re_b = \frac{\rho U_0 2h}{\mu+\kappa}$. The sum of $\mu$ and $\kappa$ represents the flow total viscosity ($\mu_t$), with $\mu$ being the molecular and $\kappa$ the micropolar one. The rest of the non-dimensional parameters are $m = \frac{\kappa}{\mu+\kappa}$ the vortex viscosity parameter; $J = \frac{j}{\delta^2}$ the dimensionless microrotation based on $\delta^2$, where $j$ is the microinertia of the fluid; and $N = \frac{\kappa\delta^2}{\gamma}$ the spin gradient viscosity parameter, where $\gamma$ is the material coefficient of the fluid. Parameter values of $N = 8.3 \times 10^4$ and $J = 1 \times 10^{-5}$ have been kept constant throughout the present study.

In the general micropolar theory, (C) is a couple stress tensor that represents the internal moments in the material. More specifically, it describes the distribution of rotational forces within the material [13]. In other words, (C) captures the resistance of the material to bending and twisting deformations and represents the moments per unit area within the material that arise due to the internal microstructure's resistance to rotation or torsion. In the specific isotropic case, such as the one that is considered here, the second-order tensor can be reduced into a scalar. From this equation, following the appropriate non-dimensionalization, the micropolar viscosity ratio (m) can be derived. A detailed derivation of the above can be found in the Appendix of Sofiadis and Sarris [14]. Different particle sizes and densities can be found in natural sediment transport and industrial processes. This diversity of internal microstructure characteristics results in different distributions of rotational forces within the material. The latter can be successfully described by micropolar theory; the couple stress tensor (C); and, subsequently, (m). All of the above is considered in an Eulerian frame of reference, which is one of the major advantages of micropolar theory.

The mean force rate is kept constant in the present simulations, which corresponds to a fixed value of mean mass flow rate and leads to a fixed $Re_b$. On the other hand, the pressure gradient is calculated and corrected in every time step. Finally, in the analysis of computed results, flow variables have been scaled according to wall coordinates. This has been achieved through the calculation of wall-shear velocity, $u_\tau = \sqrt{\frac{\tau_w}{\rho}}$, where $\tau_w = \mu_t\left(\frac{\partial <u>}{\partial y}\right)_w$ is the shear stress at the wall and <u> denotes the mean velocity value with respect to time ($t$) and space. As a consequence, the friction Reynolds number can be written as $Re_\tau = \frac{u_\tau h}{\nu}$, which in the present study corresponds to a value of $Re_\tau = 330$ for the choice of $Re_b$ = 11,200. After being non-dimensionalized, Equations (1)–(3) have been implemented in the open-source CFD platform OpenFOAM, where they have been solved directly followed by appropriate boundary conditions for the open-channel case. The OpenFOAM solver that we depart from in order to obtain the micropolar model is the solver icoFoam. The code is available upon request to the authors.

### 2.2. Numerical Schemes

The discretized governing set of Equations (1)–(3) has been advanced with up to third-order accurate schemes in space (Gauss Cubic scheme), and a second-order accurate implicit backward scheme in time. These schemes consist the standard approach of the OpenFOAM platform for the case of DNS modeling and have been previously validated against benchmark cases. Additionally, they have proven to produce accurate results along with the micropolar algorithm for the wall-bounded channel case [14,15]. The coupling of momentum with pressure has been achieved through the pressure-implicit split-operator (PISO) method. This algorithm follows an iterative procedure, where the pressure equation is solved twice and the mass conservation is enforced with an explicit correction to velocity so that momentum conservation is satisfied.

In order to obtain transient free results (statistically independent), the present paper simulations have been integrated in time for up to 10 eddy turn-over times starting from arbitrary turbulent initial conditions. A perturbation has been implemented in the initial flow field in order to create these arbitrary turbulent initial conditions. As soon as the initial criteria are satisfied, the statistics of all quantities are obtained for several eddy turn-over times until statistical stationarity occurs. The averaging period corresponds approximately to a time of 15,000 $\nu/u_\tau^2$, where $t\nu/u_\tau^2$ represents the non-dimensional time unit ($t^+ = tu_\tau^2/\nu$) [14,15]. In order to provide a standarized measure to the reader to evaluate the present model computational efficiency, the CUT (Computational Unit Time) has been utilized. This is a metric that helps with the quantification of the computational cost of a certain simulation and is calculated as CUT = (total walltime/total number of timesteps)/(total number of grid cells/number of CPU cores). In the present study, the typical CUT = 0.0013.

*2.3. Model Geometry*

The channel flow set-up (see Figure 1) is a common selection for open-channel turbulent flow investigation, especially with DNSs. The channel is typically represented by a rectangular cross-section with the dimensions and mesh size appropriately adjusted to the needs of the case. The computational domain size has been kept equal to $3\pi\delta$ and $2\pi\delta$ in the streamwise and spanwise periodic directions, respectively, while the total height (wall-normal direction) is set equal to $h = 2\delta$. The grid resolution has been selected to be $128 \times 132 \times 128$ in the streamwise, wall-normal, and spanwise directions, respectively. The aforementioned grid dimensions correspond, for the non-micropolar case (m = 0), to a uniform grid spacing at the two periodic directions, streamwise and spanwise. At these two directions, wall-scaled grid spacing is equal to $\Delta_x^+ = 12.6$ and $\Delta_z^+ = 8$. One of the most demanding tasks when solving turbulent flows with Direct Numerical Simulations (DNSs) is to capture the finer scales of the flow. This task becomes particularly important in the near wall region, where very-fast-moving small eddies reside. Thus, special attention should be paid to grid clustering very close to the wall. In the present case, the choice of appropriate grid spacing in the wall-normal direction has led to a non-uniform mesh of sizes $\Delta_{ymin}^+ = 0.6$ and $\Delta_{ymax}^+ = 4.6$ is used. For the *Re* that is investigated in the current project, the aforementioned grid resolution has been reported to be adequate in previous studies of both open and wall-bounded channels [14,15,18,19].

The boundary conditions considered in the current study include no-slip at the bottom wall and slip at the top boundary, while mean velocity has been kept constant in the streamwise direction. Periodic boundary conditions have been selected in the streamwise and spanwise directions. The velocity boundary conditions of the computational domain are summarized in Table 1.

**Table 1.** Velocity boundary conditions prescribed to the current computational domain.

| | |
|---|---|
| Top Plane | Free-Slip |
| Bottom Plane | No-Slip |
| Side Planes | Periodic |
| Inlet | Periodic |
| Outlet | Periodic |

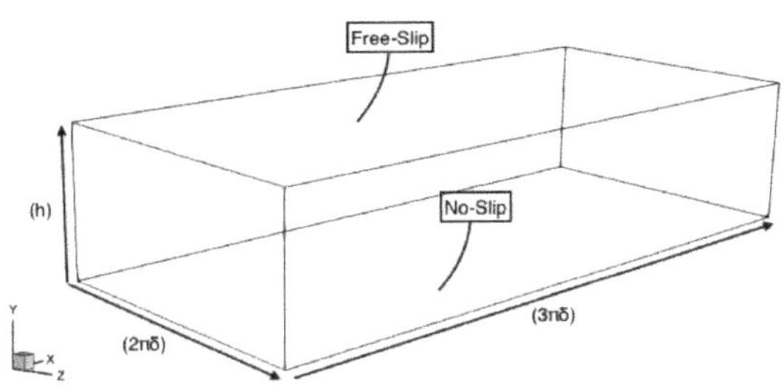


**Figure 1.** Schematic of the geometry that has been used in the present study, where $h = 2\delta$.

## 3. Results

Mean flow variables and structures have been extracted from the simulations and are presented in this section in order to analyze the flow physics.

*3.1. Mean Velocity*

At first, the normalized hydrodynamic mean velocity profile ($u^+$) has been plotted along the normalized wall-distance ($y^+$) and compared against previous experimental and DNS data. In order to calculate the hydrodynamic data, the micropolar viscosity parameter $m$ has been set equal to $m = 0$. In this case, Equation (2) reduces to the usual Navier-Stokes one, while Equation (3) does not contribute to the solution of the system. The comparison has been conducted for the hydrodynamic case of $Re_b$ = 11,200 and is presented in Figure 2. The variables to which the superscript (+) has been assigned have been scaled by wall units, i.e., $y^+ = yu_\tau/\nu$, where $\nu$ is the kinematic viscosity and $u_\tau$ the wall shear velocity.

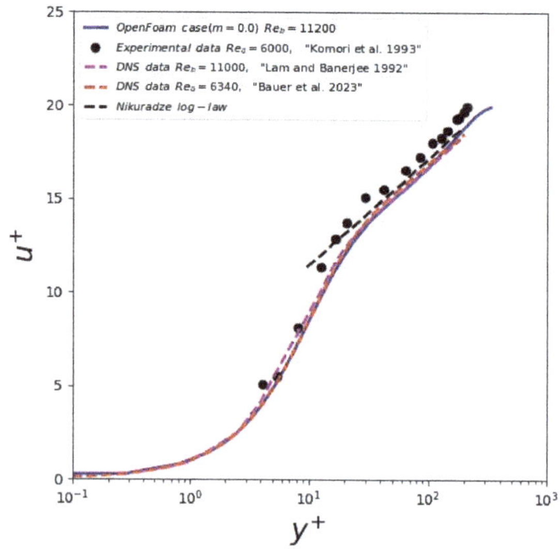


**Figure 2.** Normalized mean velocity profile of the present hydrodynamic case compared against experimental results of Komori et al. [18] and DNS results of Lam and Banerjee [19] and of Bauer et al. [20].

The mean hydrodynamic velocity profile shows excellent agreement with previous numerical and experimental data. In Figure 2, the results of the present DNS simulation seem to converge closer to the experimental values of Komori et al. [18] than the previous DNS data of Lam and Banerjee [19]. More specifically, the maximum local error between the present results and experimental ones of Komori et al. [18] can be found at $y^+ = 4$ and is $\approx$27%. The small deviation from previous data is expected as the present case of $Re_b$ = 11,200 (based on characteristic length $2h$) collapses to a value of $Re_\tau = 166$ ($Re_\tau$ based on half fluid liquid depth); in the case of Komori et al. [18], it is equal to $Re_\tau = 160$ ($Re_\delta = 6000$, based on characteristic length $\delta$); and in case of Lam and Banerjee [19], $Re_\tau = 171$ ($Re_b$ = 11,000). Moreover, the present DNS results have been also compared against the recent DNS results of Bauer et al. [20], where a larger domain of dimensions $Lx/h = 37.7142, Lz/h = 12.5664$ has been used for $Re_\delta = 6340$ and $Re_\tau = 200$. Nevertheless, the velocity profile curves are in good agreement with each other, as well as Nikuradze's log-law, without violating the physical laws. It is true that the present paper uses a smaller domain and gird resolution than the study of [20], although with no obvious drawbacks. Turbulence statistics are well converged and have adequate accuracy in order to assist the comments and discussion on the flow physics.

Apart from the hydrodynamic case, an additional set of numerical simulations was conducted, which considered the micropolar formulation of the Navier–Stokes equation. The detailed formulation of these equations has been extensively described in previous studies of Sofiadis and Sarris [14,15], where the case of a wall-bounded channel flow was examined with very interesting results. In the present case, the normalized mean velocity profile has been plotted against the normalized wall distance for three different micropolar viscosity ratios $m = 0.4, 0.7$, and 0.9, in Figure 3. The larger values of the micropolar viscosity ratio, close to 1.0, characterize stronger micropolar behavior, which has been found to be analogous to a higher volume fraction in particulate flows. The rest of the non-dimensional parameters have been kept constant. The main reason for this choice is that the micropolar viscosity ratio directly affects the flow characteristics, as can be seen through the governing Equations (1)–(3). Furthermore, since this is a first application of the micropolar model to environmental flows, a study of the effect of the rest of the parameters will be conducted in future works.

The mean velocity profiles of the various micropolar cases have been plotted along with the hydrodynamic case of the same $Re_b$ in Figure 3. The profile curves present very small deviations between each other, indicating the very small influence of the micropolar parameters. These results, although expected, are in contrast with previous findings of the velocity profile trends in the case of wall-bounded channel flow. This behavior is attributed to the existence of the slip condition on the top-wall of the geometry. Previous studies have shown that the main influence of the micropolar part of the equations can be spotted very close to the wall Sofiadis and Sarris [14,15]. Therefore, it is expected that in the present case this influence can only be strong in the lower part of the geometry (bottom wall) and will gradually decrease as we move further away from the wall.

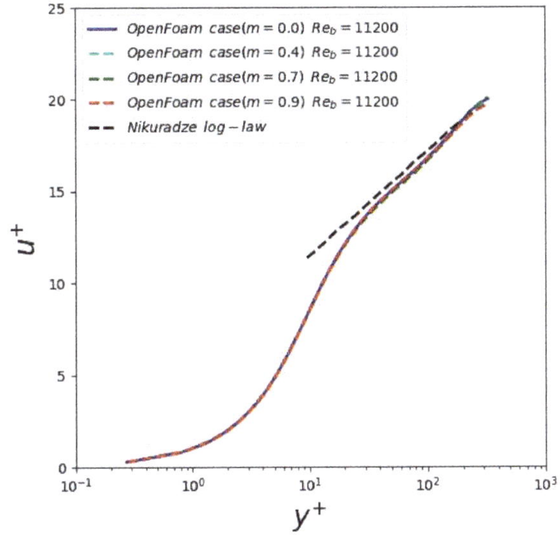


**Figure 3.** Normalized mean velocity profile of the present hydrodynamic case, compared against micropolar cases at the same $Re_b$.

The maximum difference occurs between the micropolar case of $m = 0.9$ and the hydrodynamic one, which is approximately of the order of $1 \times 10^{-2}$ . In a relevant numerical investigation [8], it has been reported that the particle secondary phase had a negligible effect on the hydrodynamic velocity profile. In addition, the difference between the single- and two-phase flow velocity profile was in that case smaller than $9.3 \times 10^{-3}$ times the maximum velocity.

On the other hand, controversial issues found in the literature regarding the effect of the sediment phase on the mean flow velocity are presented in Yu et al. [12]. It is a fact that many researchers still argue that the sediment phase leaves the mean velocity profile mainly unchanged, and, thus, the Kármán constant has approximately the same value as in the unladen case. This idea was first proposed by Coleman [21] in his original study. Nevertheless, other studies support the original findings [22], which show a clear change in the mean flow velocity profile when the sedimentation phase is present. Finally, Yu et al. [12] concluded that, when particles are added in a turbulent open-channel flow over smooth bed, the mean flow velocity profile trend changes, though in a smaller magnitude than expected. In other words, two different behaviors should be expected. In the inner region, the unladen flow presents a smaller velocity, while in the outer region this trend is reversed, with the unladen flow moving faster than the one including sediment transport. Similar behavior is also confirmed in the present study, as can be clearly seen in Figure 4. The magnitude of change is considered in this study to depend on the sediment characteristics.

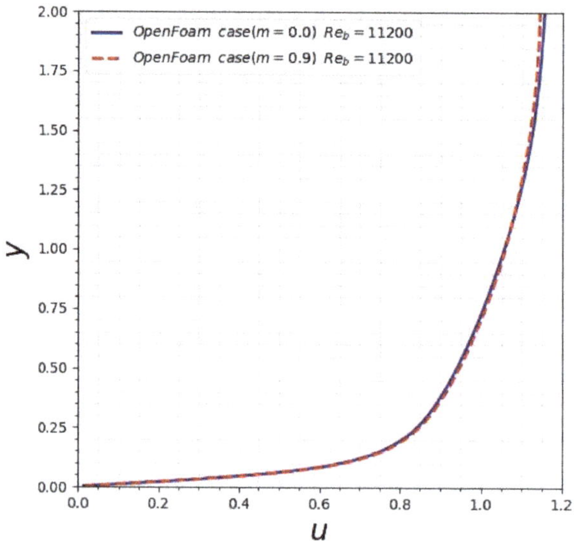


**Figure 4.** Mean velocity profile of the present hydrodynamic case, compared against the micropolar case of $m = 0.9$ at the same $Re_b$, in outer units.

In order to examine the influence attenuation of the micropolar parameters in the open-channel flow case, the spanwise micropolar velocity profile along the channel total height (y/h) is plotted in Figure 5 and is compared against the respective spanwise micropolar velocity profile of the wall-bounded channel case. The wall-bounded channel case micropolar velocity profile is plotted along the characteristic length $\delta$ ($\delta = h/2$). The reason for that is that $\delta$ and $h$ are both characteristic boundary layer thicknesses of the respective flows. Only the spanwise direction is presented for the micropolar velocity as it is the only direction in which micropolar velocity survives [14]. It is observed that the micropolar velocity presents peak values in both cases near the bottom wall ($y < 0.3$). Additionally, by examining Figure 5 it is evident that the micropolar velocity profile in the current open-channel flow case drops very quickly to zero and has no influence after $y/h > 0.1$. On the contrary, the micropolar velocity in the closed channel case "survives" at least up to $y/\delta \approx 0.3$. This observation further enhances our previous discussion about the influence attenuation of the micropolar parameters in the open-channel flow case.

Further evidence of the secondary phase presenting a strong influence only close to the bottom wall has been provided in previous experimental studies [2,23–27]. In these experiments, the particle velocity was measured as higher than the respective fluid phase one very close to the bottom wall.

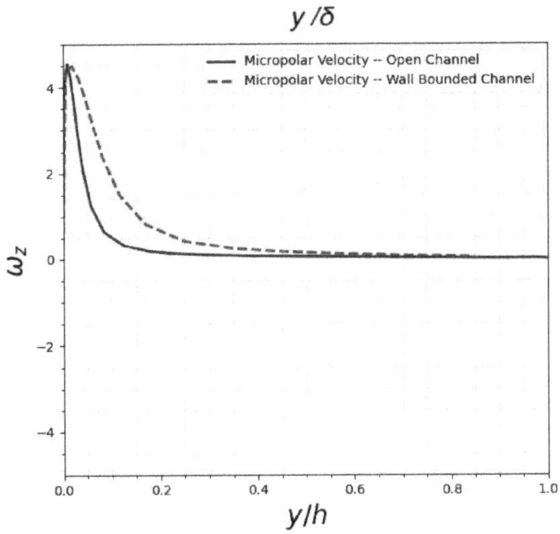


**Figure 5.** Mean spanwise micropolar velocity profile of the present open-channel case, compared against wall-bounded channel case at the same $Re_b$ = 11,200 and $m = 0.4$. The open-channel case is normalized with the total channel height (h), while the closed channel case with ($\delta = h/2$).

### 3.2. Diagnostic Functions

To answer the question of whether a logarithmic layer exists in some interval of the mean velocity profile (MVP), which has been examined in Section 3.1, we calculate a diagnostic function defined as:

$$\Xi(y^+) = y^+ \frac{du^+}{dy^+} \tag{4}$$

If a logarithmic layer exists, then $\Xi = const. = 1/\kappa$, where $\kappa$ is the von Kármán constant. In Figure 6, we graphically compare the distributions of ($\Xi$) for $m = 0.0$ and $m = 0.9$. Close to the channel bottom, $y^+ \leq 85$ ($y/h \leq 0.25$), the two curves are indistinguishable on the scale of the graph. Thus, the effect of parameter $m$ is negligible in the inner region of the turbulent boundary layer. A clear formation of a logarithmic layer is not observed for the $Re$ of this study. However, an upper bound of a "Kármán-like" constant can be estimated. Since it is found that $\Xi_{min} = 2.46$, it is speculated that in a logarithmic layer at a higher Reynolds number the value of $\kappa$ will, most probably, be lower than 0.407 (≈0.41).

Close to the free surface, $y^+ \geq 138$ ($y/h \geq 0.42$), the two ($\Xi$) curves diverge. The curve for the case of $m = 0.0$ forms an outer peak at $y^+ \approx 180$ ($y/h \approx 0.53$). In contrast, the curve of case $m = 0.9$ forms a plateau in the interval $0.4 \leq y/h \leq 0.53$ corresponding to the logarithmic behavior of the form $\frac{1}{\kappa_2} lny^+ + G$ where $\kappa_2 \approx 0.351$ ($\Xi_{mean} = 2.85$). This behavior can be marked as a characteristic of the outer layer in the micropolar turbulent case, due to the higher turbulent phenomena that are present in this case.

Next, we investigate the possibility of power law behavior in an interval of the (MVP). Here, we present the behavior of the diagnostic function $\Gamma = \frac{y^+}{u^+}\frac{du^+}{dy^+}$. If a power law ($u^+ = A(y^+)^\lambda$ ) with a constant exponent describes a "part" of the (MVP), then in that interval $\Gamma$ attains a constant value equal to $\lambda$. In Figure 7, the behavior of $\Gamma$ is presented for the cases of $m = 0.0$ and $m = 0.9$.

As expected, by examining $\Gamma$ function graphs it is confirmed that the effect of parameter ($m$) is negligible up to $y^+ \leq 125$ ($y/h \leq 0.38$). An interval where $\Gamma$ function is approximately constant is observed at $60 \leq y^+ \leq 140$ ($0.18 \leq y/h \leq 0.42$). This observation suggests that a power-law approximation of the (MVP), with an exponent of $\lambda \approx 0.162$, is satisfactory in this interval for both cases of $m = 0$ and $m = 0.9$. It is stressed here that these tentative conclusions concern the rather low turbulent Reynolds number case of $Re_b = 11{,}200$.

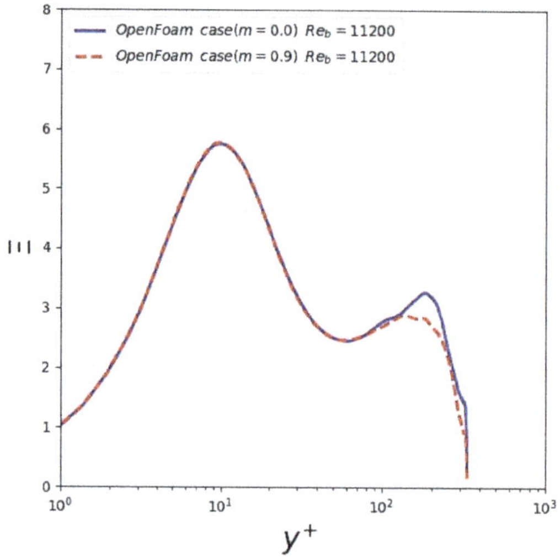


**Figure 6.** Diagnostic function ($\Xi$) plots of the present DNS data, for $Re_b = 11{,}200$ and cases of $m = 0.0$ and $m = 0.9$, plotted in inner-law units.

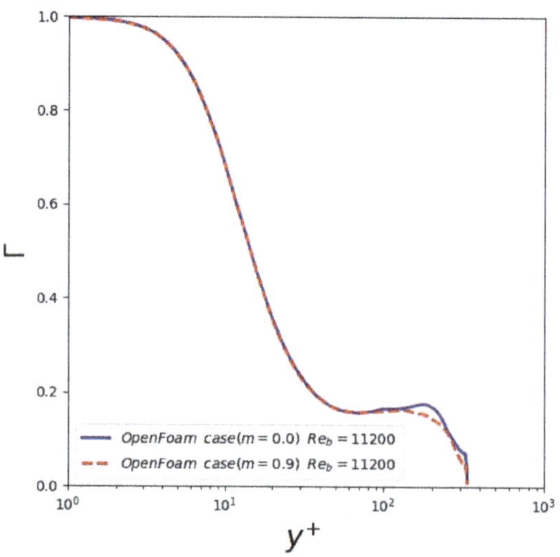


**Figure 7.** Power-law diagnostic function ($\Gamma$) based on the present DNS data, for $Re_b = 11{,}200$ and cases of $m = 0.0$ and $m = 0.9$, plotted in inner-law units.

### 3.3. Turbulence Intensity

In order to further validate our present DNS code in the specific case of the open-channel geometry, the turbulence statistics have been analyzed and compared against the usual hydrodynamic case of computational and experimental studies. Turbulence intensity has been examined through root mean square (rms) velocity profiles, which have been normalized with the shear velocity $u_\tau$ and plotted against the wall distance normalized with the channel height, $y/h$. All three rms velocity components have been plotted and are presented in Figure 8a–c, compared with the results of Komori et al. [18] and Lam and Banerjee [19].

The convergence with both the experimental and previous DNS data is satisfactory. Moreover, in some cases the present DNS data have achieved a better fit with the experimental ones, compared to literature data. More specifically, the maximum local error between the present results and previous experimental ones varies from 13 to 25% in Figure 8a–c. The turbulence intensity is accurately represented in all three components, while the peak of every case can be found close to the bottom wall. This phenomenon is more pronounced for the streamwise rms velocity profile ($u^+_{rms}$), as shown in Figure 8a. The peak of the $u^+_{rms}$ curve can be found in the area of $y/h < 0.2$ (near wall region), while for $0.2 < y/h < 1.0$, the turbulence intensity drops sharply to lower values. An additional interesting phenomenon is that the present DNS data follow the trend of the experimental ones [18], for all velocity components, closer than previous DNS data [19]. Apart from the validation of our code in terms of the mean velocity profile, this result could add to its robustness as well.

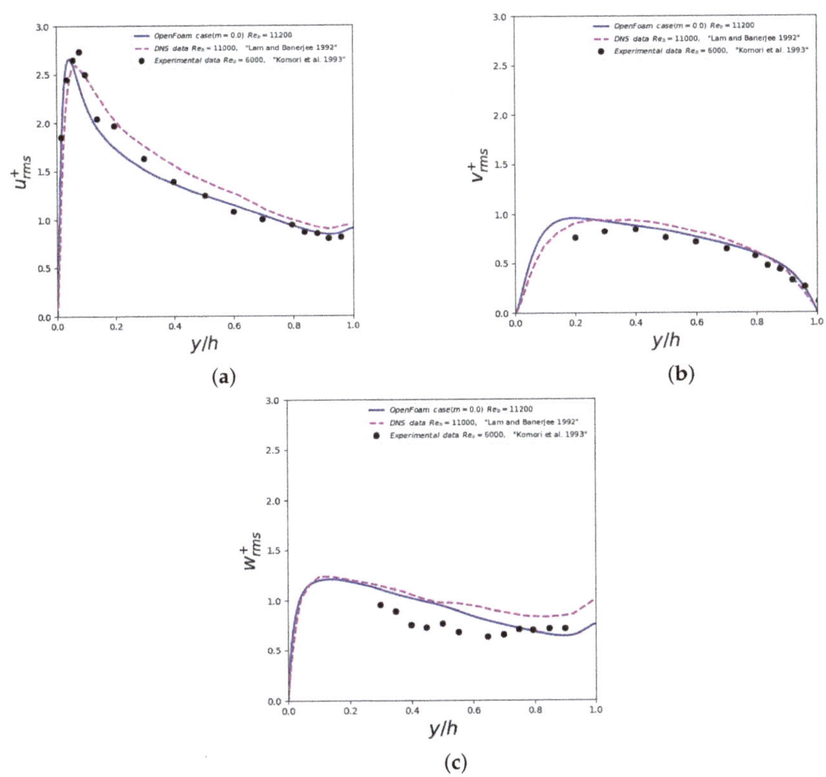


**Figure 8.** Normalized (**a**) $u_{rms}$, (**b**) $v_{rms}$, and (**c**) $w_{rms}$ velocity profiles of the hydrodynamic case ($m = 0$) plotted against global coordinates and compared with experimental results of Komori et al. [18] and DNS results of Lam and Banerjee [19].

Another point of investigation concerns the rms velocity profiles of the hydrodynamic case ($m = 0$), which are compared to the micropolar case of $m = 0.9$. Once again, the comparison is made against the "highest" micropolar case, in order for the differences to be more evident. The comparative analysis for the rms velocity between hydrodynamic and micropolar cases is presented in Figure 9a–c. Apart from the rms velocity profile curves of each case, the difference between these curves is plotted as well, as a shaded area, red, or blue, indicating the negative or positive difference between the hydrodynamic and micropolar case, respectively. The final difference values plotted have been normalized with the shear velocity so that, if the difference has a value of, for example, 0.5, it will be equal to $u_\tau/2$.

The comparative analysis presented in Figure 9a–c is particularly important as it presents different trends for the rms velocity profiles in the three directions. In the streamwise direction, the Figure 9a turbulence intensity of the micrpolar case seems to have the same behavior as the respective case shown for the streamwise velocity profile. In this sense, the micropolar turbulence intensity exceeds that of the hydrodynamic case in the inner region, while it drops to lower values in the outer region. This is an expected observation as changes in turbulence intensity are connected to the changes in the velocity profile as well. Moreover, this result could be an indication of turbulent drag reduction in the outer region.

On the contrary, in the wall-normal and spanwise directions, there seems to be a constant lag between the hydrodynamic and micropolar case along the channel height. Similar trends for the turbulence intensity profiles have been also reported in previous studies [4,23,24]. In their study, Taniere et al. [23] have found that the streamwise rms velocity profile of the particles exceeds the one of the single-phase flow very close to the wall, while the opposite trend is observed in the outer region. Interestingly, in the wall-normal direction, the rms velocity profiles of particles present a constant lag compared to that of the carrier phase, which the authors attribute to the effects of inertia and gravity in this direction. The same observations have been noted by Kiger and Pan [24] as well, who commented that flow turbulence intensity exceeds the respective unladen one only in the streamwise direction and only in the area of $y^+ < 30$. Finally, the constant lag of the laden flow in the transverse direction has been clearly shown in the results of Kulick et al. [4].

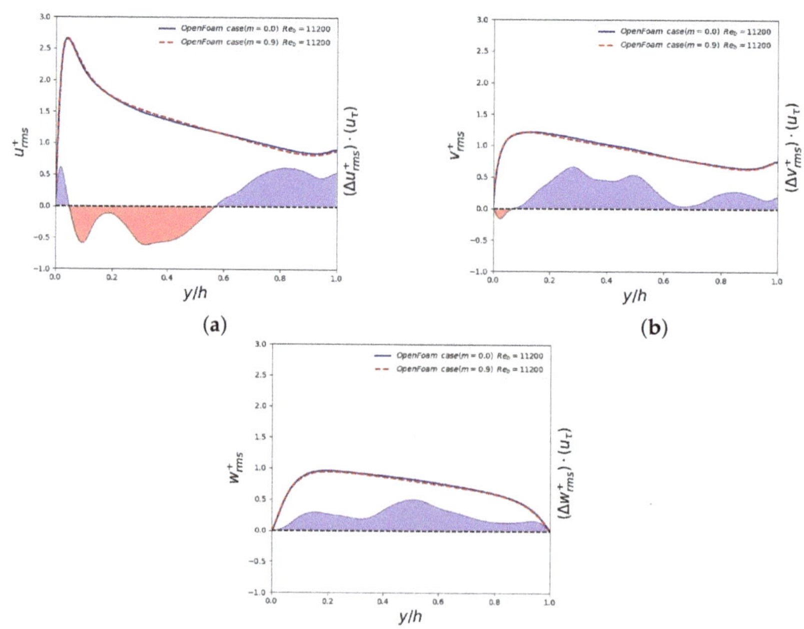


**Figure 9.** Normalized (**a**) $u_{rms}$, (**b**) $v_{rms}$, and (**c**) $w_{rms}$ velocity profiles of the hydrodynamic case ($m = 0$) compared with the micropolar case ($m = 0.9$), plotted against global coordinates. The final difference values (shaded area) have been normalized with the shear velocity so that, if the difference has a value of, for example, 0.5, it will be equal to $u_\tau/2$.

### 3.4. Shear Stresses

The shear stress profile is presented for the same $Re_b$ = 11,200 as previous results of this study and compared once again with DNS and experimental data in Figure 10, for the hydrodynamic case ($m = 0$). The shear stress is computed as:

$$\tau_s = \overline{-(u'v')} \tag{5}$$

where the overbar in Equation (5) denotes spatio-temporal averaged values. The shear stress profiles in Figure 10 have been appropriately normalized with the wall shear velocity $u_\tau$. The agreement between the present and previous DNS results is satisfying, with slight differences occurring in the near-wall region. The experimental results of Komori et al. [18] seem to reach in a peak value lower than the present and previous DNS results [19]. The maximum local error between present and previous experimental results seems to be $\approx$27%. This deviation in the shear stress peak value was also reported in the original study of Komori et al. [18], where they compared their experimental results with those of Kim et al. [28] and Lam and Banerjee [19], but they did not comment further on the source of this deviation.

Moreover, the shear stresses of the micropolar ($m = 0.9$) and hydrodynamic ($m = 0.0$) cases have been plotted in Figure 11 along with their difference normalized by $u_\tau$. The normalization of their difference follows the same approach as in the previous plots of velocity and turbulence intensity profiles. Upon closer examination of Figure 11, a similar trend with the streamwise velocity profile can be observed. This includes higher values of shear stress for the micropolar flow in the region ($y/h < 0.5$), while the hydrodynamic flow reaches higher values in the region ($y/h > 0.5$), indicating drag reduction phenomena in this region once again. These results are consistent with the findings in Figure 4, where the micropolar flow is faster than the hydrodynamic one in the region ($y/h < 0.5$) and slower when ($y/h > 0.5$) [12]. The only discrepancy in this trend occurs very close to the wall ($y/h < 0.05$), where the hydrodynamic values exceed the micropolar ones, but this result may be connected to computational error, which may be observed in regions close to the wall.

The shear stress comparison between the single- and two-phase flow presented by Kiger and Pan [24] has revealed an opposite trend to the present results, without the authors really commenting on this behavior. In this instance, differences in every case are only marginal but still pose a clear behavior of each phase. Shear stress behavior in the present study has been also discussed in Muste et al. [26]. They have also reported a linear increment of the two-phase flow shear stress towards the bed, while the single flow shear stress has been constantly collapsing to higher values across the outer region. In addition, they have reported that previous studies present mixed behaviors as far as the shear stress is concerned. There are various studies that found no difference in the shear stress values between particulate and unladen flows, such as those of Best et al. [29] and Graf and Cellino [30]. On the other hand, the experimental work of Kaftori et al. [31] has found no change in the near-wall values but instead a decrease in the outer region for the shear stress values of the particulate flow.

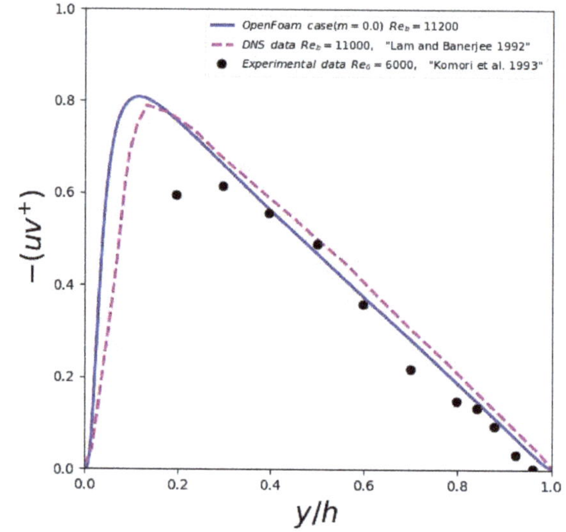


**Figure 10.** Normalized shear stress profiles of the hydrodynamic case ($m = 0$) $Re_b$ = 11,200, compared with experimental results of Komori et al. [18] and DNS results of Lam and Banerjee [19], plotted against normalized wall distance.

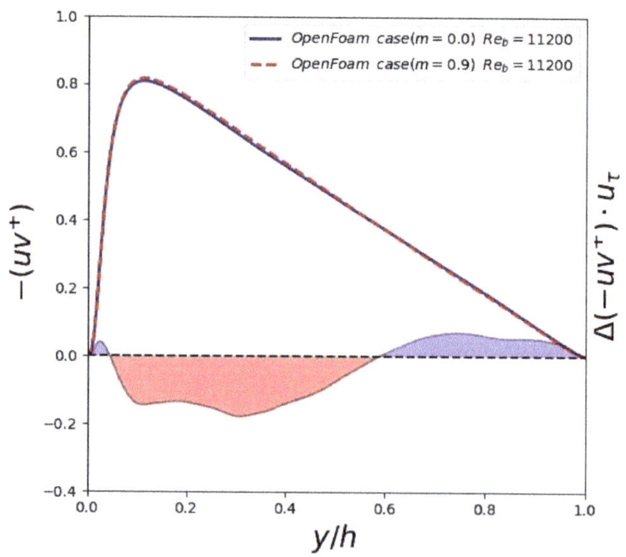


**Figure 11.** Normalized shear stress profiles of the hydrodynamic case ($m = 0$), compared with the micropolar case of ($m = 0.9$), for $Re_b$ = 11,200, plotted against normalized wall distance. The shaded areas represent the difference between the two normalized stress profiles: (red shade) $-\overline{(u'v')}_{m=0} < -\overline{(u'v')}_{m=0.9}$, (blue shade) $-\overline{(u'v')}_{m=0} > -\overline{(u'v')}_{m=0.9}$.

## 4. Conclusions

In the present paper, an Eulerian approach based on the micropolar model [13] has been employed in order to study a turbulent particulate open-channel flow, at a relatively low Reynolds number. This type of flow is of particular interest for environmental and geological applications. The micropolar model seems to be an accurate alternative for the investigation of turbulent wall-bounded flows as it can treat a particulate flow in an Eulerian frame with reduced mathematical complications and computational cost. The characteristics of the secondary phase have been introduced in the governing equations through terms that represent the spin gradient, particle viscosity, and microinertia.

The unladen flow has been successfully validated against previous experimental [18] and computational [19] results. Next, cases of varying micropolar viscosity, which correspond to varying secondary phase characteristics, have been compared against the hydrodynamic one. Differences regarding the velocity distribution between the various micropolar cases are marginal. Therefore, the case of micropolar viscosity ratio $m = 0.9$ has been compared against the hydrodynamic one, throughout the rest of the results section. One of the main disagreements in previous studies concerned the particle effect on the velocity distribution in an open-channel flow. The findings of the present study confirm the suggestions of Yu et al. [12] that the velocity distribution of the particulate flow presents mixed behavior compared to the unladen one. More specifically, it has been shown here that the velocity values of the particulate flow exceed the respective unladen ones in the inner region, while they collapse to lower values in the outer region. This may be attributed to the particles' trend of being accelerated by turbulent structures in the near-wall region.

The flow phenomena have been further examined through turbulence intensity plots. In the streamwise direction, the same trend between the particulate and single-phase flow has been observed, as in the mean velocity distribution. On the contrary, in the two other directions of the turbulence intensity, a constant lag between the two cases seems to exist, with the single-flow phase presenting slightly higher values of the order of $u_\tau/2$. Furthermore, the Reynolds shear stress profile has been plotted for the micropolar and hydrodynamic cases. The expected mixed behavior in the inner and outer regions of the flow has been confirmed. Previous studies presented controversial results, without really commenting on their observations.

To sum up, the micropolar model has shown various advantages over existing Lagrangian models when studying particulate flows, in terms of reduced computational cost and modeling complexity. Here, we have verified that this is an accurate and reliable method to analyze basic turbulent flow cases, usually found in environmental flows. Future studies should exploit the incorporation of the micropolar model in the analysis of more complex environmental and geological flows. Further DNS simulations are required for the development of composite mean velocity profile mathematical models [32], as well as accurate turbulence models very close to the wall [33].

**Author Contributions:** Conceptualization, G.S.; methodology, G.S. and A.L.; software, G.S. and F.S.; investigation, G.S. and A.L.; data curation, G.S. and A.P.; writing—original draft preparation, G.S.; writing—review and editing, A.L. and F.S.; supervision, A.L.; and funding acquisition, A.L. All authors have read and agreed to the published version of the manuscript.

**Funding:** This research project was supported by the Hellenic Foundation for Research and Innovation (H.F.R.I.) under the "2nd Call for H.F.R.I. Research Projects to Support Faculty Members and Researchers" (Project Number: 4584).

**Data Availability Statement:** Data are available to the interested reader upon request.

**Conflicts of Interest:** The authors declare no conflicts of interest.

## References


1. Rashidi, M.; Hetsroni, G.; Banerjee, S. Particle-turbulence interaction in a boundary layer. *Int. J. Multiph. Flow* **1990**, *16*, 935–949. [CrossRef]
2. Kaftori, D.; Hetsroni, G.; Banerjee, S. Particle behavior in the turbulent boundary layer. II. Velocity and distribution profiles. *Phys. Fluids* **1995**, *7*, 1107–1121. [CrossRef]
3. Nezu, I.; Azuma, R. Turbulence characteristics and interaction between particles and fluid in particle-laden open-channel flows. *J. Hydraul. Eng.* **2004**, *130*, 988–1001. [CrossRef]
4. Kulick, J.D.; Fessler, J.R.; Eaton, J.K. Particle response and turbulence modification in fully developed channel flow. *J. Fluid Mech.* **1994**, *277*, 109–134. [CrossRef]
5. Crowe, C.; Troutt, T.; Chung, J. Numerical models for two-phase turbulent flows. *Annu. Rev. Fluid Mech.* **1996**, *28*, 11–43. [CrossRef]
6. Dritselis, C.D.; Vlachos, N.S. Numerical study of educed coherent structures in the near-wall region of a particle-laden channel flow. *Phys. Fluids* **2008**, *20*, 055103. [CrossRef]
7. Balachandar, S.; Eaton, J.K. Turbulent dispersed multiphase flow. *Annu. Rev. Fluid Mech.* **2010**, *42*, 111–133. [CrossRef]
8. Kidanemariam, A.G.; Chan-Braun, C.; Doychev, T.; Uhlmann, M. Direct numerical simulation of horizontal open-channel flow with finite-size, heavy particles at low solid volume fraction. *New J. Phys.* **2013**, *15*, 025031. [CrossRef]
9. Gao, W.; Samtaney, R.; Richter, D.H. Direct numerical simulation of particle-laden flow in an open channel at. *J. Fluid Mech.* **2023**, *957*, A3. [CrossRef]
10. Rouson, D.W.; Eaton, J.K. On the preferential concentration of solid particles in turbulent channel flow. *J. Fluid Mech.* **2001**, *428*, 149–169. [CrossRef]
11. Bernardini, M.; Pirozzoli, S.; Orlandi, P. The effect of large-scale turbulent structures on particle dispersion in wall-bounded flows. *Int. J. Multiph. Flow* **2013**, *51*, 55–64. [CrossRef]
12. Yu, K.; Yoon, B.; Kim, D. Re-evaluation of change of mean velocity profiles in turbulent open-channel flows due to introducing sediment particles. *KSCE J. Civil Eng.* **2014**, *18*, 2261–2267. [CrossRef]
13. Eringen, A.C. Theory of micropolar fluids. *J. Math. Mech.* **1966**, *16* , 1–18. [CrossRef]
14. Sofiadis, G.; Sarris, I. Microrotation viscosity effect on turbulent micropolar fluid channel flow. *Phys. Fluids* **2021**, *33*, 095126. [CrossRef]
15. Sofiadis, G.; Sarris, I. Reynolds number effect of the turbulent micropolar channel flow. *Phys. Fluids* **2022**, *34*, 075126. [CrossRef]
16. Bender, J.; Koschier, D.; Kugelstadt, T.; Weiler, M. A micropolar material model for turbulent SPH fluids. In Proceedings of the ACM SIGGRAPH, Eurographics Symposium on Computer Animation, Montreal, QC, Canada, 21–23 August 2017; pp. 1–8.
17. Wang, L.; Chu, X. An extended CFD-DEM model based on micropolar fluid and its application in geological engineering. *Comput. Geotech.* **2023**, *154*, 105133. [CrossRef]
18. Komori, S.; Nagaosa, R.; Murakami, Y.; Chiba, S.; Ishii, K.; Kuwahara, K. Direct numerical simulation of three-dimensional open-channel flow with zero-shear gas–liquid interface. *Phys. Fluids A Fluid Dyn.* **1993**, *5*, 115–125. [CrossRef]
19. Lam, K.; Banerjee, S. On the condition of streak formation in a bounded turbulent flow. *Phys. Fluids A Fluid Dyn.* **1992**, *4*, 306–320. [CrossRef]
20. Bauer, C.; Sakai, Y.; Uhlmann, M. Direct numerical simulation of turbulent open-channel flow: Streamwise turbulence intensity scaling and its relation to large-scale coherent motions. In Proceedings of the 10th International Conference on Turbulence, Rome, Italy, 11–15 September 2023.
21. Coleman, N.L. Velocity profiles with suspended sediment. *J. Hydraul. Res.* **1981**, *19*, 211–229. [CrossRef]
22. Vanoni, V.A.; Nomicos, G.N. Resistance properties of sediment-laden streams. *Trans. Am. Soc. Civil Eng.* **1960**, *125*, 1140–1167. [CrossRef]
23. Taniere, A.; Oesterle, B.; Monnier, J. On the behaviour of solid particles in a horizontal boundary layer with turbulence and saltation effects. *Exp. Fluids* **1997**, *23*, 463–471. [CrossRef]
24. Kiger, K.; Pan, C. Suspension and turbulence modification effects of solid particulates on a horizontal turbulent channel flow. *J. Turbul.* **2002**, *3*, 019. [CrossRef]
25. Righetti, M.; Romano, G.P. Particle–fluid interactions in a plane near-wall turbulent flow. *J. Fluid Mech.* **2004**, *505*, 93–121. [CrossRef]
26. Muste, M.; Yu, K.; Fujita, I.; Ettema, R. Two-phase flow insights into open-channel flows with suspended particles of different densities. *Environ. Fluid Mech.* **2009**, *9*, 161–186. [CrossRef]
27. Noguchi, K.; Nezu, I. Particle–turbulence interaction and local particle concentration in sediment-laden open-channel flows. *J. Hydro-Environ. Res.* **2009**, *3*, 54–68. [CrossRef]
28. Kim, J.; Moin, P.; Moser, R. Turbulence statistics in fully developed channel flow at low Reynolds number. *J. Fluid Mech.* **1987**, *177*, 133–166. [CrossRef]
29. Best, J.; Bennett, S.; Bridge, J.; Leeder, M. Turbulence modulation and particle velocities over flat sand beds at low transport rates. *J. Hydraul. Eng.* **1997**, *123*, 1118–1129. [CrossRef]
30. Graf, W.; Cellino, M. Suspension flows in open channels; experimental study. *J. Hydraul. Res.* **2002**, *40*, 435–447. [CrossRef]
31. Kaftori, D.; Hetsroni, G.; Banerjee, S. The effect of particles on wall turbulence. *Int. J. Multiph. Flow* **1998**, *24*, 359–386. [CrossRef]

32. Liakopoulos, A. Explicit representations of the complete velocity profile in a turbulent boundary layer. *AIAA J.* **1984**, *22*, 844–846. [CrossRef]
33. Liakopoulos, A. Computation of high speed turbulent boundary-layer flows using the k−$\epsilon$ turbulence model. *Int. J. Numer. Methods Fluids* **1985**, *5*, 81–97. [CrossRef]

fluids

MDPI

Article

# Effects of Expelled Air during Filling Operations with Blocking Columns in Water Pipelines of Undulating Profiles

Vicente S. Fuertes-Miquel [1,*], Oscar E. Coronado-Hernández [2] and Alfonso Arrieta-Pastrana [2]

1 Departamento de Ingeniería Hidráulica y Medio Ambiente, Universitat Politècnica de València, 46022 Valencia, Spain

2 Instituto de Hidráulica y Saneamiento Ambiental, Universidad de Cartagena, Cartagena 130001, Colombia; ocoronadoh@unicartagena.edu.co (O.E.C.-H.); aarrietap2@unicartagena.edu.co (A.A.-P.)

* Correspondence: vfuertes@upv.es

**Abstract:** Entrapped air pockets can cause failure in water distribution systems if air valves have not been appropriately designed for expelling air during filling manoeuvres performed by water utilities. One-dimensional mathematical models recently developed for studying this phenomenon do not consider the effect of blocking columns inside water pipelines. This research presents the development of a mathematical model for analysing the filling process in a pipeline with an undulating profile with various air valves, including blocking columns during starting-up water installations. The results show how different air pocket pressure peaks can be produced over transient events, which need to be analysed to ensure a successful procedure that guarantees pipeline safety during the pressure surge occurrence. In this study, an experimental set-up is analysed to observe the behaviour of two blocking columns during filling by comparing the air pocket pressure pulses.

**Keywords:** entrapped air pocket; blocking column; filling operation; water distribution system

## 1. Introduction

Transient events caused by air pockets inside pipelines have gained attention in the last decades since pressure surges can originate catastrophic scenarios that can produce the failure of hydraulic installations [1]. Most analysed scenarios assume the worst possible conditions, so the resulting air pocket pressure peaks represent the maximum values in water pipelines [2,3], which must be considered when selecting the pipe resistance class [4]. This makes studying this problem of practical interest for water utilities [5]. To avoid issues caused by entrapped air pockets, air valves are commonly installed at various points in the system [6,7]. These valves allow air to be expelled, reducing pressure surges in installations [8]. It is crucial to note that air valves do not always provide the expected reliability and can sometimes cause severe problems when they are undersized or when filling operations are not performed according to recommendations proposed by the American Water Works Association (AWWA) [9].

Despite the advantages of using air valves [10], they also present several challenges [11]. For instance, sizing and selecting air valves must be done carefully. Estimating the amount of expelled air is also challenging [12,13]. Oversizing, especially during air expulsion, can cause high overpressures, making it dangerous to select a valve that is either too small or too large [4].

Another significant issue is the proper static and dynamic modelling of these valves. While some manufacturers provide characteristic curves in their catalogues, these data are often unreliable. Various laboratories and universities, including the Polytechnic University of Valencia, have conducted static tests on different valves, and the results generally show significant discrepancies from the manufacturers' data [14]. This can lead to critical problems due to incorrect selection based on unsuitable characteristic curves. The situation regarding dynamic characterisation is even worse since there is a lack of available information. Manufacturers overlook this aspect, and there is still much to be done from a research perspective.

Besides the difficulties in sizing air valves and the unreliability of manufacturers' characteristic curves, other issues include the "dynamic closure" of these devices (an air valve closing prematurely due to a lifting effect on the float, leaving a dangerous air pocket inside water installations), the poor correlation between the nominal diameter of an air valve (usually matching the connection diameter) and its expulsion capacity, and inadequate or non-existent maintenance, which can lead to malfunction when an air valve is needed [4]. Therefore, while air valves in hydraulic installations are highly recommended, they do not guarantee total safety. For the reasons stated above, these devices can cause more issues related to pressure surges than those they are intended to solve.

During the last decades, different mathematical models have been proposed. Martin (1976) [15] proposed the first system for analysing two-phase flows in confined pipeline systems, establishing a differential equations system composed of the continuity and momentum formulations and the thermodynamic behaviour of the air phase. Liou & Hunt (1996) [16] developed a rigid column model, applicable when high flow velocities and small pipe diameters are present. Izquierdo et al. (1999) [17] investigated scenarios involving irregular pipe profiles with entrapped air using the rigid column model. This model considers a perpendicular air-water interface (piston flow) representing rapid transients where some pipe sections are filled with air. In contrast, others are entirely occupied by water.

Wylie & Streeter (1993) [18] were the first to develop an elastic model that incorporates the elasticity of the fluid, focusing solely on single-phase flows. Zhou, Liu, and Karney (2013) [19] extended this elastic model to analyse filling processes by combining the continuity and momentum equations for the water phase with the polytropic equation governing the behaviour of the air phase. This model was developed for both simple and irregular pipe cases, accounting for entrapped air pockets within the system. More recently, Zhou et al. (2020) [12] studied, using a simple pipe, the effects occurring during rapid filling phenomena in an experimental setup measuring 8.862 m in length and 400 mm in internal diameter, with a 10-mm orifice added at the end to expel airflow.

The effects of blocking columns in confined hydraulic installations represent another critical issue that requires careful consideration. This is because the elasticity of air is significantly higher than that of water [20], leading to scenarios where water can easily compress air columns, potentially causing pressure surges in hydraulic systems. Although the effects of blocking columns have been explored regarding various entrapped air pockets using both elastic and rigid models [17,19], the literature reveals a gap in addressing these issues in systems with undulating pipe profiles, particularly when air valves or orifices sizes are installed during filling operations.

This research presents a mathematical model for studying transient events with entrapped air pockets, including air valves. This study provides a valid model for analysing transients generated by $n$ entrapped air pockets in pipelines with irregular profiles and $V$ installed air valves [21,22]. The proposed model's generalised equations, including the air valve's behaviour, are presented. The main advantage of the proposed model over previous ones is that it is suitable for considering the effects of blocking columns and air valves. Furthermore, this system of equations is particularised for a specific case: an installation with two entrapped air pockets and an air valve installed at the highest point of a water installation with an irregular profile. The problem involves the presence of a second air pocket [23]. A polytropic evolution was considered to characterise the behaviour of air pockets during transient events. A specific case is analysed using the obtained system of equations and its corresponding numerical resolution. The selected case study corresponds to a water installation located at the hydraulic laboratory at the Technical University of Valencia, Spain. Transient events are analysed considering two scenarios: first without air valves and then considering an air valve installed at the highest point in the pipeline.

The results for both situations are compared, observing the effect of air valves on the transient events. When the air pocket's location coincides with the air valve's position, the device allows free air expulsion, significantly smoothing the air compression process (reducing pressure peaks). However, when the water arrives and the air valve closes, the air pocket is again confined between two water columns with no escape to the atmosphere, leading to a new transient evolution similar to those studied without valves, generating the corresponding pressure peaks based on the conditions at the time of valve closure. Finally, this research analyses the same installation with different valve sizes. This allows for studying the effect of air valves' sizes and characteristics (mainly the expulsion coefficient and outlet size) on various transient events.

## 2. Mathematical Model

This section shows the mathematical model's development considering the air valves' behaviour. In this sense, a pipeline with an irregular profile and an air valve at the highest point in the installation is considered. At this point, an air pocket is initially trapped between two water columns (Figure 1).

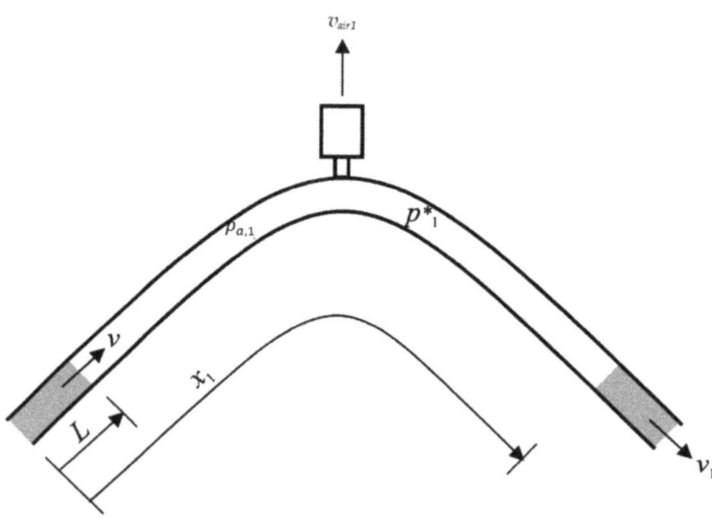


**Figure 1.** Installation of an air release valve.

The governing equations are presented below, considering the following assumptions: (i) water movement is simulated using the rigid water column model; (ii) a polytropic law is suitable for following the behaviour of entrapped air pockets; (iii) the air-water interface is considered perpendicular with the main direction of the pipe, which is applicable for pipelines with small internal diameters; (iv) a friction factor using steady flow conditions is considered; and (v) an isentropic flow is considered for the expelled air.

For a filling column, there are two equations:

1. Equation of the rigid model characterising the motion of the filling column driven by an energy source ($p_0^*$ = upstream pressure in the pipeline; $p_1^*$ = air pocket pressure)

$$\frac{dv}{dt} = \frac{p_0^* - p_1^*}{\rho L} - g\frac{\Delta z}{L} - \frac{fv|v|}{2D}, \quad (1)$$

where $v$ = water velocity, $f$ = friction factor, $D$ = internal pipe diameter, $\Delta z$ = difference elevation between two points of a water column, $\Delta z/L$ = gravity term, $\rho$ = water density, $L$ = length a water column, and $g$ = gravitational acceleration.

2. Position of the filling column

$$L = L_0 + \int_0^t v dt, \quad (2)$$

where the subscript 0 refers to an initial condition.

For an air pocket, there are three equations:

3. Evolution of the air pocket (assuming a polytropic evolution)

$$\frac{dp_1^*}{dt} = k\frac{p_1^*}{\forall_1}\frac{d\forall_1}{dt} + \frac{p_1^*}{\forall_1}\frac{k}{\rho_{a,1}}\frac{dm_{a,1}}{dt}, \quad (3)$$

where $m_{a,1}$ = air mass, $k$ = polytropic coefficient, and $\forall_1$ = air pocket volume. The subscript 1 refers to the condition of the air pocket. The polytropic coefficient varies from 1.0 (isothermal evolution) to 1.4 (adiabatic condition).

4. Continuity equation for an air pocket (considering that the air pocket density ($\rho_{a,1}$) and air density at the outlet section ($\rho_{air1}$) are equal)

$$\frac{dm_{a,1}}{dt} = -\rho_{a,1} v_{air1} A_{exp}, \quad (4)$$

where $v_{air1}$ = air velocity and $A_{exp}$ = cross-sectional area of an air valve.

Defining the air mass:

$$\frac{d(\rho_{a,1}\forall_1)}{dt} = -\rho_{a,1} v_{air1} A_{exp}, \quad (5)$$

Solving the derivate:

$$\frac{d\rho_{a,1}}{dt}\forall_1 + \rho_{a,1}\frac{d\forall_1}{dt} = -\rho_{a,1} v_{air1} A_{exp}, \quad (6)$$

Considering that:

$$\begin{cases} \forall_1 = (x_1 - L)A \\ \frac{d\forall_1}{dt} = (v_1 - v)A \end{cases} \quad (7)$$

where $A$ = cross-sectional area of the pipe, $x_1$ = position of the water column, and $v_1$ = water velocity of a blocking column.

Organising terms:

$$\frac{d\rho_{a,1}}{dt}(x_1 - L)A + \rho_{a,1}(v_1 - v)A = -\rho_{a,1} v_{air1} A_{exp}, \quad (8)$$

Solving for the air density:

$$\frac{d\rho_{a,1}}{dt} = \frac{-\rho_{a,1} v_{air1} A_{exp} - \rho_{a,1}(v_1 - v)A}{(x_1 - L)A}, \quad (9)$$

5. Equation modelling the behaviour of the air release valve

During the pipeline filling process, it is typical that air valves operate in the subsonic flow range ($p_1^* < 19.55$ m = 1.918 bar). The behaviour of expelled air in the subsonic region follows a simple expression that relates the expelled airflow $Q_{air1}$ under normal conditions (1 atm and 15 °C, that is, a density $\rho_{a,N} = 1.205\ \text{kg/m}^3$) with the pressure inside a pipeline $p_1^*$. The corresponding equation is given by:

$$Q_{air1} = c_{exp}\sqrt{(p_1^* - p_{atm}^*)p_1^*}, \quad (10)$$

where $c_{exp}$ = expelled coefficient of an air valve and $p_{atm}^*$ = atmospheric pressure.

From this, the air expulsion velocity $v_{air1}$ can be determined:

$$\rho_{a,N} Q_{air1} = \rho_{a,1} v_{air1} A_{exp}, \tag{11}$$

where $\rho_{a,N}$ = air density at atmospheric conditions (1.205 kg/m$^3$).

Equation (11) can be expressed as:

$$v_{air1} = \frac{1.205}{\rho_{a,1}} \frac{c_{exp}}{A_{exp}} \sqrt{(p_1^* - p_{atm}^*) p_1^*}, \tag{12}$$

If a filling operation were a perfectly controlled process, it would proceed slowly, and air valves would operate in the subsonic range. However, when a pipeline filling occurs more abruptly, the pressure in hydraulic installations may exceed the subsonic flow limit ($p_1^* \geq 19.55\,\text{m} = 1.918\,\text{bar}$); in this case, the expelled air is found in the sonic flow regime. The behaviour of an air valve in the sonic region fits well with a straight line, meaning the relationship between the expelled airflow under normal conditions $Q_{air1}$ and the pressure inside in a pipeline $p_1^*$ is linear. To extend the previous equation and thus have a valid expression for the sonic flow region, this line has the same slope as the one at the limit point ($p_1^* = 19.55\,\text{m}$) when the flow is subsonic. Therefore, the line should pass through the point $Q_{air1} = 13.426 c_{exp}$ and $p_1^* = 19.55$ m. The slope of this line is:

$$\text{slope} = \left. \frac{dQ_{air1}}{dp_1^*} \right|_{p_1^*=19.55} = c_{exp} \frac{2p_1^* - p_{atm}^*}{2\sqrt{(p_1^* - p_{atm}^*) p_1^*}} = 1.071, \tag{13}$$

Thus, the expression that simulates the behaviour of the expelled air when an air valve is operating in the sonic region (when $p_1^* \geq 19.55\,\text{m}$) is:

$$Q_{air1} = -7.521 c_{exp} + 1.071 c_{exp} p_1^*, \tag{14}$$

Expressing this line in terms of the air velocity $v_{air1}$ is as follows:

$$v_{air1} = \frac{1.205}{\rho_{a,1}} \frac{c_{exp}}{A_{exp}} (-7.521 + 1.071 \rho_1), \tag{15}$$

For a blocking column, there are two equations:

6. Equation of the rigid model characterizing the movement of a blocking column ($p_1$* is the pressure of the air pocket and $p_{atm}^*$ is atmospheric pressure)

$$\frac{dv_1}{dt} = \frac{p_1^* - p_{atm}^*}{\rho L_{b,1}} - g \frac{\Delta z_{b,1}}{L_{b,1}} - \frac{f v_1 |v_1|}{2D}, \tag{16}$$

where $L_{b,1}$ = length of the blocking column and $\frac{\Delta z_{b,1}}{L_{b,1}}$ = gravity term of the blocking column.

7. Position of a blocking column

$$x_1 = x_{1,0} + \int_0^1 v_1 dt \text{ or } \frac{dx_1}{dt} = v_1, \tag{17}$$

Therefore, a system composed of 7 equations needs to be solved. The 7 unknown variables for the analysed problem are:

Two unknowns for the filling column:

- Water velocity $v$ and position $L$

Three unknowns for the air pocket:

- Expulsion velocity $v_{air1}$, pressure $p_1^*$, and density $\rho_{a,1}$

Two unknowns for the blocking column:

- Velocity $v_1$ and position $x_1$

This system of equations remains valid until the filling column reaches the position of an air valve. At that moment, the water pressure will cause the float to rise, closing the valve and nullifying the air expulsion velocity ($v_{air1} = 0$).

In this new situation, the equation modelling the air valve's behaviour (Equations (12) and (15)), depending on whether it operates in the subsonic or sonic zone, respectively, is removed from the system. Then, a system of 6 equations with 6 unknowns ($v$, $Lv_1$, $x_1$, $p_1^*$, $\rho_{a,1}$) needs to be solved. The continuity equation for the air pocket (Equation (9)) can be neglected because the computation of the air density $\rho_{a,1}$ is no longer necessary; thus, Equation (3) becomes:

$$p_1^*(x_1 - L)^k = p_{1,0}^*(x_{1,0} - L_0)^k = \text{constant} \tag{18}$$

Equation (18) maintains the assumption of a polytropic evolution when an air valve closes and an air pocket is trapped. A water utility's standard procedure for cleaning and maintaining water pipelines involves starting them up. This process begins when the pipeline is at rest because the regulating valves are closed.

## 3. Description for Two Entrapped Air Pockets and an Air Valve

### 3.1. Governing Equations

The proposed model is presented for an installation with four pipe branches of constant slope ($m = 4$) and two entrapped air pockets ($n = 2$) where there is an air valve ($V = 1$) at the highest point of the profile (see Figure 2).

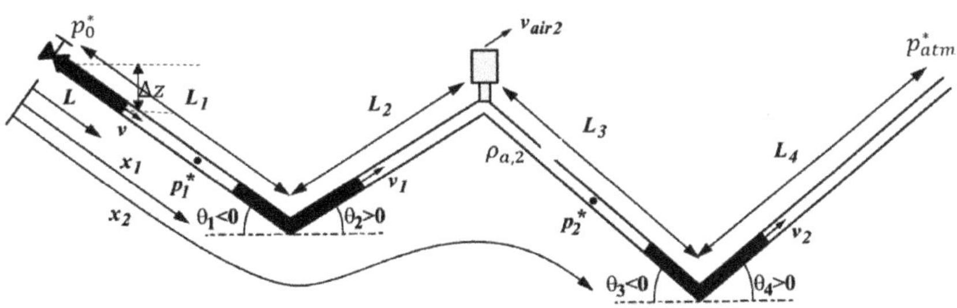


**Figure 2.** Installation with two entrapped air pockets and an air release valve (Phase A).

Ten equations must be used to model the pipeline filling for an installation with two entrapped air pockets and an air valve. The different phases of this process are presented below.

#### 3.1.1. Phase A

The system starts at rest ($t = 0$), as depicted in Figure 2. The system of equations for modeling this problem are described as follows:

- For the filling column:

1. Equation of the rigid model to consider the movement of the filling column ($p_0^*$ is the upstream pressure in the pipeline and $p_1^*$ is the pressure of the first air pocket)

$$\frac{dv}{dt} = \frac{p_0^* - p_1^*}{\rho L} - g\frac{\Delta z}{L} - \frac{fv|v|}{2D}, \tag{19}$$

2. Position of the filling column

$$L = L_0 + \int_0^t v dt \ \left(\frac{dL}{dt} = v\right), \tag{20}$$

- For the first air pocket, there is 1 equation:

3. Evolution of air pocket 1 (polytropic law)

$$p_1^*(x_1 - L)^k = p_{1,0}^*(x_{1,0} - L_0)^k = \text{constant}. \tag{21}$$

- For the first blocking column, there are 2 equations:

4. Equation of the rigid model that characterizes the movement of blocking column 1 ($p_1^*$ is the pressure of the first air pocket and $p_2^*$ is the pressure of the second air pocket)

$$\frac{dv_1}{dt} = \frac{p_1^* - p_2^*}{\rho L_{b,1}} - g\frac{\Delta z_{b,1}}{L_{b,1}} - \frac{f v_1 |v_1|}{2D}, \tag{22}$$

5. Position of blocking column 1

$$x_1 = x_{1,0} + \int_0^t v_1 dt \quad \left(\frac{dx_1}{dt} = v_1\right), \tag{23}$$

For the last air pocket, there are 3 equations:
6. Evolution of air pocket 2 (assuming a polytropic evolution)

$$\frac{dp_2^*}{dt} = k\frac{p_2^*}{\forall_2}\frac{d\forall_2}{dt} + \frac{p_2^*}{\forall_2}\frac{k}{\rho_{a,2}}\frac{dm_{a,2}}{dt}, \tag{24}$$

7. Continuity equation for air pocket 2 (applying a similar procedure described in Section 2)

$$\frac{dm_{a,2}}{dt} = -\rho_{a,2} v_{air2} A_{exp}, \tag{25}$$

$$\frac{d(\rho_{a,2}\forall_2)}{dt} = -\rho_{a,2} v_{air2} A_{exp}, \tag{26}$$

$$\frac{d\rho_{a,2}}{dt}\forall_2 + \rho_{a,2}\frac{d\forall_2}{dt} = -\rho_{a,2} v_{air2} A_{exp}, \tag{27}$$

$$\forall_2 = (x_2 - x_1 - L_{b,1})A, \tag{28}$$

$$\frac{d\forall_2}{dt} = (v_2 - v_1)A, \tag{29}$$

$$\frac{d\rho_{a,2}}{dt}(x_2 - x_1 - L_{b,1})A + \rho_{a,2}(v_2 - v_1)A = -\rho_{a,2} v_{air2} A_{exp}, \tag{30}$$

$$\frac{d\rho_{a,2}}{dt} = \frac{-\rho_{a,2} v_{air2} A_{exp} - \rho_{a,2}(v_2 - v_1)A}{(x_2 - x_1 - L_{b,1})A}, \tag{31}$$

where subscript 2 refers to the second air pocket.
8. Equation for the air valve characterisation
Subsonic flow region ($p_2^* < 19.55\,\text{m} = 1.918\,\text{bar}$):

$$Q_{air2} = c_{exp}\sqrt{(p_2^* - p_{atm}^*)p_2^*}, \tag{32}$$

$$v_{air2} = \frac{1.205}{\rho_{a,2}}\frac{c_{exp}}{A_{exp}}\sqrt{(p_2^* - p_{atm}^*)p_2^*}, \tag{33}$$

Sonic flow region ($p_2^* \geq 19.55\,\text{m} = 1.918\,\text{bar}$):

$$Q_{air2} = -7.521 c_{exp} + 1.071 c_{exp} p_2^*, \tag{34}$$

$$v_{air2} = \frac{1.205}{\rho_{a,2}}\frac{c_{exp}}{A_{exp}}(-7.521 + 1.071 p_2^*), \tag{35}$$

For the last blocking column, there are 2 equations:

9. Equation of the rigid model characterizing the movement of blocking column 2 ($p_2^*$ is the pressure of the second air pocket and $p_{atm}^*$ is the downstream pressure of the pipeline)

$$\frac{dv_2}{dt} = \frac{p_2^* - p_{atm}^*}{\rho L_{b,2}} - g\frac{\Delta z_{b,2}}{L_{b,2}} - \frac{f v_2 |v_2|}{2D}, \tag{36}$$

10. Position of blocking column 2

$$x_2 = x_{2,0} + \int_0^t v_2 dt \quad \left(\frac{dx_2}{dt} = v_2\right), \tag{37}$$

where $L_{b,2}$ = length of the blocking column 2, $\frac{\Delta z_{b,2}}{L_{b,2}}$ = gravity term of the blocking column 2, and $v_2$ = water velocity of the blocking column 2.

This results in a system of 10 equations, the resolution of which, along with the initial conditions ($t = 0$) and the corresponding boundary conditions ($p_0^*$ and $p_{atm}^*$), are needed to compute the 10 unknowns of the problem, which are:

- Two unknowns for the filling column: $v$, $L$
- One unknown variable for the air pocket 1: $p_1^*$
- Two unknowns for the blocking column 1: $v_1$, $x_1$
- Three unknowns for the air pocket 2: $v_{air2}$, $p_2^*$, $\rho_{a,2}$
- Two unknowns for the blocking column 2: $v_2$, $x_2$

### 3.1.2. Phase B

The previous system of equations is only valid while the air valve is expelling air, that is, until the first blocking column reaches the position of the valve ($x_1 + L_{b,1} = x_{v1}$). The air valve closes from this moment and the air expulsion velocity will become zero ($v_{air2} = 0$).

Since the air valve no longer functions in this phase (see Figure 3), the corresponding equation that models the air valve's behaviour can be neglected (Equations (33) and (35)), depending on whether the flow is subsonic or sonic.

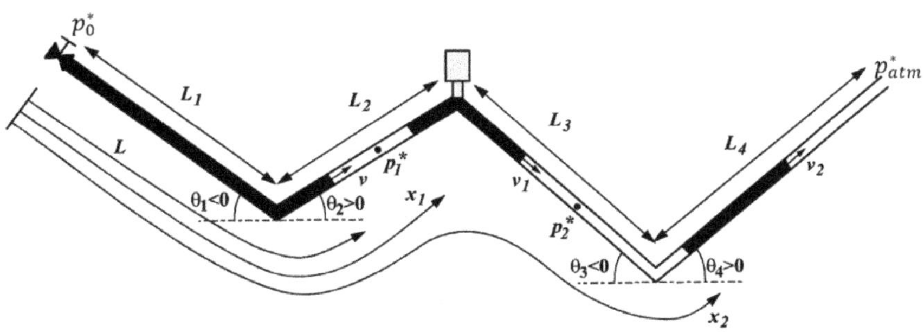


**Figure 3.** Phase B of the transient event for the two entrapped air pockets and an air valve.

Thus, for this phase of the transient, there is a system composed of 9 equations with 9 unknown variables ($v$, $L$, $v_1$, $x_1$, $p_1^*$, $v_2$, $x_2$, $p_2^*$, $\rho_{a,2}$). This system can be simplified by omitting the continuity equation for the second air pocket (Equation (31)), as it is no longer necessary to calculate the air density $\rho_2$ to solve the problem. Then, the system is reduced to 8 equations with 8 unknown variables ($v$, $L$, $v_1$, $x_1$, $p_1^*$, $v_2$, $x_2$, $p_2^*$). Equation (25) can be written as follows:

$$p_2^*(x_2 - x_1 - L_{b,1})^k = p_{2,0}^*(x_{2,0} - x_{1,0} - L)^k = \text{constant}, \tag{38}$$

3.1.3. Phase C

This new situation remains until the first air pocket reaches the position of the air valve ($x_1 = x_{v1}$). The air valve opens by expelling air, as shown in Figure 4.

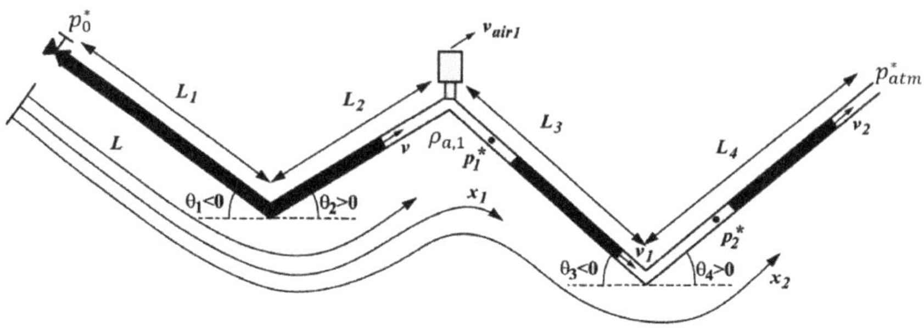


**Figure 4.** Phase C of the transient event for the two entrapped air pockets and an air valve.

As in the first phase, a system of 10 equations with ten unknowns is obtained:
For the filling column, there are 2 equations:
1. Equation of the rigid model characterizing the movement of the filling column

$$\frac{dv_2}{dt} = \frac{p_0^* - p_1^*}{\rho L} - g\frac{\Delta z}{L} - \frac{fv|v|}{2D}, \tag{39}$$

2. Position of the filling column

$$L = L_0 + \int_0^t vdt \;\; \frac{dL}{dt} = v, \tag{40}$$

For the first air pocket, there are 3 equations:
3. Evolution of air pocket 1 (considering a polytropic evolution)

$$\frac{dp_1^*}{dt} = k\frac{p_1^*}{\forall_1}\frac{d\forall_1}{dt} + \frac{p_1^*}{\forall_1}\frac{k}{\rho_{a,1}}\frac{dm_{a,1}}{dt}, \tag{41}$$

4. Continuity equation for air pocket 1

$$\frac{d\rho_{a,1}}{dt} = \frac{-\rho_{a,1}v_{air1}A_{exp} - \rho_{a,1}(v_1 - v)A}{(x_1 - L)A}, \tag{42}$$

5. Equation for modelling the behaviour of the air valve
Subsonic flow region ($p_1^* < 19.55\,\text{m} = 1.918\,\text{bar}$)

$$Q_{air1} = c_{exp}\sqrt{(p_1^* - p_{atm})p_1^*}, \tag{43}$$

$$v_{air1} = \frac{1.205}{\rho_{a,1}}\frac{c_{exp}}{A_{exp}}\sqrt{(p_1^* - p_{atm})p_1^*}, \tag{44}$$

Sonic flow region ($p_1^* \geq 19.55\,\text{m} = 1.918\,\text{bar}$)

$$Q_{air1} = -7.521c_{exp} + 1.071c_{exp}p_1^*, \tag{45}$$

$$v_{air1} = \frac{1.205}{\rho_{a,1}}\frac{c_{exp}}{A_{exp}}(-7.521 + 1.071p_1^*), \tag{46}$$

For the first blocking column, there are 2 equations:

6. Equation of the rigid model characterizing the movement of blocking column 1

$$\frac{dv_2}{dt} = \frac{p_1^* - p_2^*}{\rho L_{b,1}} - g\frac{\Delta z_{b,1}}{L_{b,1}} - \frac{fv_1|v_1|}{2D}, \tag{47}$$

7. Position of blocking column 1

$$x_1 = x_{1,0} + \int_0^t v_1 dt \left(\frac{dx_1}{dt} = v_1\right), \tag{48}$$

8. Evolution of air pocket 1

$$p_2^*(x_2 - x_1 - L_{b,1})^k = p_{2,0}^*(x_{2,0} - x_{1,0} - L_{b,1})^k = \text{constant}, \tag{49}$$

For the last blocking column, there are 2 equations:
9. Equation of the rigid model characterizing the movement of blocking column 2

$$\frac{dv_2}{dt} = \frac{p_2^* - p_{atm}^*}{\rho L_{b,2}} - g\frac{\Delta z_{b,2}}{L_{b,2}} - \frac{fv_2|v_2|}{2D}, \tag{50}$$

10. Position of blocking column 2

$$x_2 = x_{2,0} + \int_0^t v_2 dt \left(\frac{dx_2}{dt} = v_2\right), \tag{51}$$

The solution of this system composed of 10 equations, along with the initial conditions ($t = 0$) and boundary conditions ($p_0^*$ y $p_{atm}^*$), must be solved to determine the 10 unknowns of the problem posed:

- Two unknowns for the filling column: $v$, $L$
- Three unknowns for air pocket 1: $v_{aire1}$, $p_1^*$, $\rho_{a,1}$
- Two unknowns for blockage column 1: $v_1$, $x_1$
- One unknown for air pocket 2: $p_2^*$
- Two unknowns for blockage column 2: $v_2$, $x_2$

### 3.1.4. Phase D

This system of equations is valid only until the filling column reaches the position of the air release valve ($L = x_{v1}$). The air valve closes again at that moment and the air expulsion velocity is null ($v_{aire1} = 0$).

Since the air valve is inactive during this transient phase (Figure 5), the equation is neglected.

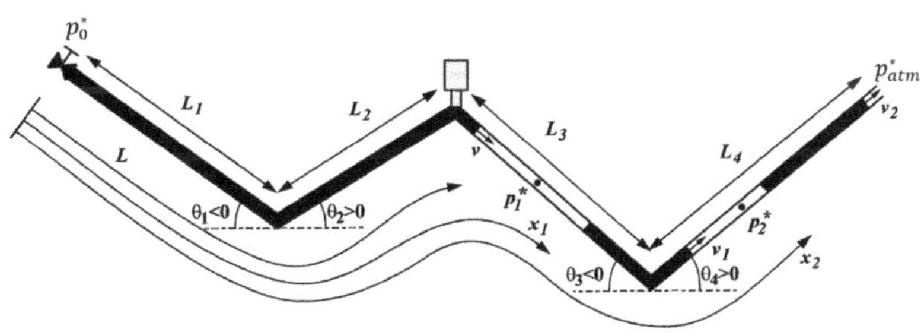


**Figure 5.** Phase D of the transient event for two entrapped air pockets and an air valve.

Therefore, the system that simulates the problem consists of a total of 9 equations with 9 unknown variables ($v$, $L$, $v_1$, $x_1$, $p_1^*$, $\rho_{a,1}$, $v_2$, $x_2$, $p_2^*$). Equation (4) can be omitted since there is no need to calculate the air density $\rho_{a,1}$. Thus, the system is reduced to 8 equations with 8 unknown variables ($v$, $L$, $v_1$, $x_1$, $p_1^*$, $v_2$, $x_2$, $p_2^*$), where Equation (41) becomes:

$$p_1^*(x_1 - L)^k = p_{1,0}^*(x_{1,0} - L_0)^k = \text{constant,} \tag{52}$$

### 3.1.5. Phase E

The previous situation persists as long as there are two air pockets and two blocking columns. When the last water column completely leaves the water installation ($x_2 = L_{total}$), a new phase begins. There is only one blocking column and air pocket (Figure 6).

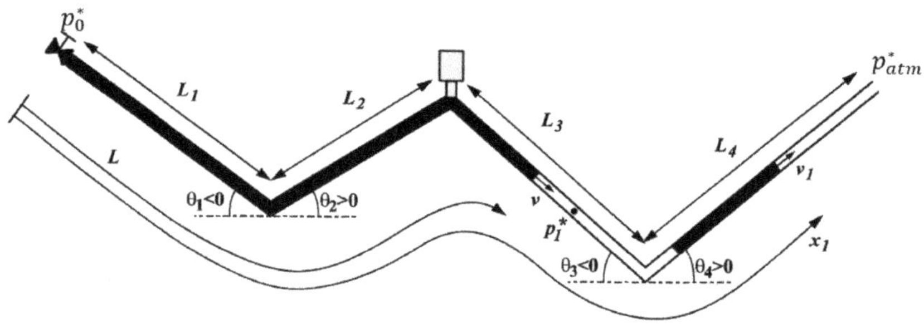

**Figure 6.** Phase E of the transient event for two entrapped air pockets and an air valve.

In this transient phase, all equations corresponding to the last blocking column and the last air pocket disappear. The downstream pressure of the blocking column becomes ($p_2^* = p_{atm}^*$) at the atmospheric conditions, resulting in the following system of equations:

$$\frac{dv}{dt} = \frac{p_0^* - p_1^*}{\rho L} - g\frac{\Delta z}{L} - \frac{fv|v|}{2D}, \tag{53}$$

$$L = L_0 + \int_0^t v dt \quad \left(\frac{dL}{dt} = v\right), \tag{54}$$

$$p_1^*(x_1 - L)^k = p_{1,0}^*(x_{1,0} - L_0)^k = \text{constant,} \tag{55}$$

$$\frac{dv_1}{dt} = \frac{p_1^* - p_{atm}^*}{\rho L_{b,1}} - g\frac{\Delta z_{b,1}}{L_{b,1}} - \frac{fv_1|v_1|}{2D}, \tag{56}$$

$$x_1 = x_{1,0} + \int_0^t v_1 dt \left(\frac{dx_1}{dt} = v_1\right), \tag{57}$$

The system of equations modelling the installation during this phase of the filling procedure, where there is only one entrapped air pocket, has been reduced to 5 equations with 5 unknown variables ($v$, $L$, $v_1$, $x_1$, $p_1^*$).

### 3.1.6. Phase F

The previous phase ends when the only remaining blocking column completely comes out the installation ($x_1 = L_{total}$). At this point, the existing air pocket disappears, leaving only the filling column, which partially occupies the installation (Figure 7).

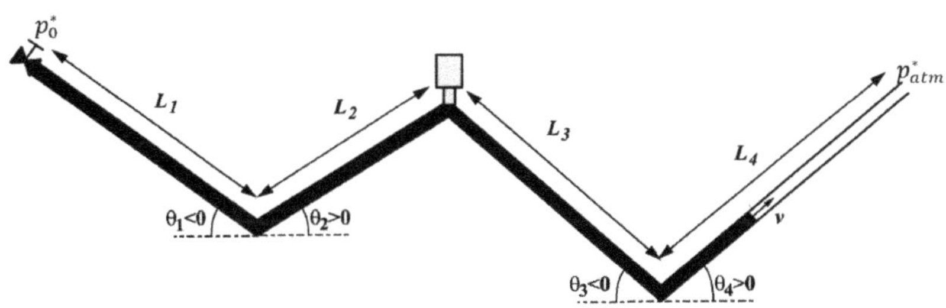


**Figure 7.** Phase F of the transient event for two entrapped air pockets and an air valve.

Under these conditions, the equations corresponding to the first blocking column and air pocket disappear. The filling column's downstream pressure becomes atmospheric ($p_1^* = p_{atm}^*$), reducing the system to two equations:

$$\frac{dv}{dt} = \frac{p_0^* - p_{atm}^*}{\rho L} - g\frac{\Delta z}{L} - \frac{fv|v|}{2D}, \tag{58}$$

$$L = L_0 + \int_0^t v dt \quad \left(\frac{dL}{dt} = v\right), \tag{59}$$

Transient modelling during the filling process can be carried out in this phase where there are no longer any air pockets, using a system of 2 equations with 2 unknown variables ($v$, $L$).

3.1.7. Phase G

Finally, when the filling column reaches the end of the hydraulic installation and begins to drain ($L = L_{total}$), the last phase of the transient event begins. The length of the filling column remains constant ($L = constant$) and the rigid model can be used to characterise the movement of the water column:

$$\frac{dv}{dt} = \frac{p_0^* - p_{atm}^*}{\rho L} - g\frac{\Delta z}{L} - \frac{fv|v|}{2D}, \tag{60}$$

This differential equation models the final stage of the transient until it reaches the steady state.

### 3.2. Numerical Resolution

The most general case involves the presence of entrapped air pockets and various installed air valves along hydraulic pipelines. This scenario requires solving a system of $2 + 3n + 2V$ equations that determine the behaviour of the filling column (length $L$ and velocity $v$) and three unknowns associated with each filling column and entrapped air pockets (upstream boundary position $x_i$, trapped air pocket pressure $p_i^*$, and blocking column velocity $v_i$), as well as two unknown variables for each installed air valve (air expulsion velocity $v_{air,r}$ and density of the air pocket exiting through the device $\rho_{a,r}$).

There is a system of $2 + 3n + 2V$ differential equations that must be solved. It is a nonlinear initial value problem, yet it meets the necessary assumptions to ensure the existence of a mathematical solution.

The resolution can be obtained using an appropriate numerical method. An adaptive step method was selected. The primary reason is that the transient phases involve abrupt changes in variables (requiring very small integration steps), while other phases are much smoother (allowing for larger integration steps). Therefore, the adaptive fifth-order Runge-Kutta method is suitable for this problem.

It is essential to highlight that the filling process constitutes a transient phenomenon characterised by distinct phases governed by different equations. In summary, the numerical resolution of the phenomenon involves solving a sequence of initial value problems with a decreasing number of differential equations. The initial condition for each issue is the final state of the previous system, with variables no longer holding physical significance.

## 4. Analysis of Results and Discussion

### 4.1. Practical Application

The proposed model is applied to study the filling of a system with four pipe branches of constant slope ($m = 4$), two air pockets ($n = 2$), and one air valve ($V = 1$) installed at the highest point in the pipeline of irregular profile.

An experimental facility was configured at the Technical University of Valencia, Spain (see Figure 8). This installation is fed using a pump, which was tested in the experimental facility. The pump curve is $H_b(\text{m}) = 38.68 - 19.76\ Q\ (\text{l/s})^2$. It discharges into the atmosphere in a reservoir. The pipeline has a total length $(L_{total})$ of 8.62 m, a diameter $(D)$ of 18.8 mm, and a friction factor $(f)$ of 0.020. Once a steady-state regime is established, the flow rate is $Q_{\text{steady}} = 0.954 \frac{l}{s}$ (with a velocity of $v_{\text{steady}} = 3.437 \frac{m}{s}$). The steady state was reached at the end of the transient event, which was considered by numerical and experimental results.

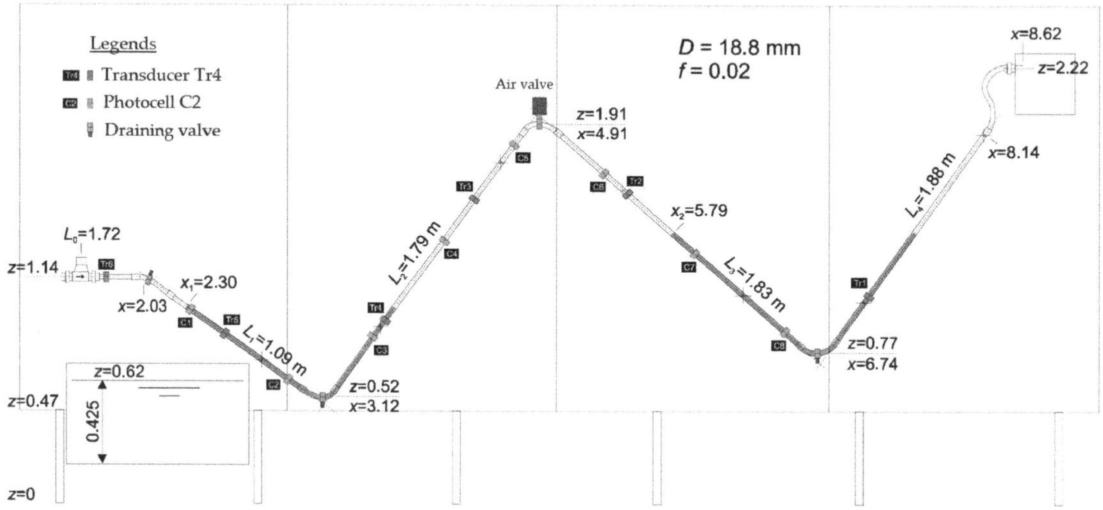


**Figure 8.** Experimental facility (installation with two entrapped air pockets and an air valve).

For the practical application, the initial positions of the blocking columns and, consequently, the initial sizes of the air pockets are:

- $x_{1,0} = 2.30\,\text{m}$ ($L_{b,1} = 1.443\,\text{m}$, $L_{air1,0} = 0.58\,\text{m}$, $\forall_{1,0} = 0.161\,\text{l}$)
- $x_{2,0} = 5.79\,\text{m}$ ($L_{b,2} = 1.725\,\text{m}$, $L_{air2,0} = 2.047\,\text{m}$, $\forall_{2,0} = 0.5682\,\text{l}$)

Once the pump is turned on, the discharge electro-valve is opened, thus beginning the pipeline's filling operation.

### 4.2. Installation without Air Valves

Solving the corresponding system of equations yields the results depicted graphically in Figure 9. This figure presents the pressure evolution of the two air pockets throughout the transient process. To measure the pressure pulses of the air pockets, transducers Tr3, Tr4, and Tr5 were used [20]. At the onset of the transient event, air pocket 1 is positioned at the location of Tr5, and the peak value predicted by the proposed model is 18.577 m at 0.19 s, which closely matches the experimental measurement of 17.484 m. A similar correspondence is observed with air pocket 2 and transducer Tr3. It is crucial to emphasise that monitoring air pockets should consider the different transducers used. In this sense, the proposed model was used. Figure 9 shows, in purple colour, the difference between the computed (continuous line) and measured (discontinuous line) position of transducer Tr5, where the proposed model can follow the air pocket pulses during the transient event. The proposed model is suitable for predicting extreme pressure pulses for both air pockets. Similarly, Figure 10 shows the water velocity evolution of the three water columns.

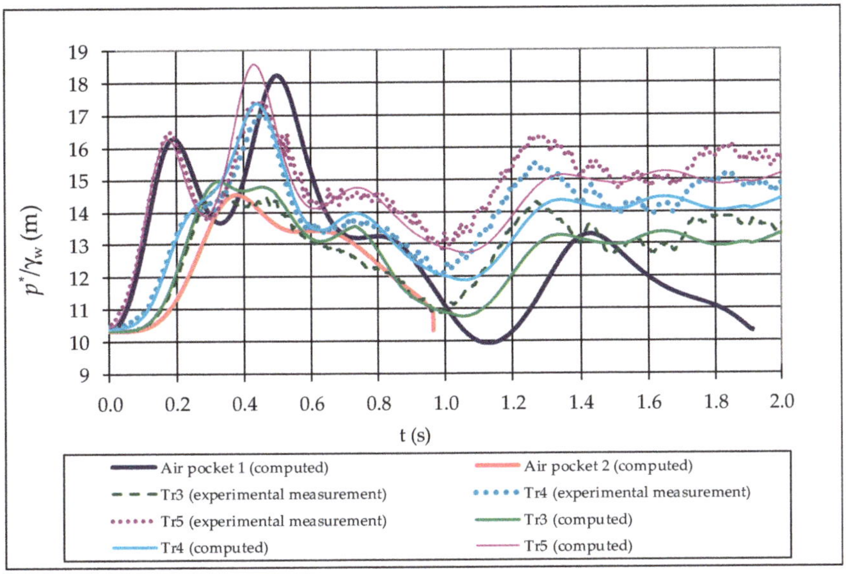


**Figure 9.** Pressure oscillations during the transient without air valves.

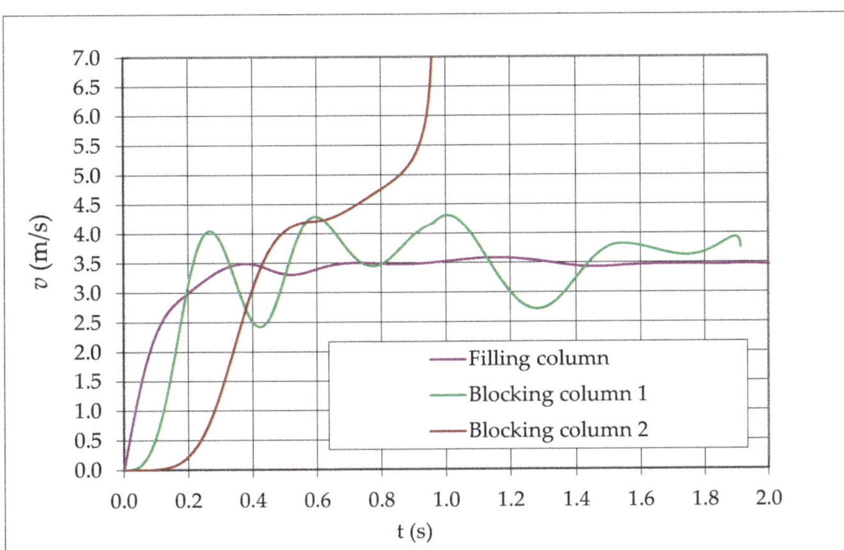


**Figure 10.** Water velocity pulses during the transient without air valves.

Due to the small size of the studied installation, the transient process lasts only a few seconds. An adiabatic evolution was considered during the experimental stage since entrapped air pockets are rapidly compressed during the transient event, considering that there is no air expulsion. Table 1 presents the times at which the three water columns reach the downstream end of the pipeline.

**Table 1.** Characteristic times of the transient event without air valves.

| Water Column | Time (s) |
|---|---|
| Blocking column 2 comes out of the installation | 0.963 |
| Blocking column 1 comes out of the installation | 1.912 |
| The filling column reaches the final point of the installation | 2.080 |

During the initial instant of the transient event, the air pockets compress notably (especially the first one) and reach their maximum pressures. Table 2 presents these maximum values and their corresponding time occurrences.

**Table 2.** Maximum air pocket pressure heads without air valves.

| Type of Peak | Pressure Head (m) | Time (s) |
|---|---|---|
| First peak in air pocket 1 | $p_1^* = 16.286$ | 0.193 |
| Second peak in air pocket 1 | $p_1^* = 18.230$ | 0.500 |
| First peak in air pocket 2 | $p_2^* = 14.540$ | 0.379 |

*4.3. Installation Using an Air Valve*

This section presents the results considering an air valve located at the highest point of the installation ($x_{V1} = 4.91$ m), which is suitable for expelling air. Table 3 presents the main characteristics of the air valve. An isothermal process was assumed for this analysis, as the air release valves expel air, reducing the likelihood of abrupt changes in the entrapped air pockets.

**Table 3.** Characteristics of the selected air valve.

| Parameter | Value and Units |
|---|---|
| Expulsion coefficient | $c_{exp} = 0.00028\ (\mathrm{Nm^3/s})/\mathrm{m}$ |
| Cross-sectional area | $A_{exp} = 9.8\ \mathrm{mm^2}$ |
| Internal diameter | $D_{exp} = 3.5\ \mathrm{mm}$ |

The air valve has a coefficient $c_{exp} = 0.00028 \frac{\left(\frac{\mathrm{Nm^3}}{\mathrm{s}}\right)}{\mathrm{m}}$, which is suitable of expelling an airflow rate of approximately 8.9 $\frac{\mathrm{Nm^3}}{\mathrm{h}}$ ($Q_{air} = 2.48 \times 10^{-3}\ \frac{\mathrm{Nm^3}}{\mathrm{s}}$) with a differential pressure of 0.5 bar ($\Delta p = 5.1$ m).

The numerical resolution yields the transient evolution shown in Figure 11 (pressure of the two air pockets) and Figure 12 (water velocities of the filling column and the two blocking columns).

It is observed that with the presence of the air vent, the second blocking column comes out of the installation at time $t = 1.259$ s (30% later than filling without the air valve).

The maximum values of air pocket pressure heads considering the installation of an air valve are presented in Table 4.

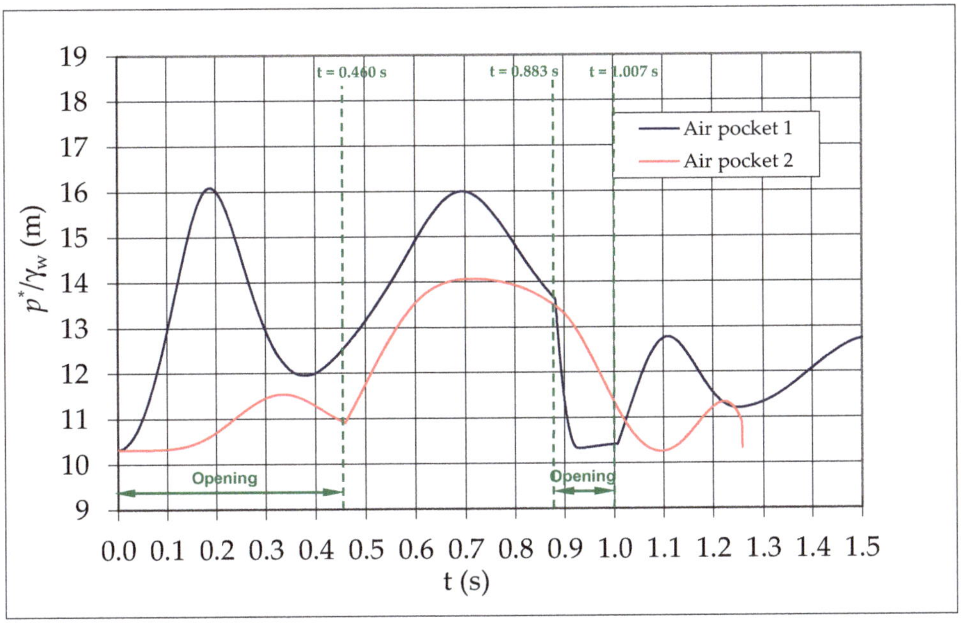


**Figure 11.** Pressure oscillations during the transient using the air valve.

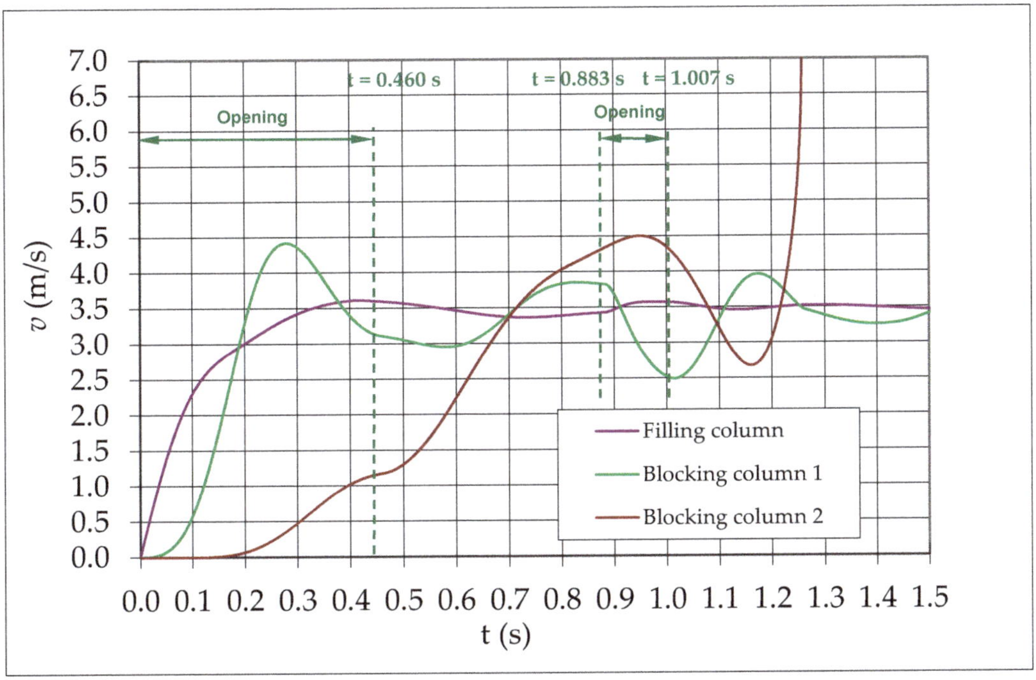


**Figure 12.** Water velocity pulses during the transient using the air valve.

**Table 4.** Maximum air pocket pressure heads using the air valve.

| Type of Peak | Pressure Head (m) | Time (s) |
|---|---|---|
| First peak in air pocket 1 | $p_1^* = 16.093$ | 0.189 |
| Second peak in air pocket 1 | $p_1^* = 15.997$ | 0.695 |
| First peak in air pocket 2 | $p_2^* = 11.538$ | 0.335 |
| Second peak in air pocket 2 | $p_2^* = 14.070$ | 0.714 |

Initially, the position of the air valve coincides with the second air pocket, so the device is opened and begins expelling air until the corresponding blocking column arrives ($t = 0.460$ s). Throughout this time, the presence of the air valve produces a lower peak of pressure heads compared to when it is neglected. The explanation is straightforward: as the installation fills, the air pocket compresses, causing high-pressure peaks when no air valves are present. The air valve allows the expelling of airflow, preventing the air pocket from compressing as much.

When the water column arrives, the air valve closes and remains closed until another air pocket arrives ($t = 0.883$ s). The air pockets compress, usually with corresponding pressure peaks, since there is no air outlet through the air valve during this period. From the moment the air valve opens again, coinciding with the position of the first air pocket, the pressure in this pocket quickly decreases. The effect of this new opening on the other air pockets is relatively small. The pressure evolution in the first air pocket is smooth until the filling column reaches the pair valve position and closes again ($t = 1.007$ s).

Figure 13 depicts the flow rate and velocity of air expelled through the air valve during the filling procedure. It is observed that during the first opening of the air valve, the maximum expelled flow rate is $Q_{air} = 3.78\frac{\text{Nm}^3}{\text{h}}$. When the maximum pressure reaches the second air pocket, the maximum flow rate is simultaneously expelled. On the other hand, when the air valve opens for the second time, the pressure in the first air pocket is much higher, consequently increasing the expelled airflow as well ($Q_{air} = 6.66\frac{\text{Nm}^3}{\text{h}}$ with $v_{air}$ = 146.7 m/s).

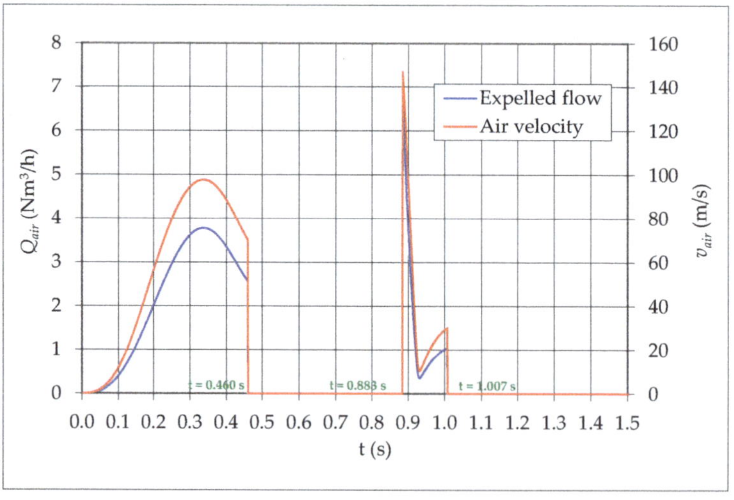


**Figure 13.** Evolution of expelled air during the transient event.

### *4.4. Comparison between the Transient Event Considering and Neglecting the Air Valve*

After separately analysing the pipeline's filling without air valves and considering them, a comparison was performed, as shown in Figures 14 and 15. Table 5 shows the maximum value of air pocket pressure reached for the two analysed scenarios.

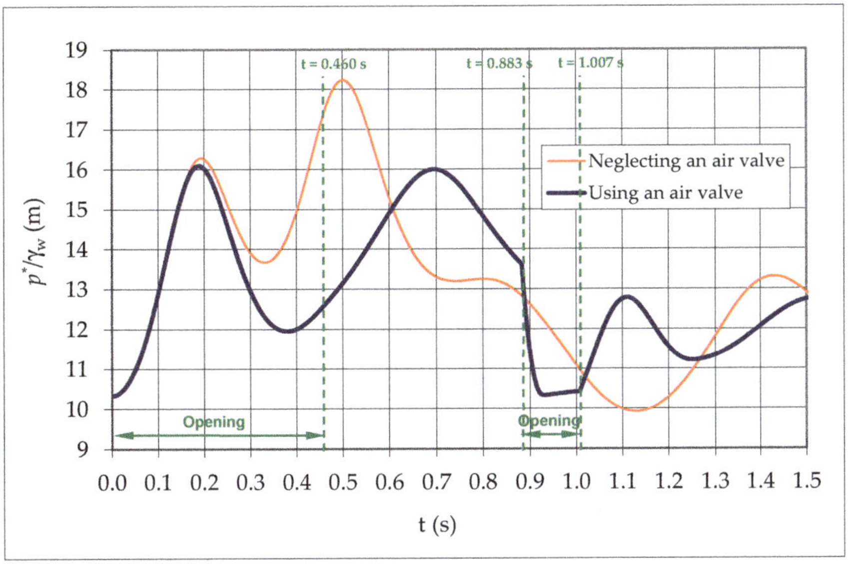


**Figure 14.** The pressure of the first air pocket $p_1^*$ with and without the air valve.

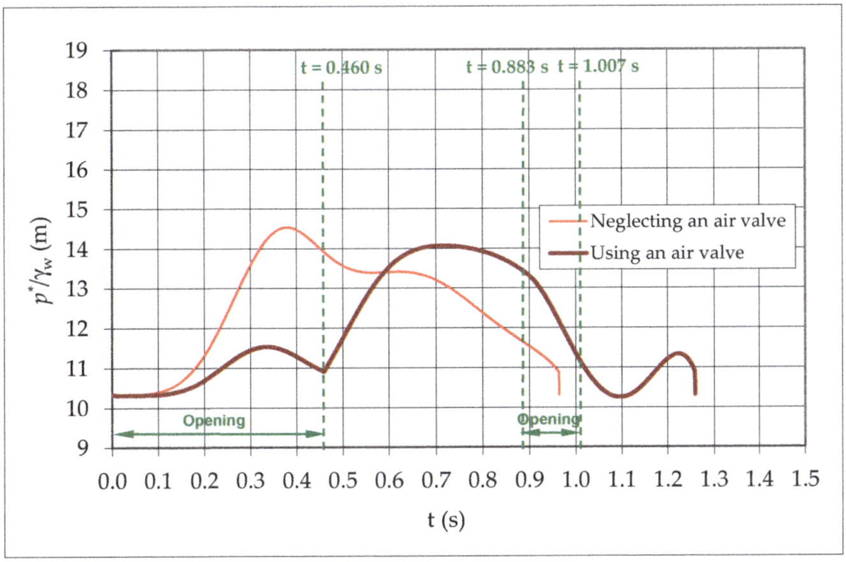


**Figure 15.** The pressure of the second air pocket $p_2^*$ with and without the air valve.

**Table 5.** Maximum pressure heads for the first air pocket.

| Type of Peak | Without the Air Valve | Considering the Air Valve |
|---|---|---|
| First peak in air pocket 1 | $p_1^* = 16.286$ m ($t = 0.193$ s) | $p_1^* = 16.093$ m ($t = 0.189$ s) |
| Second peak in air pocket 1 | $p_1^* = 18.230$ m ($t = 0.500$ s) | $p_1^* = 15.997$ m ($t = 0.695$ s) |

As expected, the first pressure peak is slightly reduced with the installation of the air valve (16.093 m versus 16.286 m). It should be noted that when the hydraulic installation starts, the filling column abruptly compresses the first air pocket. Still, the first water column reacts much more slowly due to its greater inertia. Therefore, the air valve has almost no effect on the pressure of the first air pocket during these times. Significant differences appear when the first blocking column moves significantly, compressing the second air pocket and causing air to be expelled through the air valve. The second pressure peak is much smaller, with 15.997 and 18.230 m values occurring at 0.695 and 0.500 s, respectively. When the air pocket reaches the air valve position and opens, the pressure drops to nearly atmospheric values. The air pocket begins to compress once more when the air valve closes again as the filling column reaches its position.

Table 6 presents the maximum pressures reached for the second air pocket in the two analysed situations (Figure 15).

**Table 6.** Maximum pressure heads for the second air pocket.

| Type of Peak | Without the Air Valve | Considering the Air Valve |
|---|---|---|
| First peak in air pocket 2 | $p_2^* = 14.540$ m ($t = 0.379$ s) | $p_2^* = 11.538$ m ($t = 0.335$ s) |
| Second peak in air pocket 2 | $p_2^* = 13.420$ m ($t = 0.620$ s) | $p_2^* = 14.070$ m ($t = 0.714$ s) |

The effect of the air valve's installation is evident from the beginning. From the start of the transient event until t = 0.460 s (the period during which the air valve is expelling air), the evolution of the air pocket pressure is much smoother, reaching a maximum value of 11.538 m, which is significantly lower than when the installation fills without the air valve ($p_2^* = 14.540$ m). Once the air valve closes, the air pocket compresses again, generating the corresponding pressure peaks.

Figures 16–18 present the analysis of the evolution of the water velocities for the water columns.

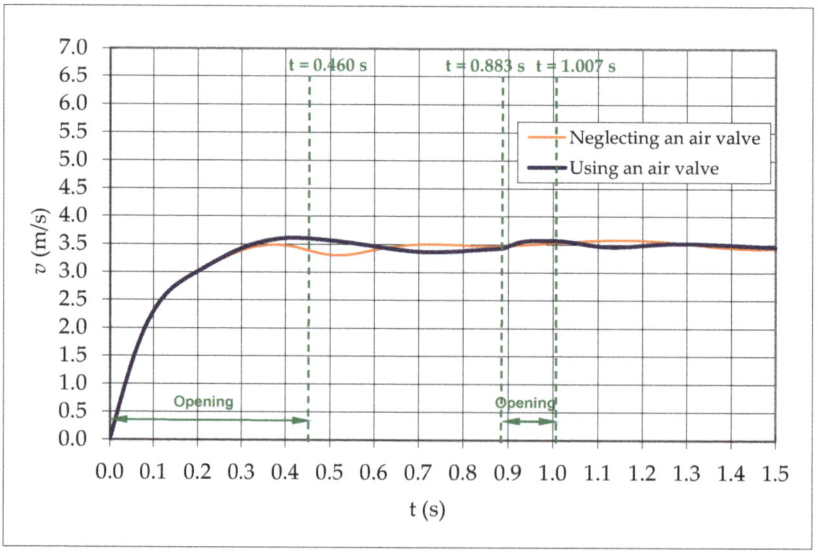


**Figure 16.** The velocity of the filling column with and without the air valve.

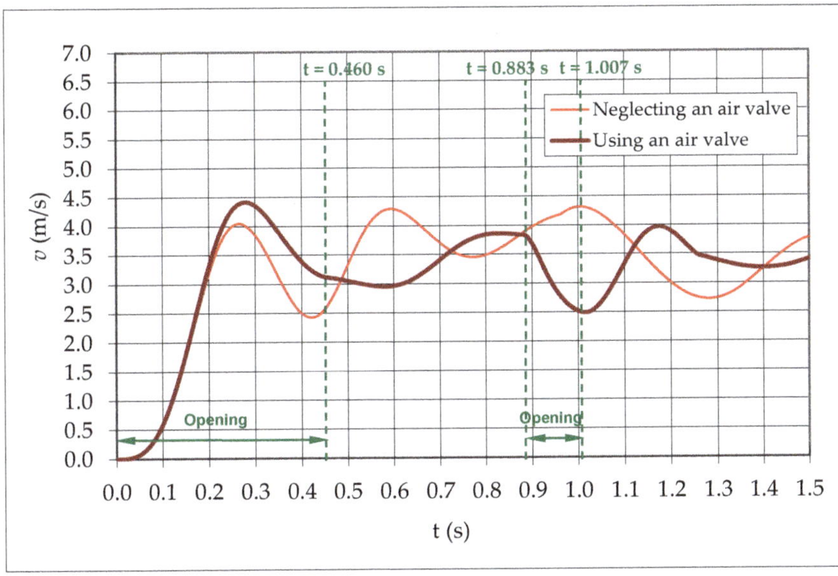


**Figure 17.** The velocity of the first blocking column $v_1$ with and without the air valve.

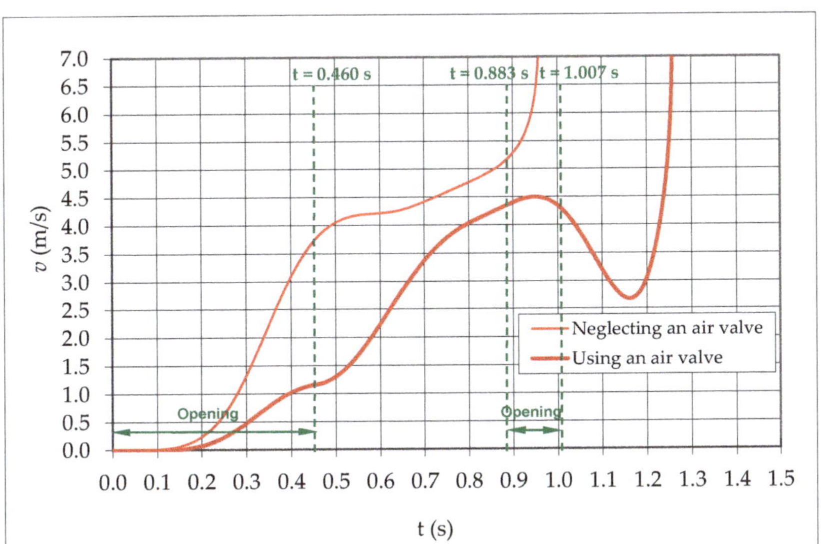


**Figure 18.** The velocity of the second blocking column $v_2$ with and without the air valve.

The installation of the air valve affects the velocity of the filling column. The first blocking column's velocity slows down slightly after the beginning of the hydraulic event. However, the movement of the second blocking column is significantly slower when the air valve is installed, as illustrated in Figure 18.

### *4.5. Comparison of Various Air Valve Sizes*

The hydraulic event was analysed, considering the characteristics of the air valve presented in Table 3. The influence of various characteristic curves of air valves is performed.

Table 7 shows three different air valves: the one already studied, a smaller one, and a larger one.

**Table 7.** Characteristics of the air valves.

| Parameter | Air Valve A | Air Valve B | Air Valve C |
|---|---|---|---|
| $c_{exp}$ [(Nm$^3$/s)/m] | 0.00014 | 0.00028 | 0.00056 |
| $A_{exp}$ [mm$^2$] | 4.9 | 9.8 | 19.6 |
| $D_{exp}$ [mm] | 2.5 | 3.5 | 5.0 |

With these expulsion coefficients, for a gauge pressure in the pipe of 0.5 bar, air valve A can expel an airflow rate of $Q_{air} = 4.5\frac{\text{Nm}^3}{\text{h}}$, air valve B can expel an airflow rate of $Q_{air} = 8.9\frac{\text{Nm}^3}{\text{h}}$, and air valve C can expel an airflow rate of $Q_{air} = 17.9\frac{\text{Nm}^3}{\text{h}}$.

Due to their significantly different expulsion capacities, the development of the transient generated will vary notably from one scenario to another. Table 8 shows when each air valve opens and closes during the conduit's filling process.

**Table 8.** Characteristic times for the different air valves.

| Situation | Air Valve A | Air Valve B | Air Valve C |
|---|---|---|---|
| Closure | $t = 0.479$ s | $t = 0.460$ s | $t = 0.445$ s |
| Opening | $t = 0.886$ s | $t = 0.883$ s | $t = 0.880$ s |
| Closure | $t = 1.009$ s | $t = 1.007$ s | $t = 1.007$ s |

Figure 19 illustrates the pressure evolution of the first air pocket, depicted for the four analysed scenarios: installation without an air valve and considering the three air valves. As observed, during the initial instants of the hydraulic event, the influence of the air valve is practically negligible. When the air valve closes, a compression process of the air pocket begins, which varies for each scenario depending on the specific conditions of the closure's occurrence. For instance, with the most oversized air valve, a higher second peak pressure is reached ($p_2^*$ = 18.280 m), slightly exceeding the pressure obtained without air valves.

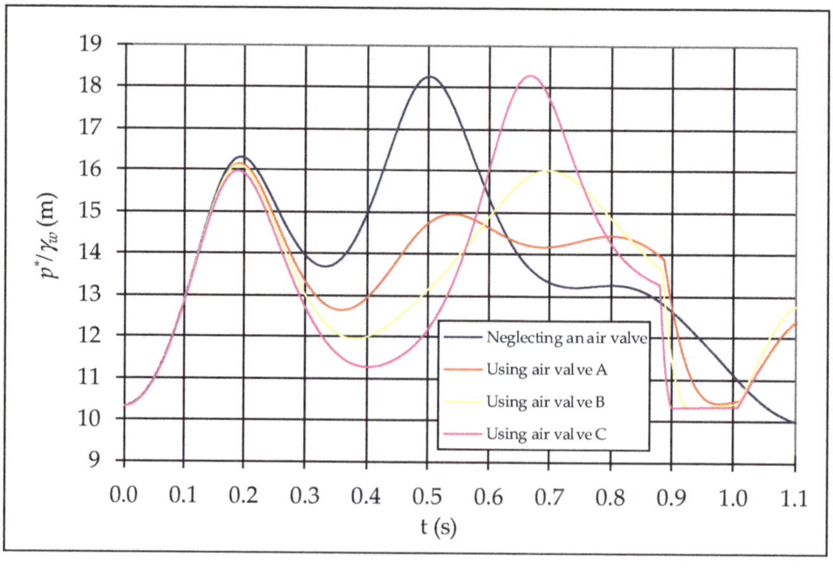


**Figure 19.** The pressure pulses of the first air pocket $p_1^*$ for different air valves.

Figure 20 presents the effect of the air valve size on the transient occurring during the filling manoeuvre. It shows the pressure evolution of the second air pocket. Table 9 shows the value of the first pressure peak, the time when the maximum pressure is reached, and the time for the first closure of the air valve.

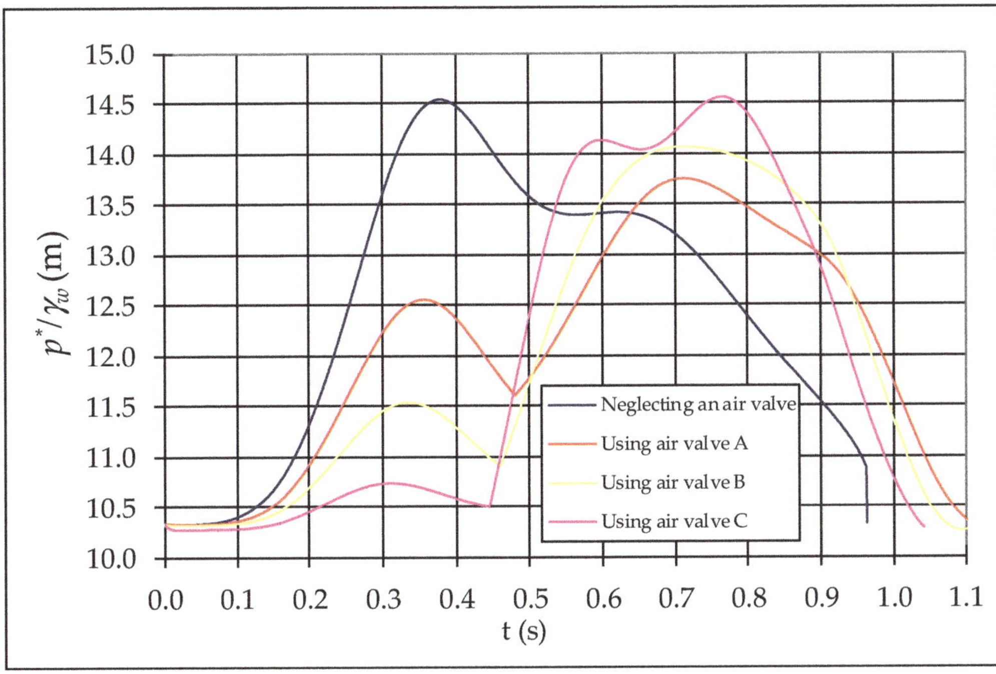


**Figure 20.** The pressure pulses of the second air pocket $p_2^*$ for different air valves.

**Table 9.** Results for the different air valves (second air pocket).

| Variable | | Air Valve A | Air Valve B | Air Valve C |
|---|---|---|---|---|
| Maximum air pocket pressure | Value (m) | 12.551 | 11.538 | 10.739 |
| | Time (s) | 0.356 | 0.335 | 0.311 |
| Time closure (s) | | 0.479 | 0.460 | 0.445 |

Table 9 shows the influence of the air valve size on the hydraulic event. The greater the air valve's expulsion capacity, the lower the pressure peak. This peak is reached earlier and the air valve closes sooner due to the arrival of the water column.

When the air valve closure is produced, the air pockets compress, reaching their corresponding pressure peaks. Considering the air valve closure, the highest-pressure peak in the hydraulic installation occurs using the most oversized air valve (see Figure 20); since the air valve is not active, the air cannot be expelled, reducing the pressure peaks. Figure 21 illustrates the expelled flow rate of the three air valves during the transient event, while Figure 22 shows the air velocity through the outlet orifice.

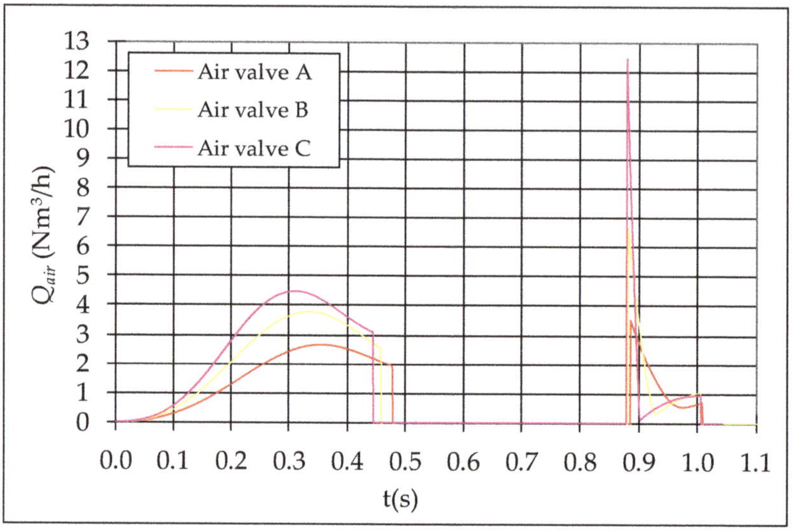


**Figure 21.** Expelled airflow rate $Q_{air}$ through the different air valves.

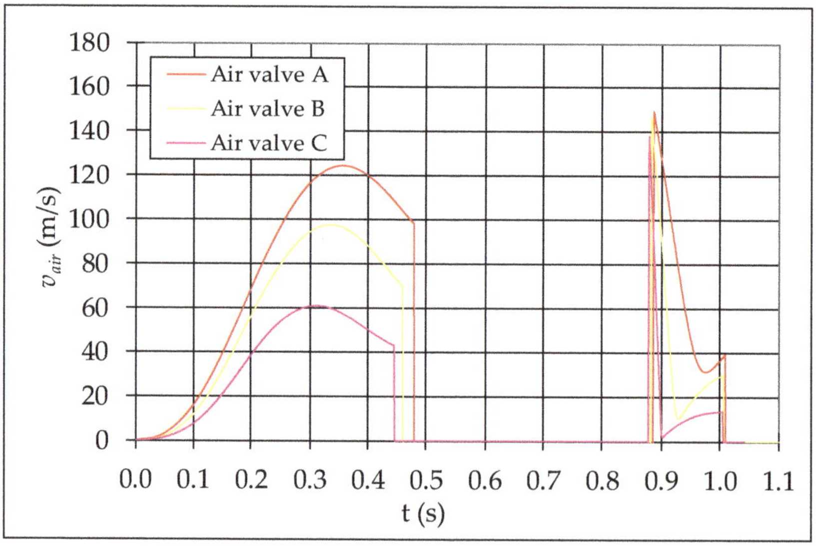


**Figure 22.** Air velocity $v_{air}$ through the different air valves.

## 5. Conclusions

A general mathematical model based on the Rigid Column Model has been developed to analyse transient events generated by $n$ entrapped with air pockets in water pipelines with irregular profiles, incorporating various air valves. The proposed model's main advantage over the previous models is that it considers the effect of blocking columns, even considering the impact of expelled air. The findings from the analysed small experimental facility cannot be directly extrapolated to actual pipelines.

Based on this research, it is crucial to consider the following:

- Model Applicability: The model proposed in this research can serve as a starting point for water utilities to compute the maximum pressure of entrapped air pockets during filling operations in water installations.
- Avoiding Blocking Columns: Blocking columns should be avoided in actual pipelines as they can generate hazardous pressure surges by compressing trapped air pockets. Water utilities must ensure that pipelines are wholly drained before filling operations commence. The placement of draining valves should guarantee complete drainage before starting up the pipelines.
- Controlled Filling: Pipeline filling should be conducted slowly to facilitate the expulsion of air through the valves. Care must be taken when filling pipelines with trapped air, as even with air valves installed, complete system safety cannot be guaranteed.
- Air Valve Selection: It is advisable to choose a reliable air valve manufacturer for sizing water installations to ensure proper characterisation of the differential pressure-versus-airflow curve.
- Polytropic Evolution: Selecting the appropriate polytropic process remains challenging for engineers and designers, as determining a specific value for real-world water installations is complex. This research recommends that an adiabatic behaviour should be considered for rapid transient flows, while an isothermal behaviour is more appropriate for slower events. Intermediate values may be adopted depending on the scenarios under consideration.
- Numerical Resolution: The filling procedure involves solving a system of differential-algebraic equations, which was addressed using an adaptive fifth-order Runge-Kutta method.

**Author Contributions:** Conceptualization, V.S.F.-M., A.A.-P. and O.E.C.-H.; methodology, A.A.-P. and O.E.C.-H.; formal analysis, O.E.C.-H.; validation, A.A.-P. and V.S.F.-M.; writing—original draft preparation, A.A.-P. and O.E.C.-H.; writing—review and editing, V.S.F.-M. All authors have read and agreed to the published version of the manuscript.

**Funding:** This research received no external funding.

**Data Availability Statement:** Databases are available in this article.

**Conflicts of Interest:** The authors declare no conflicts of interest.

## References

1. Pozos-Estrada, O.; Pothof, I.; Fuentes-Mariles, O.A.; Dominguez-Mora, R.; Pedrozo-Acuña, A.; Meli, R.; Peña, F. Failure of a Drainage Tunnel Caused by an Entrapped Air Pocket. *Urban Water J.* **2015**, *12*, 446–454. [CrossRef]
2. Wang, L.; Wang, F.; Karney, B.; Malekpour, A. Numerical Investigation of Rapid Filling in Bypass Pipelines. *J. Hydraul. Res.* **2017**, *55*, 647–656. [CrossRef]
3. Ferreira, J.P.; Ferràs, D.; Covas, D.I.C.; van der Werf, J.A.; Kapelan, Z. Air Entrapment Modelling during Pipe Filling Based on SWMM. *J. Hydraul. Res.* **2024**, *62*, 39–57. [CrossRef]
4. Fuertes-Miquel, V.S.; Coronado-Hernández, O.E.; Mora-Meliá, D.; Iglesias-Rey, P.L. Hydraulic Modeling during Filling and Emptying Processes in Pressurized Pipelines: A Literature Review. *Urban Water J.* **2019**, *16*, 299–311. [CrossRef]
5. Ferreira, J.P.; Ferràs, D.; Covas, D.I.C.; Kapelan, Z. Air Entrapment Modelling in Water Supply Networks during Pipe Filling Events. *Urban Water J.* **2024**, *21*, 685–697. [CrossRef]
6. Tasca, E.S.A.; Karney, B. Improved Air Valve Selection through Better Device Characterization and Modeling. *J. Hydraul. Eng.* **2023**, *149*, 4023019. [CrossRef]
7. Apollonio, C.; Balacco, G.; Fontana, N.; Giugni, M.; Marini, G.; Piccinni, A.F. Hydraulic Transients Caused by Air Expulsion During Rapid Filling of Undulating Pipelines. *Water* **2016**, *8*, 25. [CrossRef]
8. Fontana, N.; Galdiero, E.; Giugni, M. Pressure Surges Caused by Air Release in Water Pipelines. *J. Hydraul. Res.* **2016**, *54*, 461–472. [CrossRef]
9. American Water Works Association (AWWA). *Air-Release, Air/Vacuum, and Combination Air Valves: M51*; American Water Works Association: Denver, CO, USA, 2001.
10. Tran, P.D. Pressure Transients Caused by Air-Valve Closure While Filling Pipelines. *J. Hydraul. Eng.* **2017**, *143*, 4016082. [CrossRef]
11. Pozos, O.; Gonzalez, C.A.; Giesecke, J.; Marx, W.; Rodal, E.A. Air Entrapped in Gravity Pipeline Systems. *J. Hydraul. Res.* **2010**, *48*, 338–347. [CrossRef]

12. Zhou, L.; Cao, Y.; Karney, B.; Bergant, A.; Tijsseling, A.S.; Liu, D.; Wang, P. Expulsion of Entrapped Air in a Rapidly Filling Horizontal Pipe. *J. Hydraul. Eng.* **2020**, *146*, 4020047. [CrossRef]
13. Huang, B.; Fan, M.; Liu, J.; Zhu, D.Z. CFD Simulation of Air–Water Interactions in Rapidly Filling Horizontal Pipe with Entrapped Air. In Proceedings of the World Environmental and Water Resources Congress, Online, 7–11 June 2021; pp. 495–507. [CrossRef]
14. Iglesias-Rey, P.L.; Fuertes-Miquel, V.S.; García-Mares, F.J.; Martínez-Solano, J.J. Comparative Study of Intake and Exhaust Air Flows of Different Commercial Air Valves. *Procedia Eng.* **2014**, *89*, 1412–1419. [CrossRef]
15. Martin, C.S. Entrapped Air in Pipelines. In Proceedings of the 2nd International Conference on Pressure Surges, British Hydromechanics Research Assoc., London, UK, 22–24 September 1976; pp. 5–28.
16. Liou, C.P.; Hunt, W.A. Filling of Pipelines with Undulating Elevation Profiles. *J. Hydraul. Eng.* **1996**, *122*, 534–539. [CrossRef]
17. Izquierdo, J.; Fuertes, V.S.; Cabrera, E.; Iglesias, P.L.; Garcia-Serra, J. Pipeline Start-up with Entrapped Air. *J. Hydraul. Res.* **1999**, *37*, 579–590. [CrossRef]
18. Wylie, E.B.; Streeter, V.L. *Fluid Transients*; Advanced Book Program; McGraw-Hill International Book Company: New York, NY, USA, 1978.
19. Zhou, L.; Liu, D.; Karney, B. Investigation of Hydraulic Transients of Two Entrapped Air Pockets in a Water Pipeline. *J. Hydraul. Eng.* **2013**, *139*, 949–959. [CrossRef]
20. Fuertes-Miquel, V.S. Hydraulic Transients with Entrapped Air Pockets. Ph.D. Thesis, Polytechnic University of Valencia, Valencia, Spain, 2001.
21. Tijsseling, A.S.; Hou, Q.; Bozkuş, Z.; Laanearu, J. Improved One-Dimensional Models for Rapid Emptying and Filling of Pipelines. *J. Press. Vessel Technol.* **2015**, *138*, 031301. [CrossRef]
22. Zhou, L.; Liu, D.; Karney, B.; Wang, P. Phenomenon of White Mist in Pipelines Rapidly Filling with Water with Entrapped Air Pockets. *J. Hydraul. Eng.* **2013**, *139*, 1041–1051. [CrossRef]
23. Zhou, L.; Liu, D.; Karney, B.; Zhang, Q. Influence of Entrapped Air Pockets on Hydraulic Transients in Water Pipelines. *J. Hydraul. Eng.* **2011**, *137*, 1686–1692. [CrossRef]

fluids

MDPI

*Article*

# On the Modelling of Asymptotic Wavefronts in Long Ducts with Chambers

**Alan E. Vardy**

University of Dundee and Dundee Tunnel Research, Dundee DD1 4HN, UK; a.e.vardy@dundee.ac.uk

**Abstract:** A novel method of determining the possible shapes of pressure wavefronts in ducts after they have travelled sufficient distances to evolve to asymptotic states is introduced. Although it is possible in principle to achieve the same outcome by simulating complete flow histories from the time of the creation of the wavefronts, this can be impracticable. It is especially unsuitable to use such methods when extremely small grid lengths are needed to represent the final outcome adequately. The new method does not simulate the propagation phase at all. Instead, it explores what final end states are possible, but gives no information about the initiating disturbance or the wavefront evolution towards the assessed asymptotic state. Accordingly, the two methods do not overlap, but instead are complementary to each other. A typical case in which the new capability has high potential is described and used to illustrate the purpose and use of the methodology. However, the primary focus is on the presentation and assessment of the method, not on any particular phenomenon. It is shown that the required computational resources are far smaller than those needed for conventional unsteady flow simulations of propagating wavefronts. The potential numerical limitations of the method are highlighted and, with one exception, are shown to be either of no consequence or easily reduced to acceptable levels. Special attention is paid to the one exception because it cannot be proven to be unimportant and, indeed, it would be unsafe to use it in general analyses of wave propagation. However, strong evidence is presented of its acceptability for the study of asymptotic wavefronts.

**Keywords:** 1D unsteady flow; wavefront propagation; method of characteristics; micro-pressure waves; ultra-steep wavefronts; air chambers

## 1. Introduction

In many, perhaps most, studies of unsteady fluid flows in a pipe or duct, it is sufficient to treat the whole flow as moving almost as a solid body. This is also true for individual pipes in networks of connected pipes. However, when studying the consequences of disturbances induced suddenly—e.g., rapid valve closures—account needs to be taken of fluid compressibility, which enables rapid accelerations to occur without inducing excessive pressure changes (Fox 1977 [1], Wylie and Streeter 1993 [2], Chaudhry 2014 [3]). In such cases—commonly known as 'water hammer', even when the fluid is not water—the initiating event first influences only the fluid closest to it. This then influences the adjacent fluid before that influences its adjacent fluid, and so on. That is, the region of flow that has been influenced by the event continually increases in length. The interface between this region and the region of flow that has not yet been affected typically propagates along a pipe at approximately the speed of sound. In this paper, this interface is at the leading part of a well-defined wavefront causing a finite increase in pressure. The trailing part of such wavefronts is usually less clearly defined. Loosely, however, it is where the region of rapid change ceases and a subsequent region of much slower change begins.

Although pressure wavefronts typically propagate at approximately the speed of sound, they usually change in shape as they do so. This is because the *local* speed of propagation depends upon the local flow state, which necessarily varies between the leading tip and the trailing edge. There are two primary reasons why this dependency exists. Attention is commonly drawn first to the fact that the speed of sound is itself a variable. It depends not only on the intrinsic properties of the fluid, but also on the local pressure and/or temperature. This dependence is more pronounced in gases than in liquids, but it exists for both. In practical flows, the speed of sound increases with increasing pressure; so, if the initiating event causes increasing pressure, then, in the absence of any other influence, the higher-pressure regions of the wavefront will tend to travel faster than its toe. Given sufficient time, they will catch up to the toe, and the wavefront will then be a single, discontinuous step, namely a shock. In contrast, if the initiating event causes decreasing pressure, the wavefront will lengthen indefinitely.

A second—and usually more important—reason for the differing speeds of propagation is that waves move at the speed of sound *relative to the fluid*, not relative to a fixed reference such as a pipe. Now, the initiating event necessarily causes changes in velocity, and these must be allowed for when estimating speeds of propagation. Again, therefore, the consequence will be a tendency for the wavefront to either lengthen or shorten, potentially into a shock.

Wavefront shortening and lengthening is rarely of importance in the case of liquid flows, except in special cases such as cavitating flows. However, it can be of critical importance in gas flows. Figure 1 shows pressure histories measured at successive locations along a railway tunnel in Germany. The graphs are offset in time to enable the shapes of the individual histories to be shown clearly. There are strong secondary effects before and after the main part of the wavefront, but the shortening effect is demonstrated clearly.

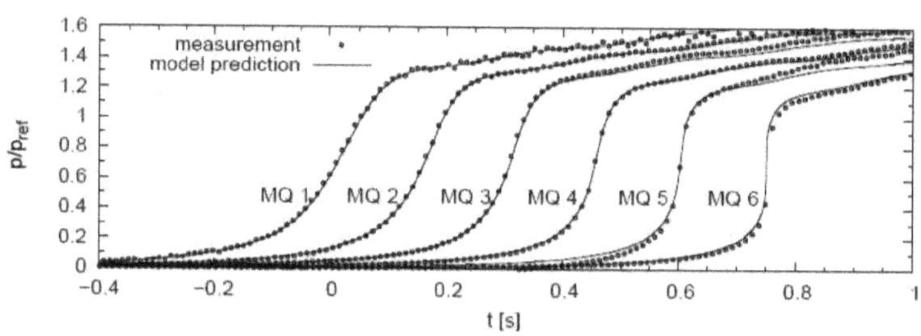


**Figure 1.** Wavefront shortening/steepening during propagation along a 7.7 km long railway tunnel. (Reproduced with kind permission of Deutsche Bahn AG, DB Systemtechnik). NB: The graphs are offset in time to compensate for times of travel between successive measurement locations. The 'time' axis should be used only to infer the scales of individual graphs.

The challenge for engineers is to ensure that wavefronts do not become so steep that subsequent reflections at tunnel portals will cause unacceptably large audible phenomena in the surrounding environment. To do this, the first task is to assess whether the secondary effects will constrain the steepening process sufficiently during propagation. This depends strongly on the type of track. Historically, all track was ballasted, but slab track construction is increasing chosen because of its less demanding maintenance requirements. Reports of analyses and experimental measurements of wavefront propagation in these different types of tunnel include Mashimo et al. (1997) [4], Fukuda et al. (2006) [5], Nakamura et al. (2011) [6], Wang et al. (2018) [7], Fukuda et al. (2020) [8] and Iyer et al. (2021) [9]. When such investigations do not confirm that the resulting steepness will be acceptable, the most common design response is a modification of the tunnel entrance region to reduce the initial steepness of train-entry wavefronts. Alternatively, a reduction in the train-entry speed can have the same effect. Another possibility, so far explored only theoretically, involves utilising unused space beneath tunnel walkways to create air chambers that mimic a key effect of ballast tracks (Liu et al. 2021 [10]). The methodology developed below has been devised to address a difficulty encountered when assessing the final shapes of such wavefronts after propagating long distances in such a tunnel. The method is potentially of use for studying other asymptotic wavefronts, but the only specific application discussed below is the particular case of air chambers.

### 1.1. Practical Flows

In real flows of finite amplitude, the characteristics of an evolving wavefront depend on the environment in which it is propagating, as well as upon the initiating event and the inertial behaviour discussed so far. An almost universal complication arises as a consequence of fluid viscosity and of the so-called 'no-slip' condition that causes fluid velocities on a solid surface to remain the same as the velocity of the surface itself. This causes frictional resistance to flow that influences the pressure–flowrate dependency and hence changes the evolution of wavefronts. This phenomenon is quite complex because its consequences can depend upon rates of change of flow, as well as upon absolute values of these changes. Broadly speaking, however, long regions of slowly varying flow are influenced primarily by low-frequency behaviour of friction, whereas short wavefronts are influenced primarily by high-frequency behaviour.

Other phenomena that influence some flows include (i) a non-uniform pipe area, (ii) heat transfers with pipe walls and (iii) the leakage of mass through the pipe walls (whether inflow or outflow). The author expects the generic methodology described herein to be used for a range of purposes, and this is reflected in the presentation given below. However, in the particular example presented in Section 5 to illustrate the use of the methodology, all influences of friction, non-uniform area and heat transfer are neglected. Attention is focussed on compressive wavefronts and on a particular form of leakage that, in principle, has the potential to exactly counterbalance the inertial-shortening process illustrated in Figure 1 and thereby prevent the adverse consequences associated with very short wavefronts.

The chosen geometrical configuration is a duct that has wall holes through which air exchange with adjacent air chambers is possible (see Figure 2). The chambers are isolated from one another; so, although they contribute to the effective bulk compressibility of air in the duct, they do not provide a path for axial flows along the duct. This configuration is being investigated to assess its potential for countering wavefront steepening in railway tunnels, thereby reducing the likelihood of sudden pressure disturbances inside a tunnel causing unacceptable disturbances outside it. In a nutshell, the challenge for tunnel designers is to prevent wavefronts becoming steeper than a prescribed upper limiting value.

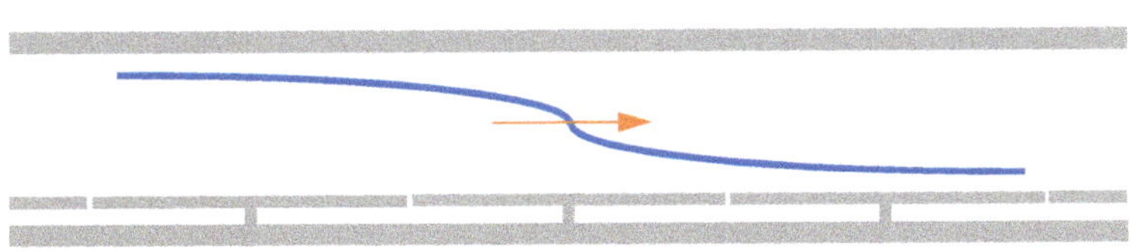

**Figure 2.** Wavefront propagating in a duct with air chambers.

### 1.2. Structure of the Paper

The remainder of this paper begins with a qualitative discussion of some characteristics of wavefronts, with special attention being paid to a condition that might be approached asymptotically in long ducts. Then, two sections are devoted to describing the numerical method used herein to investigate this behaviour. Unusually, instead of prescribing a cause and predicting its effect, the method assesses (the characteristics of) a possible effect without making any assumption about its possible cause. This enables the asymptotic behaviour to be studied in fine detail with high accuracy, without the complications that usually arise in studies of wavefront propagation in long ducts. The method is then used to demonstrate a particular characteristic of asymptotic wavefronts, and the paper closes with a short section giving a brief summary and highlighting the key conclusions.

## 2. Wavefront Shapes

In most water hammer-like applications, wavefronts are quite long, and this is recognised implicitly in the terminology used to define the phenomenon. For example, in the classical reservoir pipe valve case, any valve closure that causes a wavefront length to be shorter than two pipe lengths is termed 'rapid'. Then, neglecting the influences of other causes of change, the maximum pressure reached at the valve will be almost independent of the detailed sequence of the closure. In contrast, if the valve closure time exceeds that required for the tip of the initial disturbance to travel all the way along the pipe and back to the valve before the closure is complete, the maximum pressure will be smaller. The amount of the reduction will depend on the closure sequence and hence upon the shape of the wavefront—e.g., 'steep near the tip and less steep afterwards' or 'shallow near the tip and steeper afterwards', as illustrated in Figure 3.

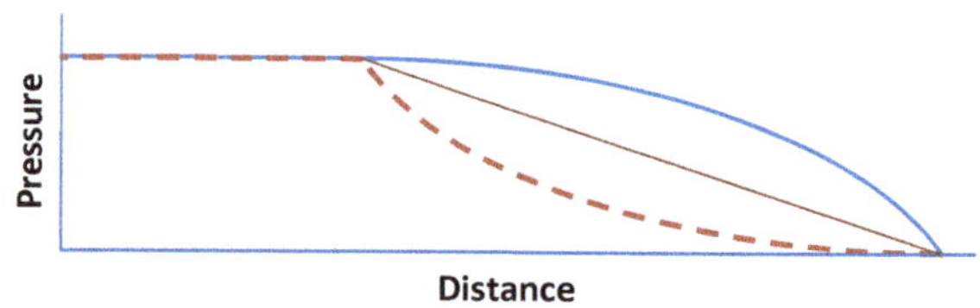


**Figure 3.** Alternative, nominally possible initial wavefront shapes.

An important feature of this illustration is that the shape of the wavefront exerts its influence only when the wavefront reaches a boundary. Another example of this is described by Tijsseling (2021) [11] in relation to the reflection of a wavefront at an open end of a pipe connected to a reservoir. During such reflections, most of the internal energy of the fluid is 'converted' to kinetic energy, and the reflected wavefront travels upstream to induce this change. However, a small amount of energy is released into the reservoir during the reflection process, causing waves to radiate outwards, and, in this context, the shape of the wavefront is important. The amplitudes of the radiated waves depend upon the variation in steepness along the wavefront. That is, they depend upon the *rates of change* of pressure and velocity. Counter-intuitively, the amplitude of the incident wavefront is not important in its own right. Instead, what matters is the rate at which the change occurs.

Waves radiated into water reservoirs during wavefront reflections are of little practical significance except, perhaps, in highly specialist activities such as sonic surveillance. However, the same behaviour can be hugely important in air. An obvious instance is the design of musical instruments that exploit the phenomenon to delightful effect. Unfortunately, this contrasts sharply with the railway tunnel example cited above. That process is essentially the same, but the consequence can be far from delightful. The most common initial cause of unacceptable wavefronts occurs when the nose of a train enters a tunnel at high speed. Initially, the length of the wavefront might be in the order of, say, ten tunnel diameters, but it can shorten greatly as it propagates. If the tunnel is sufficiently long, the wavefront can be very steep when it reaches the exit portal of the tunnel. In extreme cases, a loud bang can be emitted from the tunnel. It is neither musical nor pleasant, so it has to be prevented. A YouTube link [12] to an example of this phenomenon is given in the references at the end of the paper.

Various counter-measures exist to reduce the amplitudes of the emitted disturbances. Most are designed to reduce the abruptness of the process of nose-entry to the tunnel. These include, for example, reducing the train speed (unpopular with operators), tapering the entrance region, allowing some air to escape to the atmosphere through holes in the walls of tunnel extensions (Winslow and Howe 2005 [13], Zhang et al., 2018) [14] or extending train noses (Kikuchi et al. 2011 [15], Miyachi et al. 2022 [16]). Perhaps unexpectedly, these methods are more reliable and more effective than similar measures at the tunnel exit, although the latter have received attention—even including active measures, as well as passive ones analogous to those at tunnel entrances (Nishimura et al., 1994 [17],

Aoki et al., 1999, [18], Wang et al., 2015 [19]). A third approach is to implement measures along the tunnel to prevent the unacceptable steepness of wavefronts arriving at the exit portal (Hieke et al., 2009 [20], Liu et al., 2021) [10]. The last of these investigates the influence of air chambers by studying pressure wave propagation over large distances. However, the time steps used in its numerical calculations are considered to be too coarse for high confidence to be placed in the detailed shapes of the leading parts of the resulting wavefronts, even though useful conclusions could be drawn.

*Asymptotic Wavefronts*

As indicated above, inertial forces tend to cause compressive wavefronts to shorten, and hence steepen, as they propagate, whereas other phenomena tend to have the opposite effect. For some time after its initiation, the *shape* of a wavefront will depend primarily on the initiating event itself—e.g., on rates of change of flow during a valve closure. Gradually, however, the shape will depend increasingly on the environment in which the wavefront is propagating and less and less on its initial shape. In sufficiently long ducts, an approximate balance will gradually develop between the steepening and flattening influences, and, thereafter, the wavefront will propagate in an almost unchanged form. Thus, for example, all three wavefronts in Figure 3 would tend to evolve towards the same eventual shape. An important corollary of this fact is that it is not possible to infer the initial shapes from a knowledge of the asymptotic states. However, the characteristics of the asymptotic conditions are of interest in their own right, independently of how they were created and irrespective of particular practical applications.

In principle, it is possible to study asymptotic wavefronts by simulating the whole process of propagation until an asymptotic condition is reached. Indeed, this is exactly how the present author has studied them over a period spanning several decades. However, that method has both practical and technical disadvantages. The most obvious practical one arises from the need to use small grid lengths to represent conditions along steep wavefronts in appropriate detail. Unfortunately, this same grid size must be used along the whole length of the duct. Furthermore, the corresponding integration time steps must also be small. Another big disadvantage is that current understanding of pressure wave damping during propagation over long distances inside ducts is weak. Furthermore, there is a risk of the accumulation of numerical errors at the wavefronts themselves. At best, these difficulties reduce the level of confidence that can realistically be justified in predicting the detailed shapes of wavefronts when they reach the exit.

Vardy and Brown (2000) [21] reduced some of the disadvantages by simulating only within a limited band of $(x,t)$ space moving with the wavefront. This greatly reduced CPU requirements, but it also broke the link between the wavefront and the upstream flows, thereby partially negating the apparent benefit. More important, in all of the solutions, the leading region of the simulated wavefronts developed a ramp-like shape that the authors were unable to explain with confidence. They hypothesised that it was probably a numerical consequence of interpolation, but were unable to rule out the possibility of it being a true physical phenomenon.

In the following sections, problems associated with simulating wavefront propagation per se are avoided completely. Instead of asking what final state will result from prescribed initial conditions, the analysis focusses exclusively on possible end results. The aim is to see what final *shapes* are possible for wavefronts that have travelled sufficiently far along ducts/tunnels of a prescribed geometry. As indicated above, the answer to this question does not depend upon the process that initiated the wavefront—although this will influence the length of duct required for the asymptotic state to be approached closely.

This approach might seem like 'putting the cart before the horse', but it is potentially valuable because it enables highly accurate predictions of the range of possible asymptotic states. At the very least, it will enable the outcomes of more conventional analyses to be assessed with greater confidence than hitherto. Importantly, the method of achieving this benefit requires only trivial CPU resources, so large numbers of cases could be studied rapidly.

For completeness, it is emphasised that the analysis cannot be extended backwards in time to identify what initial conditions could have led to the eventual outcome. The primary reason for this is that the method enforces asymptotic behaviour everywhere, regardless of how far back in time (or how far back in distance from the wavefront tip), the analysis is extended. Conversely, of course, analyses that begin with the initiating event cannot stipulate the characteristics of the resulting wavefronts a priori.

## 3. Numerical Grid

This numerical simulation is a particular form of the Method of Characteristics (MoC), which is widely used for the simulation of waves in pipelines. It is especially popular in studies of low Mach number flows in which strong changes occur over short timescales. It has a well-known inability to satisfy conservation laws exactly, but, for the purposes of studying details of highly localised wavefronts, that is far less is less important than the spurious oscillations that can arise when using the control volume or finite element methods. Furthermore, the most important aspect of the following development is not the MoC equations themselves, but the method of specifying the numerical grid. Accordingly, this is discussed first, and the presentation of the relevant equations is deferred to the following Section.

Figure 4a depicts the profile of an isolated, asymptotic wavefront at three instants as it travels at a steady velocity $U_w$ (*NB: all points on the wavefront are travelling at the same speed; otherwise, asymptotic conditions would not exist*). At the time indicated by the line A-A, the wavefront has almost passed location C, but its tip has not yet reached location D. At the second instant, the whole wavefront is between these locations, and nominally steady conditions exist at both of them. Ambient conditions still prevail at D, whereas, at C, the pressure and velocity have changed in response to the wavefront. At the third instant, depicted by the line B-B, the wavefront has begun to disturb conditions at D. Importantly, because the wavefront is prescribed as being asymptotic, all three curves in the figure are identical. That is, the spatial profiles at successive instants differ only by a uniform distance. Likewise, the temporal histories at different locations differ only by a uniform time interval. Moreover, these statements are true regardless of whether the intervals are small or large. The chosen numerical grid takes advantage of this unique characteristic behaviour of the asymptotic state.

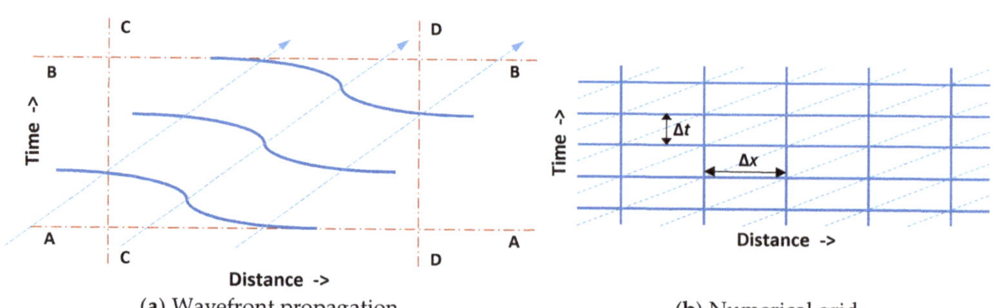


(**a**) Wavefront propagation (**b**) Numerical grid

**Figure 4.** Using the asymptotic behaviour to tailor the numerical grid structure.

Figure 4b shows a fixed rectangular grid. It is designed to ensure that the ratio $\Delta x/\Delta t$ is exactly equal to the speed of the wavefront, $U_w$. As a consequence, valid solutions at every grid point will necessarily be identical except for offsets in time and space. The methodology described in the following section enforces this behaviour. It is worth emphasising that this grid structure is possible even though the wavefronts under investigation are of finite amplitude. However, it would not be possible if the wavefront were not in an asymptotic state or if other disturbances were travelling in the opposite direction. Either of these complications would lead to different physical behaviours at different locations, just as they do in more general simulations of pressure surges in pipelines. The only other instance in which a fixed rectangular grid can exactly match the physical propagation of waves is with vanishingly small-amplitude disturbances. In that case, there is no need to stipulate either asymptotic behaviour or an absence of waves travelling in the opposite direction. However, when these restrictions are removed, it is no longer the case that identical solutions exist at every location in space and time.

The use of a numerical grid structure satisfying $\Delta x/\Delta t = U_w$ has two important benefits and one important disbenefit. These are discussed fully below (Section 4.2), but they can be summarised as follows:

*Benefit-1*: It enables the numerical methodology to ensure that all points along the whole of the finite-amplitude wavefront propagate at an identical, constant, physically-correct speed. A loose analogy can be drawn with the widely accepted practice of forcing all points on general wavefronts in low Mach number, transient flows to propagate at exactly the same speed (usually chosen as the ambient speed of sound). However, that is an approximation to the true physics. It is made for simplicity and is justified only when the resulting errors are acceptable.

*Benefit-2*: It enables the speed of propagation of the wavefront to be chosen a priori by the software user. Indeed, it enforces that action. Use is made of this below (Section 5.3) when illustrating a practical application of the methodology. It is worth noting here that the chosen speed must be below the speed of sound. Propagation at a greater speed would imply that the wavefront is an abrupt discontinuity, namely a normal shock, whereas the analysis used herein is valid only in the absence of discontinuities.

*Disbenefit*: The downside of choosing the numerical grid to satisfy $\Delta x/\Delta t = U_w$ is that this makes the use of extrapolation inevitable—because the gradients of the MoC characteristic lines are greater than the speed of sound. This is a serious matter; numerical methods that involve extrapolation are usually inherently unstable. Accordingly, repeated attention is drawn to this concern below, including an assessment in which the potential errors are magnified tenfold.

In the particular software used to obtain the outcomes presented below, the numerical solution is obtained at successive times at a single spatial location, with use being made of the asymptotic property to infer values at adjacent locations and times. Hence, the numerical grid needs to include only this location and grid points very close to it. Nevertheless, the analysis has to include every time step in which the wavefront has an influence at the chosen location, and the number of such increments could be quite large when a fine grid pattern is used. However, in the author's software, the grid exists only at time steps close to the current solution instant. This is made possible by nesting the whole solution at the end of each time step. As a result, the complete solution is obtained using only $4 \times 4$ arrays. It is emphasised, however, that this choice has been made solely for programming convenience. It has no influence whatsoever on the resulting solution, so little purpose would be served by expanding on the relevant programming details herein.

## 4. Numerical Method

The cross-sectional-averaged continuity and momentum equations for 1D unsteady flow in the presence of mass addition/subtraction (through the duct walls) and skin friction (on the wall surfaces) can be combined and expressed in characteristic form as

$$\text{1D MoC}: \frac{dp}{dt} \pm \rho c \frac{dU}{dt} = -c^2 \frac{\dot{m}_y}{A} \mp c \left\{ -\frac{\dot{m}_y}{A} (U_{yx} - U) - \frac{\tau_w \updownarrow}{A} \right\} \tag{1}$$

in which all symbols are defined in the nomenclature.

In general, the wall shear stress $\tau_w$ can have both steady-flow and unsteady-flow components, but, for simplicity, it is treated as zero herein because the principal purpose is to introduce a new methodology, not to study particular practical applications comprehensively. However, since $\tau_w = 0$, the lateral mass flowrate $\dot{m}_y$ cannot also be zero. If it were, the right-hand side of Equation (1) would be zero and nothing would exist to counter the inevitable long-term consequence of inertial steepening, namely the evolution of the wavefront into a normal shock. At the other extreme, if the holes in the duct wall led directly to the atmosphere, the cumulative mass flowrate would be unconstrained and, in effect, the wavefront would evolve in a dispersive manner. Indeed, assuming ambient conditions prevail indefinitely on the other side of the wall, the amplitude of the wavefront would decrease indefinitely with an increasing distance from its source. Instead, therefore, the wall holes lead to chambers of finite volume. Flows into these cause their internal pressure to increase, imposing increased resistance to further flows into them. After the wavefront has passed, the pressures in the tunnel and chamber are equal.

To preserve generality, at least provisionally, the mass flowrate per unit length of wall is represented generically by

$$\text{Mass flowrate}: \ \dot{m}_y = \dot{m}_y\{a_y, V_{CH}, \rho_A, \rho_{CH}, p_A, p_{CH}, K_y\} \tag{2}$$

where $a_y$ is the area of holes per unit length of the duct, $V_{CH}$ is the chamber volume per unit length and $K_y$ is a resistance coefficient for flows through the wall. Details of the mechanism and its analysis are given by Liu et al. (2021) [10] and, using a different representation of the flow, by Vardy et al. (2022) [22]. In both cases, the importance of the resistance characteristics of the wall openings is highlighted. NB: If there were no resistance, the chambers would behave like Helmholtz resonators.

### 4.1. Illustration

Figure 5 shows indicative pressure histories at an arbitrary location during the passage of an asymptotic wavefront along a duct with air chambers. This example is discussed in greater detail in Section 5, which is focussed on the application of the methodology in the context of railway tunnels, but the qualitative behaviour is of general relevance in its own right. The example is included here for reference purposes when discussing the numerical solution process. It is seen that the pressure in the chambers lags behind that in the open duct. Indeed, a lag is inevitable whenever *any* wavefront propagates along such a duct, regardless of whether asymptotic conditions have developed. The magnitude of the lag depends upon the time required for air to flow into the chambers in response to the evolving pressure. The resulting wavefront can be asymptotic only if the *total* rate of mass flow into the chambers during the wavefront passage is exactly equal to the difference between the longitudinal mass flowrates upstream and downstream of the wavefront. If the lateral rate is insufficient, the wavefront will shorten. If it is too great, the wavefront will lengthen. The stated balance needs to exist relative to axes moving with the wavefront itself, not relative to axes that are stationary relative to the tunnel.

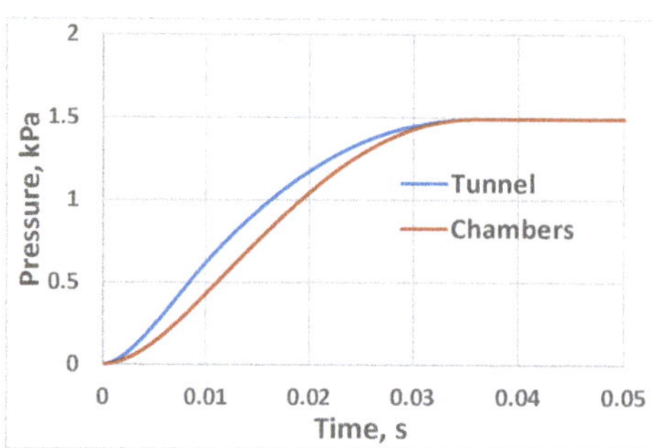


**Figure 5.** Indicative form of an asymptotic wavefront in a duct with air chambers (*Pressures in tunnel and chambers*).

### 4.2. Details of the Numerical Integration Process

The focus now moves to the numerical integration process. This is the heart of the solution methodology, so it is described in detail. The methodology has many features in common with conventional MoC solutions using a fixed-grid approach. In principle, it is a time-stepping process in which the known solution up to some instant $t_{old}$ is used, together with boundary conditions, to deduce the corresponding solution at a later instant $t_{new}$. In wider use, either implicit or explicit schemes may be used. With the former, the solution is obtained simultaneously at every grid point at $t_{new}$. With explicit schemes, it is obtained independently at each grid point. The present scheme is nominally explicit. Furthermore, since the whole solution needs to be obtained at only one spatial location, only one solution point exists at any instant $t_{new}$. In Figure 6, the point 'A' is the one at which a solution is to be obtained. The remaining points are as follows:

- M,N: Grid points at the most recent solution instant. The values of the flow parameters at these location are known from solutions obtained in the previous two time steps;
- B,C: Identical locations along the wavefront as A, but at different times. The flow parameters at these points are not known, but it is known that they are identical to the values at A (because, as stipulated in Section 3, $\Delta x/\Delta t = U_w$);
- L,R: Locations on characteristic wave paths projected backwards in time from A;
- L2: The same spatial location as L, but at the previous time step;
- L′,L2′: The same locations along the wavefront as L and L2, but one time step later. The flow parameters at L′ & L2′ are identical to those at L and L2.

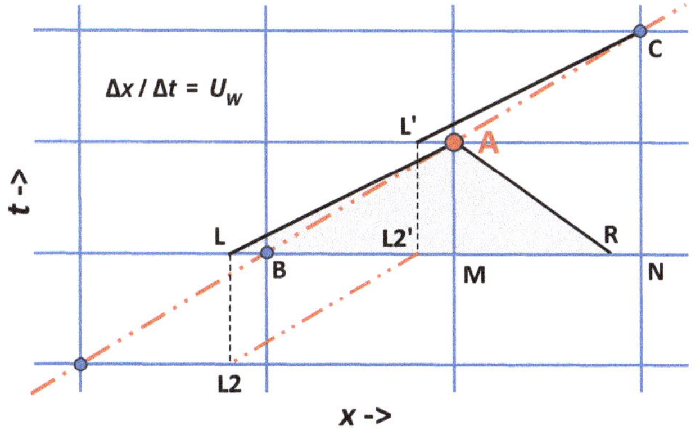


**Figure 6.** Paths of asymptotic wavefront and natural grid characteristics (*Shaded zone = domain of dependence at solution point A*).

The shaded zone in the figure includes all points in (x,$t$) space in the current time interval on which the *physical* solution at A depends. Notice, in particular, locations outside the range $x_L < x < x_R$ at the instant $t = t_M$ are too remote for waves originating at them to reach $x_A$ in one time step. This is a crucially important complication to which reference is made frequently below, even though its practical influence on the presented solutions is small.

Solution Sequence

The iterative solution sequence leading to values of the flow parameters at point A can be summarised as follows:

1. Make an initial estimate of the pressure at A. Designate it $p_{AG}$;
2. Infer the corresponding value of $c_A$, assuming isentropic change between A and the tip of the wavefront;

3. Estimate the location of R and interpolate between M and N to obtain implied values of $p_R$, $U_R$ and $\dot{m}_{y,R}$;
4. Solve Equation (1) along RA and Equation (2) along MA simultaneously. Hence, deduce values of $U_{AR}$, $\dot{m}_{y,A}$ and $p_{CH,A}$ consistent with the estimated $p_{AG}$. NB: The velocity at A is given the suffix AR to distinguish it from a second estimate of $U_A$ obtained independently below;
5. It is now possible to make an improved estimate of the location of R implied by the estimated value of $p_{AG}$. Hence, repeats steps 2 and 3 until an acceptable convergence is achieved. Then proceed to the following steps;
6. Using a suitable equation for the gradient of an MoC wave path, derive a relationship between $x_L$ and the local gradient of $(U + c)_L$. In the author's software, the gradient of the wave path is evaluated as ½[$(U + c)_L + (U + c)_A$}, with $U_A$ equal to the current estimate for $U_{AR}$ obtained in step 4 above;
7. Obtain a second relationship between $x_L$ and the local gradient of $(U + c)_L$ by extrapolation from M and B, noting that all values at B are the same as at A;
8. Solve these two equations to give estimates of $x_L$ and $(U + c)_L$. Use this solution to improve the estimated gradient of the MoC wave path, and repeat the process until convergence is achieved. Then extrapolate for the value of $\dot{m}_{y,L}$.
   NB: The use of extrapolation in this step will justifiably ring warning bells in the minds of readers with appropriate experience. A detailed discussion of it is given later;
9. It is necessary to know $U_L$ and $c_L$ independently, whereas the preceding step has yielded only their sum. Solve this together with a standard relationship for a simple wave-like disturbance between L and B. Then use an isentropic relationship to infer $p_L$ from $c_L$;
10. Interpolate between B and M to obtain implied values of $p_{L2'}$, $u_{L2'}$ and $\dot{m}_{y,L2'}$ and note that identical values are applicable at L2;
11. Solve Equation (2) between L2 and L to obtain the implied value of $p_{CH,L}$;
12. Solve Equation (1) along RA to obtain a second estimate of the velocity at A, namely $U_{AL}$;
13. At this stage, except for the dual value of $U_A$, everything is consistent with the chosen pressure $p_{AG}$. Accordingly, any difference between the two values of $U_A$ will exist only because of an error in the estimated value of $p_A$. Therefore, a unique solution would be found if steps 1–12 were repeated with the true value of $P_A$. In practice, this value is found by repeating the whole process with successively improved estimates of $p_{AG}$. Convergence is achieved fairly rapidly because it is known a priori that increases in $p_{AG}$ will increase $U_{AR}$ but will decrease $U_{AL}$, and vice versa. Hence the sign of $(U_{AR} - U_{AL})$ determines whether the estimated $p_{AG}$ needs to be increased or decreased;
14. After completing step 13, the final solution at A is known, and the whole sequence is repeated at successive grid times until the whole of the wavefront has passed the chosen location.

It should be noted that some of the above steps have been left somewhat open in that they could be undertaken in any of a number of ways. However, provided that suitable choices are made, accurate solutions will be achievable by simply reducing the grid size until further reductions have a negligible influence. Unlike many computational studies, computer resources are of almost no consequence for the present purposes. As an example, the CPU time for each of the examples given in Section 5 was in the order of one second on a simple laptop computer.

### 4.3. Numerical Limitations

Before illustrating the use of the methodology, five possible (or apparent) limitations are raised and methods of minimising or eliminating potential adverse consequences are described.

#### 4.3.1. Asymptotic Wave Speed $U_w$

Perhaps the most obvious apparent limitation is that the whole process begins by stipulating that the grid structure must satisfy $\Delta x/\Delta t = U_w$ exactly. This appears to conflict with the inevitable fact that designers will never know a priori what wave speed will exist in any particular application. However, this is misleading, because the whole basis of the analysis is to choose an outcome and then to explore how it could arise, not vice versa. As a consequence, we are completely free to choose whatever value of the wave speed we wish and to use the analysis to determine the remaining properties of wavefronts that could cause it. Naturally, it makes sense to focus on ranges of values that are expected to be physically meaningful, but that is a practical matter, not a formal requirement.

#### 4.3.2. Wavefront Amplitude

Likewise, the overall *amplitude* of the wavefront cannot be known a priori, but again, it is misleading to regard this as a limitation. Unlike the wavefront speed, we are not free to choose it a priori, but there is no need to do so. Indeed, it is properly regarded as an output of the methodology, not as an input into it. This is illustrated by the role of the parameter $p_{AG}$ in the detailed procedure presented above for the solution in a typical timestep. From this perspective, the final pressure after the whole wavefront has passed is simply the end result of a sequence in which the pressures at all intermediate times have been determined.

#### 4.3.3. Interpolation

The third potential limitation, namely the need for interpolation to determine the location of, and parameter values at, the point R at the base of the MoC characteristic line RA in Figure 6, is real, not merely apparent. First, since the parameter values need to be inferred from values at the adjacent grid points M and N, they cannot always be exactly correct. Second, even if exactly correct values could be deduced, the location of point R itself would be subject to error because it has to be inferred from data at points A and R alone, even though, physically, it depends on local values everywhere along the true path between R and A. This issue is present in any MoC analysis based on numerical discretisation. Its influence cannot be eliminated completely, but it can be reduced greatly by using sufficiently small values of $\Delta x$ and $\Delta t$. It is usually assumed to be very small if a series of simulations with successively smaller integration steps converge to a common solution. This is the approach adopted in the illustrative solutions presented below.

#### 4.3.4. Extrapolation

Although errors caused by interpolation can reasonably be expected to reduce with reducing the grid size, this is not necessarily also true for errors resulting from extrapolation. Therefore, special attention needs be paid to the use of extrapolation to locate point L and to infer parameter values at it. In more general MoC analyses, it is usual to use one of two ways to avoid this problem. One includes regarding $c$ as a constant and disregarding convective terms in the analytical equations, thereby yielding wave paths satisfying $\mathrm{d}x/\mathrm{d}t = c$ instead of $\mathrm{d}x/\mathrm{d}t = U \pm c$. However, that would be a nonsense in the present context because it would prevent the numerical simulation of the inertial steepening that the chambers are used to counteract.. The other common way of avoiding extrapolation involves using greater time steps than those illustrated in Figure 6, even though this simultaneously increases the influence of interpolation at points such as R. However, this approach is not possible herein because the whole solution process is applicable only for grids that satisfy $\Delta x/\Delta t = U_w$. Instead, therefore, a more direct method of assessing the influence of extrapolation is used below.

Specifically, solutions obtained using simple linear extrapolation are compared with alternative solutions in which the changes implied by linear extrapolation are deliberately magnified. This is a demanding test in its own right, but the risk of serious error is further reduced by using extrapolation only for two parameters that can be expected to vary only slowly, namely $\dot{m}_y$ and the sum $(U + c)$. The assessment process is illustrated in Section 5.2

below, where it is seen that differences between solutions obtained with and without magnification by factors of ±10 are small in comparison with the physical effect under study. It is not claimed that the test demonstrates that extrapolation will be safe in other contexts, but it is a strong indicator that its influence for the present purposes is weak, indeed probably negligible. It is certainly tiny in comparison with the physically important differences illustrated in Section 5.3.

#### 4.3.5. Triggering the Integration

The final limitation considered in this section is more academic than practical, but it nevertheless merits attention. It relates to the definition of the leading tip of the wavefront. So far, the discussion has been based on an implicit assumption that the whole of the wavefront can propagate at a uniform speed. However, this cannot be exactly true, especially for the examples considered herein because, in all cases, the assumed wave speed is smaller than the speed of sound. In reality, therefore, there must be a leading portion between the tip of the simulated wavefront and a more distant location that will have been reached by an infinitesimal disturbance travelling at the speed of sound. Herein, this issue is circumvented by defining the location of the leading tip of the asymptotic wavefront to be at a position where the pressure has already increased by a small finite value. One of the examples in Section 5.2 compares outcomes obtained using different values of this increase.

## 5. Numerical Examples

The use of the methodology is now illustrated. First, features of an asymptotic wavefront corresponding to particular geometrical parameters are described in detail. Then, the sensitivity of the numerical processes to key factors is assessed and is shown to be acceptably small. Having established this, the method is used to investigate the physical relationship between the asymptotic wavefront and the amplitude of the upstream pressure.

### *5.1. Base Case Asymptote*

Figure 7 shows predicted histories of the pressure, velocity and lateral mass flowrate during the passage of an asymptotic wavefront along a tunnel with air chambers. The origin of the time axis is at the leading tip of the wavefront, so these graphs are valid at all axial locations. Also, since the whole wavefront is propagating at a uniform speed, $U_w$, it would be equally acceptable to interpret the graphs as spatial profiles (*of a wavefront travelling from right to left*). The only change required for this is a scaling of the horizontal axis by $U_w$.

The cross-sectional area of the tunnel is 100 m$^2$, and the volume of the uniformly distributed chambers is equal to 3% of the corresponding volume of the tunnel. The rate of flow through connections between the tunnel and the chambers is prescribed to vary as the square of the flowrate. The wavefront speed is stipulated to be 2 m/s slower than the ambient speed of sound, so $U_w$ = 338.4 m/s. Therefore, since the duration of the period when the rate of change of pressure becomes negligible (Figure 7c) is approximately 0.036 s, the length of the wavefront itself may be regarded as approximately 12 m. This is only slightly greater than the tunnel diameter, and hence the use of a one-dimensional methodology is open to question. However, this will depend upon the geometry of the flowpaths between the tunnel and the chambers. Although shown as discrete holes in Figure 2 above, it is expected that continuous slots will be preferable for practical reasons. Also, it is important to bear in mind that the asymptotic condition is the cumulative consequence of large numbers of chambers. Each individual chamber has only a small influence.

In this particular example, the pressure rises along the wavefront to a maximum of approximately 1.5 kPa, and the same maximum necessarily applies in the chambers too. Inevitably, the increase in the chambers lags behind that in the tunnel—because flow into the chambers can exist only whilst the pressure inside them is smaller than that in the tunnel. By inspection, in this instance, the duration of the period required for the pressures to almost equalise is slightly more than 10% of the duration of the overall pressure change caused by the wavefront itself. The graphs of the axial velocity (inside the tunnel) and the lateral mass flowrate per unit length into the chambers are included primarily for completeness. However, it is noteworthy that the maximum rate of flow into the chambers occurs long after the instant corresponding to the maximum rate of change of pressure in the tunnel.

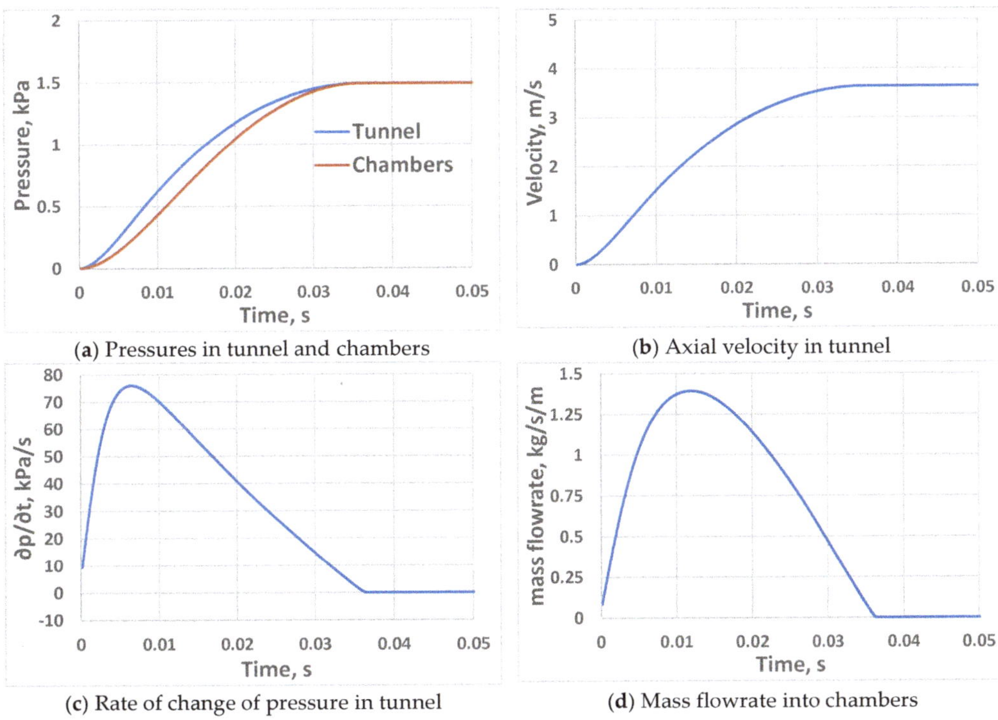


**Figure 7.** Asymptotic wavefront in a tunnel with lateral air chambers.

The most obvious practical significance of the chambers is that they facilitate the existence of the asymptotic behaviour. That is, they prevent the wavefront steepening into a shock. Even so, the maximum steepness of the wavefront in this example, namely about 76 kPa/s, is significantly greater than would be acceptable for a railway tunnel portal close to buildings. It follows that chambers with the particular properties chosen here would not be suitable in a practical application, so other combinations of chamber volume and connection properties would need to be assessed. However, since the simulations require such small CPU resources, large numbers of such simulations can be undertaken in a short time. This is a huge practical benefit, because the alternative of exploring options by means of simulations involving propagation along the whole tunnel requires CPU resources that are many thousands of times greater (sometimes millions of times). For the avoidance of doubt, it is not claimed that the need for any such simulations is completely eliminated. Some will be necessary for a range of other reasons. Nevertheless, the required number of them will be greatly reduced.

### *5.2. Numerical Accuracy*

Figure 8 illustrates the sensitivity of the base case solution to various numerical factors. This is done by comparing rates of change of pressure, not the pressure itself. There are two reasons for this. The more important of these for the theoretical modelling purposes of this paper is that the rates of change are more sensitive than the absolute values, so errors can be seen more easily. However, it is also true that the physical application that triggered the research depends far more strongly on rates of change of pressure than on the pressure itself.

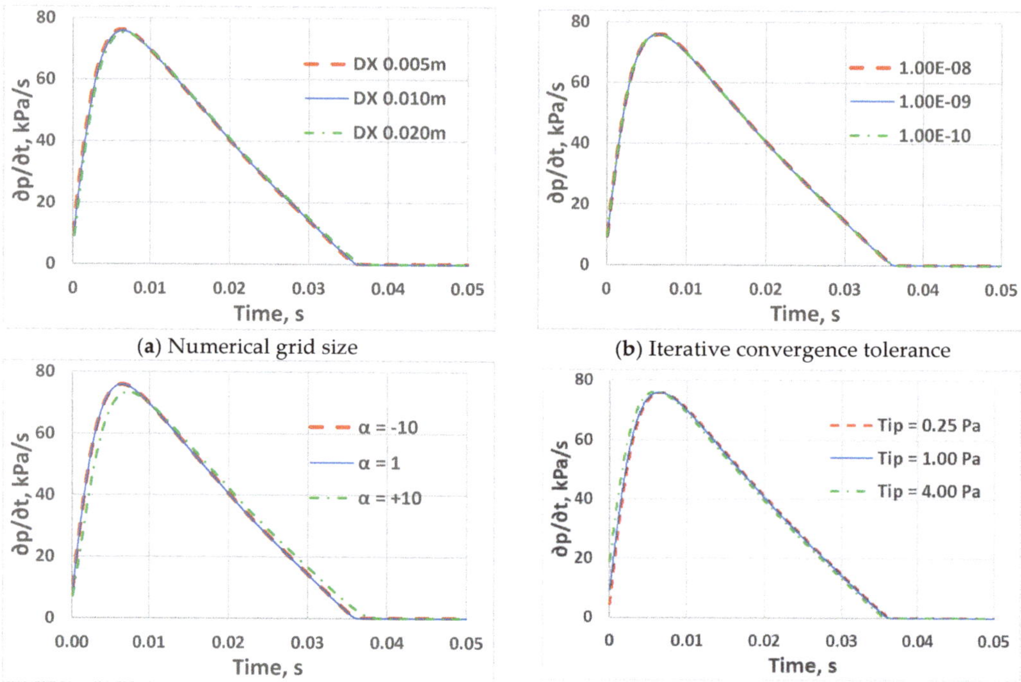


(**a**) Numerical grid size (**b**) Iterative convergence tolerance

(**c**) Influence of multiples of linear extrapolation (**d**) Pressure at tip of wavefront

**Figure 8.** Numerical sensitivity of the solution methodology. The parameter α in the legend of (**c**) is a numerical amplifier of linear extrapolation (see Equation (3)).

Each of the four boxes in the figure shows three graphs, and, in each box, the continuous line denotes a base case that is common to all four boxes. The broken lines show corresponding solutions with greater and smaller values of the parameter under consideration. The upper row shows the influence of (a) the grid size and (b) the convergence tolerance used to control the inbuilt iterative processes. By inspection, numerical errors due to these causes are so small that it is difficult to distinguish between the curves.

The same is true for only two of the curves in Figure 8c, which assesses the sensitivity to the extrapolation process to which attention has been drawn above. In the base case, a simple linear extrapolation of the relevant flow parameter from values at points M and B in Figure 6 is used to estimate the required value at point L. In the other two cases, the prescribed difference between the values at B and L is 10 times greater than the difference corresponding to linear extrapolation. Using $\varphi$ to denote any extrapolated parameter, the values used at the extrapolated point L in Figure 6 above satisfy

$$\text{Extrapolation}: \ \varphi_L - \varphi_B = \alpha[\ (\varphi_B - \varphi_M)\ \{(x_L - x_B)/(x_B - x_M)\}] \tag{3}$$

in which $\alpha$ is a numerical amplifier. For simple linear extrapolation, $\alpha = 1$, and for the cases used to assess the consequences of extrapolation, $\alpha = +10$ and $\alpha = -10$. The author considers these differences to be implausibly large and expects the base case to be the closest of the three to the true numerical solution of the equations.

Less qualitative arguments in support of approximately linear extrapolation are possible. One is based on the simple observation that the dominant process is inertial and that the lateral flows merely modulate it. This is important because the amplified extrapolations magnify the whole change, including the dominant inertial contribution. Another is that errors in the assumed values of the lateral mass flowrate have a self-correcting tendency. Overestimation in one time step will cause the pressure in the chambers to increase too much, and hence the driving force in the next time interval will be smaller than it would have been if the overestimation had not occurred. Notwithstanding these observations, it is not possible to provide a rigorous proof that the extrapolation is 100% safe. Instead, readers must decide for themselves whether the author's confidence in the presented results is justified.

The final box in Figure 8 illustrates the influence of the numerical method of kick-starting the whole process. Ahead of the wavefront, ambient conditions prevail, and these would persist indefinitely unless some method of triggering a disturbance existed. In the software, this is done by arbitrarily deeming a small pressure change to have occurred and then tracking its consequences upstream. Of necessity, this initial step will not be physically meaningful, and so the chosen value must be acceptably small. Figure 8d compares predicted profiles based on three trigger values. By inspection, differences can be seen, even though the trigger values are in the order of 1000 times smaller than the final amplitude of the wavefront. Nevertheless, the differences are small, and, in particular, the variation in the predicted maximum rate of change of pressure is small. Attention is drawn to this because, as explained in Section 2 above, $(\partial p/\partial t)_{max}$ is the parameter of greatest practical importance in the context of pressure disturbances emanating from railway tunnel portals.

### 5.3. Practical Use of the Methodology

Although the primary purpose of this paper is to introduce the new methodology in its own right, there is merit in presenting a simple example to illustrate its potential practical value. Figure 9 shows predicted histories of the pressure and its rates of change for three cases with identical values of all prescribed values except the asymptotic wave speed. The continuous blue line is the base case used above. For it, the prescribed wave speed is 2 m/s slower than the ambient speed of sound. The other two cases correspond to 1 m/s and 3 m/s deficits, respectively. By inspection, the smaller the prescribed speed difference, the greater the pressure that can be sustained indefinitely. For practical design purposes, however, it is more useful to rephrase this to reflect the parameters that designers and operators can influence directly. At the outset, they can choose a cross-sectional area of the tunnel to limit the maximum pressure amplitude. Later, adjustments can be made to the train speed for the same purpose. Then, graphs such as those in Figure 9 can be used to estimate the maximum rate of change of pressure at the tunnel exit portal. The exact speed of the wavefront is of no direct practical interest in this context. However, it is the input parameter used in the analysis to assess the influence of pressure on the properties of the asymptotic wavefronts.

For completeness, it is important to recognise that large differences may exist between the amplitude of the initial disturbance and the subsequent amplitude of the wavefront as it nears the tunnel exit. Many factors can influence changes during propagation along tunnels, including, for instance, skin friction, the track geometry and airshafts. To allow for these, designers will still need to undertake simulations of wave propagation along the whole tunnel. However, for the required purpose of estimating the resulting pressure amplitude, it will be sufficient for such analyses to use a coarse grid size—perhaps a few tunnel diameters. Such simulations necessitate vastly smaller CPU demands than simulations designed to give accurate predictions of wavefront *shapes*—and hence of rates of change of pressure. Instead, the method introduced herein can assess the likelihood of a wavefront with the expected final amplitude being sufficiently steep to causes an unacceptable disturbance outside the tunnel. This complementary use of the two types of analysis has the potential to reduce greatly the resources needed to obtain reliable predictions, especially in the case of long tunnels.

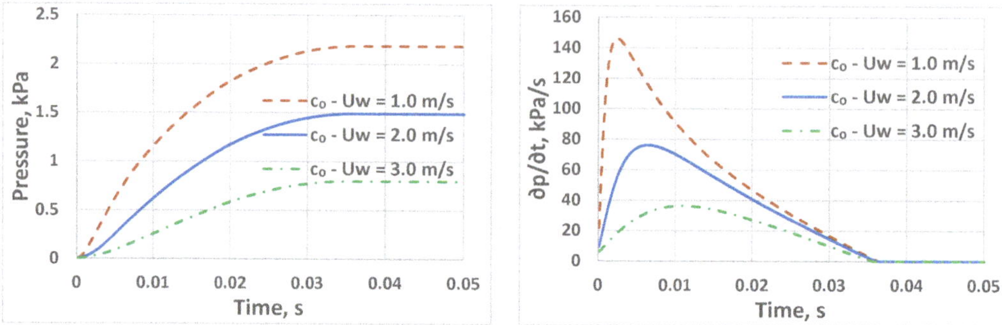


(**a**) Pressure histories (**b**) Rates of change of pressure

**Figure 9.** Relationship between the wavefront speed and pressure amplitude.

**6. Summary and Conclusions**

The possibility of wavefronts evolving towards an asymptotic state as they propagate along a duct has been demonstrated, and a method of predicting the range of possible detailed forms of such asymptotes has been developed. The method, which requires minimal CPU resources, is unusual in that it predicts possible outcomes without any need to know what has caused them. Indeed, since the method is tailored expressly to mimic asymptotic conditions, it is wholly unsuitable for studying wavefronts during preceding periods of evolution before asymptotic forms have developed. In this sense, the method complements conventional analyses in which prescribed boundary conditions define initial 'causes' and simulations are used to deduce the resulting 'effects'. In such methods, the continual accumulation of numerical errors during propagation makes them poorly suited to modelling eventual asymptotic states.

The new analysis needs to simulate only short lengths of duct—e.g., a few duct diameters—and only the small times needed for wavefronts to travel such distances. It needs only trivial CPU resources, even when tiny grid sizes are used to achieve high accuracy—e.g., one-thousandth of the duct diameter. The use of such grid sizes in simulations along complete ducts would be impracticable.

The methodology is inherently simple, and it has been shown to be highly stable numerically. Importantly, this is true even for the method's only obvious drawback, namely a need for two parameters to be determined by means of extrapolation, not interpolation. The consequences of this drawback have been assessed in detail and shown to be minimal, even when the influence of the extrapolations has been magnified by factors of $\pm 10$.

The use of this method has been illustrated for the particular case of a duct that has small, enclosed air chambers into which air can flow, thereby modifying the dynamics of airflow along the duct. This configuration has been studied previously for a particular practical purpose, but the simulations have been of uncertain accuracy because of the CPU limitations cited above. It has been found that one-to-one correlations exist between the pressure amplitude of an asymptotic wavefront and (i) its speed and (ii) its maximum steepness. Unique relations such as these do not exist in wavefronts that have not evolved to an asymptotic state.

**Funding:** This research received no external funding.

**Data Availability Statement:** Not applicable.

**Conflicts of Interest:** The author declares no conflict of interest.

## Nomenclature

| | |
|---|---|
| 1D | one-dimensional |
| $A$ | cross-sectional area of duct/tunnel, m$^2$ |
| $a_y$ | area of wall holes per unit length, m |
| $c$ | speed of sound, m/s |
| $K_y$ | resistance coefficient, flow through wall |
| $\ell$ | duct perimeter length, m |
| $\dot{m}_y$ | mass flowrate per unit length through wall (positive = outflow), kg·s$^{-1}$m$^{-1}$ |
| $p$ | pressure, Pa |
| $t$ | time coordinate, s |
| $U$ | axial velocity (cross-sectional average), m/s |
| $U_{yx}$ | axial velocity component of flow through wall, m/s |
| $U_w$ | axial velocity of asymptotic wavefront |
| $V_{CH}$ | volume of chambers per unit length, m$^2$ |
| $x$ | spatial coordinate along tunnel, m |
| Greek characters | |
| $\alpha$ | numerical amplifier of extrapolated variations |
| $\rho$ | density, kg/m$^3$ |
| $\tau_w$ | wall shear stress, Pa |
| $\varphi$ | typical parameter subject to extrapolation |
| Subscripts | |
| A,B,... | grid points defined in Figure 6 |
| $CH$ | chamber |
| $w$ | asymptotic wavefront |

## References


1. Fox, J.A. *Hydraulic Analysis of Unsteady Flow in Pipe Networks*; Macmillan: London, UK, 1977.
2. Wylie, E.B.; Streeter, V.L. *Fluid Transients in Systems*; Prentice Hall: Hoboken, NJ, USA, 1993; 463p.
3. Chaudhry, M.H. *Applied Hydraulic Transients*, 3rd ed.; Springer: New York, NY, USA, 2014; 583p.
4. Mashimo, S.; Nakatsu, E.; Aoki, T.; Matsuo, K. Attenuation and distortion of a compression wave propagating in a high-speed railway tunnel. *JSME Int. J. Ser. B Fluids Therm. Eng.* **1997**, *40*, 51–57. [CrossRef]
5. Fukuda, T.; Ozawa, S.; Iida, M.; Takasaki, T. Distortion of Compression Wave Propagating through Very Long Tunnel with Slab Tracks. *JSME Int. J. Ser. B Fluids Therm. Eng.* **2006**, *49*, 1156–1164. [CrossRef]
6. Nakamura, S.; Sasa, D.; Aoki, T. Attenuation and distortion of a compression waves propagating in very long tube. *J. Therm. Sci.* **2011**, *20*, 53–57. [CrossRef]
7. Wang, H.; Lei, B.; Bi, H.; Yu, T. Wavefront evolution of compression waves propagating in high speed railway tunnels. *J. Sound Vib.* **2018**, *431*, 105–121. [CrossRef]
8. Fukuda, T.; Nakamura, S.; Miyachi, T.; Saito, S.; Kimura, N.; Matsunuma, M. Influence of ballast quantity on compression wavefront steepening in railway tunnels. *Proc. Inst. Mech. Eng. Part F J. Rail Rapid Transit* **2020**, *234*, 607–615. [CrossRef]
9. Iyer, R.S.; Kim, D.H.; Kim, H.D. Propagation characteristics of compression wave in a high-speed railway tunnel. *Phys. Fluids* **2021**, *33*, 086104. [CrossRef]
10. Liu, F.; Vardy, A.E.; Pokrajac, D. Influence of air chambers on wavefront steepening in railway tunnels. *Tunn. Undergr. Space Technol.* **2021**, *117*, 104120. [CrossRef]
11. Tijsseling, A.S. Discussion of "Effect of boundary on water hammer wave attenuation and shape" by Huade Cao, Ioan Nistor, and Magdi Mohareb. *J. Hydraul. Eng.* **2021**, *147*, 07021011. [CrossRef]
12. YouTube Video Posted by Korea Railroad Research Institute (The Boom is Heard Soon after 1 min 15 s into the 3 min 8 s Video). Available online: https://www.youtube.com/watch?v=euFn8eOsnDI&t=91s (accessed on 7 October 2024).
13. Winslow, A.; Howe, M.S. Influence of a scarfed portal on the compression wave generated by a high-speed train entering a tunnel. *J. Low Freq. Noise Vib. Act. Control* **2005**, *24*, 203–217. [CrossRef]
14. Zhang, L.; Thurow, K.; Stoll, N.; Liu, H. Influence of the geometry of equal-transect oblique tunnel poral on compression wave and micro-pressure wave generated by high-speed trains entering tunnels. *J Wind. Eng. Ind. Aerodyn.* **2018**, *178*, 1–17. [CrossRef]
15. Kikuchi, K.; Iida, M.; Fukuda, T. Optimisation of train nose shape for reding micro-pressure wave radiation from tunnel exit. *J. Low Freq. Noise Vib. Act. Control* **2011**, *30*, 1–19. [CrossRef]
16. Miyachi, T.; Kikuchi, K.; Hieke, M. Multistep train nose for reducing micro-pressure waves. *J. Sound Vib.* **2022**, *520*, 116665. [CrossRef]

17. Nishimura, M.; Kondo, M.; Iwamoto, K.; Tsujimoto, Y. Active attenuation of impulsive noise from a railway tunnel exit. In Proceedings of the Inter-Noise 94, Yokohama, Japan, 29–31 August 1994; pp. 1383–1388.
18. Aoki, T.; Vardy, A.E.; Brown, J.M.B. Passive alleviation of micro-pressure waves from tunnel portals. *J. Sound Vib.* **1999**, *220*, 921–940. [CrossRef]
19. Wang, H.; Vardy, A.E.; Pokrajac, D. Pressure radiation from a perforated duct exit region. *J. Sound Vib.* **2015**, *351*, 29–42. [CrossRef]
20. Hieke, M.; Kaltenbach, H.-J.; Tielkes, T. Prediction of micro-pressure wave emissions from high-speed railway tunnels. In Proceedings of the 13th Int Symp on Aerodynamics and Ventilation of Vehicle Tunnels, New Brunswick, NJ, USA, 13–15 May 2009; Sweetland, I., Ed.; BHR Group: Cranfield, UK, 2009; pp. 487–501.
21. Vardy, A.E.; Brown, J.M.B. Influence of ballast on wave steepening in tunnels. *J. Sound Vib.* **2000**, *238*, 595–615. [CrossRef]
22. Vardy, A.E.; Liu, F.; Pokrajac, D. Large air chambers for rail tunnel exit regions. In Proceedings of the 19th International Symposium on Aerodynamics, Ventilation and Fire in Tunnels, Brighton, UK, 28–30 September 2022; Tabarra, M., Ed.; BHR Group: Cranfield, UK, 2022; pp. 185–200.

fluids

MDPI

Article

# Mesh Sensitivity Analysis of Axisymmetric Models for Smooth–Turbulent Transient Flows

**Pedro Leite Ferreira [1] and Dídia Isabel Cameira Covas [2,*]**

[1] Department of Civil Engineering, Polytechnic of Porto, 4249-015 Porto, Portugal; plf@isep.ipp.pt
[2] CERIS, Instituto Superior Técnico, Universidade de Lisboa, 1049-001 Lisbon, Portugal
* Correspondence: didia.covas@tecnico.ulisboa.pt

**Abstract:** The current paper focuses on the assessment of radial mesh influence on the description of the transient event obtained by an axisymmetric model. The objective is to reduce computational effort while accurately calculating hydraulic transients in smooth–turbulent pressurized pipes. The analyzed pipe system has a reservoir–pipe–valve configuration with an inner diameter of 0.02 m and a total length of 14.96 m, with the initial discharge being equal to $120 \times 10^{-3}$ L/s (Re = 7638). An extensive study is carried out with 80 geometric sequence meshes by varying the total number of cylinders, the geometric common ratio, and the pipe axial discretization. The benefit of increasing the geometric common ratio is highlighted. A detailed comparison between two meshes is presented, in which the best mesh (i.e., the one with the lowest computational effort) has a three-fold higher value of the geometric common ratio. The two meshes show small differences for the instantaneous valve closure, limited to a time interval immediately after the arrival of the pressure surge and only during the first pressure wave. The dynamic characterization of the transient phenomenon demonstrates the in-depth consistency between the model results and the hydraulic transients' phenomenon in terms of the piezometric head, the wall shear stress, and the mean velocity time-history, in comparison to the results obtained with the shear stress, lateral velocity, and axial velocity profiles.

**Keywords:** hydraulic transient; unsteady friction; axisymmetric model; mesh optimization; smooth–turbulent transient flows

## 1. Introduction

Transient flows occur when pressure and flow rate changes are created by manoeuvres in valves or in turbomachines. The numerical modelling of these flows is a complex engineering task that should ideally consider secondary phenomena that are not described by the Joukowsky formulation, such as, fluid–structure interaction, cavitation, frequency-dependent energy dissipation, and pipe wall viscoelasticity [1,2].

A hydraulic transient is not a one-dimensional (1D) phenomenon and has complex radial velocity profile dynamics. It has a different behaviour between the pipe core and the wall region, with the latter having higher velocity gradients. Hence, the unsteady energy dissipation is not in phase with the mean velocity. Therefore, there is no exact and completely satisfying solution for the calculation of unsteady friction in fast transient flows in 1D models [3,4]. The 1D instantaneous acceleration-based formulations lack generality and accuracy [5–9]. The 1D convolution integral-based models require a significant computational effort, and the simplified versions lack both accuracy and generality [10–12]. Conversely, the three-dimensional computational fluid dynamics (3D CFD) models are highly accurate in portraying the 3D nature of the transient flow. However, these models are extremely time-consuming for engineering practice [13]. Moreover, the inconsistencies between the outcomes of the various turbulence models raise questions about the rationale for their application [11].

The axisymmetric models, hereafter referred as quasi-2D (Q2D), have proven to be consistent with the phenomenon dynamics, leading to highly accurate results [14]. Their main advantage is to describe solely the most essential features of the transient. The uniform cross-sectional pressure and the flow radial symmetry assumptions ensure the "simplicity" of these mathematical models. These simplifications eliminate the radial and circumferential momentum equations and neglect the circumferential velocity component in the continuity equation [15,16]. Additionally, turbulent models can be adapted for axisymmetric flows, extending beyond the frozen-viscosity approximation, as is the case with 1D convolutional models [17,18]. The drawback of their use is also the numerical effort, resulting in longer computational times than the simplest 1D models. Numerically more efficient versions have been proposed in the past [19–21].

Still, Q2D models can be considerably improved with the development and investigation of new optimized meshes. The mesh definition significantly affects the computational effort, but also, more importantly, affects the accuracy of the results. Indeed, the accuracy of a numerical simulation is dependent on the mesh size and geometry [22]. A second important aspect for grid definition refers to the adaption of the mesh to the flow physics to ensure that the velocity variation is of the same order of magnitude along the grid. In practice, in pressurized transient analysis, a non-uniform grid must be generated due to the strong gradients near the wall; finer meshes are needed near the wall to better describe high velocity gradients, whereas coarser meshes are sufficient in the pipe core [23].

This paper aims at the assessment of radial mesh influence on the computation of unsteady energy dissipation in smooth–turbulent flows using a Q2D model. An extensive numerical analysis of the effect of the radial mesh on the computation of transient flows in pressurized pipes is carried out. A compromise between accuracy and computational time is established. A detailed analysis and comparison of the obtained Q2D results for the two best meshes for a theoretical instantaneous valve closure are developed.

The current paper innovatively presents the following: (i) the identification of the necessary improvements in the Q2D model radial meshes to accurately simulate steady and unsteady flows with less computation effort; (ii) guidelines to the definition of radial meshes based on transient simulation for 80 different meshes; (iii) in-depth analysis of the results for two different approaches to radial mesh construction; and (iv) the comprehensive characterization and analysis of transient flow simulation for the best identified mesh.

The numerical simulations consider a pressurized pipe system with a reservoir–pipe–valve configuration. The pipe has an inner diameter, $D$, of 20 mm, a wall thickness, $e$, of 0.001 m, a total length, $L$, of 15.2 m, and a wave velocity, $c$, of 1250 m/s (experimental and theoretical values of wave velocity yielded similar results). The numerical study considered an initial discharge of $120 \times 10^{-3}$ L/s, a kinematic viscosity of $1.01 \times 10^{-6}$ $m^2/s$, corresponding to a Reynolds number of 7638, and the downstream valve's theoretical instantaneous closure. It corresponds to the experimental facility assembled at the Laboratory of Hydraulics, Water Resources and Environment at Instituto Superior Técnico, which has been used for several experimental and numerical studies with different 1D, Q2D, and 3D models [13,24,25].

## 2. Numerical Models

Traditional axisymmetric equations, assumptions, and numerical schemes are used. The governing equations constitute a system of two partial differential equations with three unknown parameters ($p$, $u$, $v$) with the following form [14,17]:

$$\frac{\partial H}{\partial t} + \frac{c^2}{g}\frac{\partial u}{\partial x} = -\frac{c^2}{gr}\frac{\partial (rv)}{\partial r} \tag{1}$$

$$\frac{1}{g}\frac{\partial u}{\partial t} + \frac{\partial H}{\partial x} = \frac{1}{r\rho}\frac{\partial r\tau}{\partial r} \tag{2}$$

$$\tau = \rho\nu\frac{\partial u}{\partial r} - \rho\overline{u'v'} \tag{3}$$

where $x$ is the distance along the pipe; $r$ is the distance from the axis in the radial direction; $t$ is time; $H$ is the piezometric head; $u(x, r, t)$ is the local axial velocity; $v(x, r, t)$ is the local radial velocity; $g$ is the gravitational acceleration; $c$ is the wave velocity; $\rho$ is the liquid density; $\tau$ is the shear stress; $\nu$ is the fluid kinematic viscosity; and $\rho\overline{u'v'}$ represents the Reynolds stress term. The Reynolds stress calculation used the five-layer algebraic and time-constant turbulence model [14,26].

The main code was developed in VBA and some parts in C++, the parts corresponding to a higher number of computations. The implemented numerical scheme was the one proposed by Zhao and Ghidaoui [19], which allows to obtain the solution in $N_C$ operations instead of the $N_C{}^3$ necessary for solving the original Vardy and Hwang [14] model, with $N_C$ being the total number of cylinders considered in the radial discretization. The numerical scheme, the mesh setup definition, the steady- and unsteady-state flow calculations, the boundary conditions, the truncation error, and the implemented five-region turbulence model are described in detail in Ferreira and Covas [22].

The radial mesh (or grid) generation used a geometric sequence (GS) defined by fixing the first cylinder thickness near the pipe wall, $\Delta r_{j=N_C-1}$, and the geometric sequence common ratio, $C_R$. The thickness of cylinder $j$ is calculated as follows:

$$\Delta r_j = \Delta r_{j=N_C-1} \cdot (1 + C_R)^{N_C - j} \tag{4}$$

A particular GS mesh is referred to as "GS $_{CR}$ $N_C$". For example, a GS$_{5\%}$40 mesh, depicted in Figure 1, corresponds to a radial grid comprising a total of 40 cylinders ($N_C$ = 40), with a 5% increase in cylinder thickness from the wall to the centre of the pipe ($C_R$ = 5%).

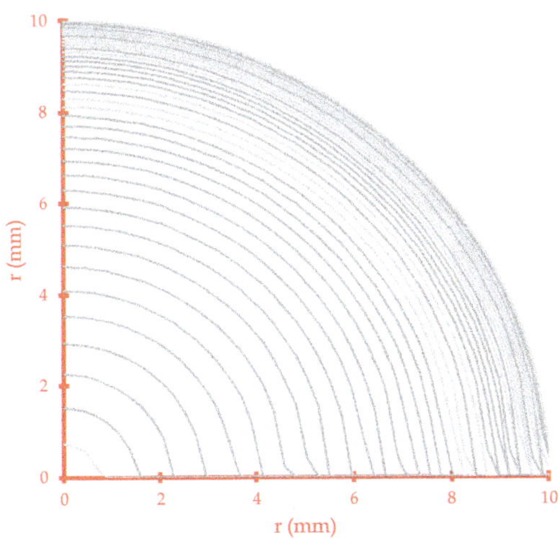


**Figure 1.** Geometric sequence mesh for $N_C$ = 40 and $C_R$ = 5%, referred to as GS$_{5\%}$40.

## 3. Mesh Analysis

The most efficient mesh in the context of the current research stands for the one that ensures the best compromise between accuracy and computational time. For this purpose, a mesh sensitivity analysis is conducted, including two components as follows: (i) the steady-state analysis, which should assure a minimum initial mean velocity error; and (ii) the unsteady-state analysis, focusing on the unsteady wall shear stress computation.

### 3.1. Steady-State Mesh Assessment

The implemented Q2D model uses the method of characteristics to solve the transient flow propagation along the pipe axis and a finite difference numerical scheme for the shear stress and volume transfer calculation between adjacent cylinders in each pipe cross section.

The latter numerical scheme has a truncation error proportional to $\Delta r^2.\partial^3 ru / \partial^3 r$, which is proportional to $\Delta r^2$, $O\left(\Delta x^2\right)$, and to the third derivative of the axial velocity.

The performance of the used finite difference scheme is analyzed based on the capacity to correctly represent the shear stress variation (i.e., high-gradient areas corresponding to higher mesh resolutions). In the case of a uniform shear stress variation, as in laminar flow conditions, an equal area mesh has been demonstrated to be a good approximation [22]. However, in turbulent flows, the difference between the total viscosity (i.e., the summation of the molecular or kinematic viscosity and the turbulent viscosity) near the wall and in the pipe core implies a steeper velocity gradient near the pipe wall. The high velocity gradient in this region requires a higher concentration of grid points near the wall, which can be achieved by using the geometric sequence (GS) mesh.

Since the refinement in the wall region is crucial, the adopted GS mesh is assessed for different numbers of cylinders, $N_C$, and of cylinder increase ratio values, $C_R$. The variation in the relative error of the mean velocity obtained by the GS mesh as a function of the $N_C$ values is shown in Figure 2, corresponding each line to different $C_R$ values. Note that the reference velocity, $U_{ref}$, corresponds to the mean velocity to which the velocity converges (when increasing the $N_C$) for each $C_R$ value.

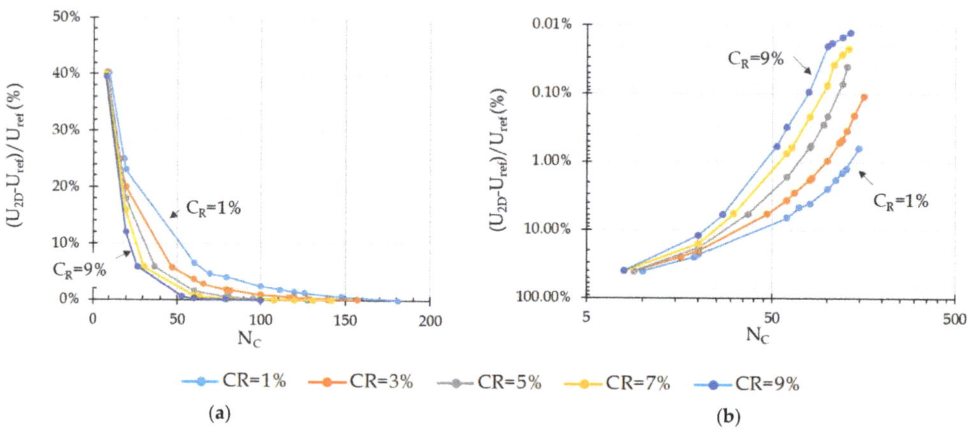


**Figure 2.** GS grids' accuracy variation with $N_C$ and $C_R$ values represented in (**a**) a linear scale and (**b**) a logarithmic scale.

Two features should be highlighted. The first is that the mesh error decreases exponentially with $N_C$ and the decay variation increases with $C_R$. For low $N_C$ meshes (e.g., $N_C = 10$), the accuracy is almost $C_R$-independent. For intermediate $N_C$ meshes, the accuracy increases with the $C_R$ value. For $N_C = 60$, the $C_R = 9\%$ mesh attains a high accuracy with an error of 0.3%, while the $C_R = 1\%$ mesh leads to a much higher error of 6.65%. To achieve the same accuracy with the lowest $C_R$ value (1%), a mesh with $N_C = 120$ should be considered (i.e., twice the radial grid points and computation time). The second feature is that, after reaching a certain accuracy value ($N_C = 100$), higher $C_R$ meshes ($C_R = 9\%$ and $C_R = 7\%$) present a significant decrease in the accuracy increase with $N_C$. However, low $C_R$ value meshes are far from this pattern for the analyzed $N_C$ range (Figure 2b).

Figure 3 depicts the dimensionless velocity, $u^+(=U/u_\tau)$, with the dimensionless distance from the wall, $y^+$ $(=u_\tau y/\nu)$, for two meshes, both with 60 cylinders: the first mesh with a $C_R = 1\%$ (Figure 3a) and the second with $C_R = 9\%$ (Figure 3b). The second mesh ($C_R = 9\%$) allows to reduce the thickness of the first cylinder by almost two orders of magnitude (compared to $C_R = 1\%$) by increasing the number of mesh points near the wall. This can be observed in Figure 3a,b by comparing the first mesh point near the pipe wall in both meshes. Results for $C_R$ = 9% have a better fitting with the theoretical law-of-the-wall $u^+ = y^+$ than those for $C_R$ = 1%.

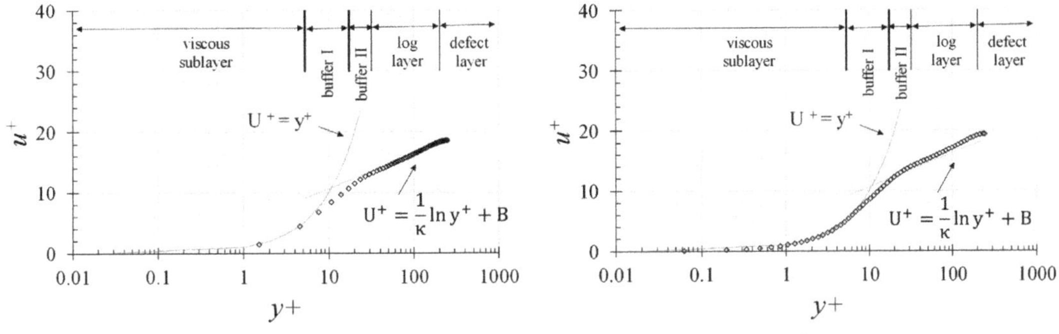


**Figure 3.** Axial velocity profile according with the turbulent layers for two GS mesh with $N_C$ = 60: (**a**) $C_R$ = 1%; (**b**) $C_R$ = 9%.

The previous examples justify that the error increases with the $C_R$ decrease, due to the fact that the grid interval undergoes a notable expansion near the wall, where higher velocity gradients are observed. The effect of the steady-state velocity error is more significant in the calculation of the first pressure wave, in which the energy dissipation is small and less significant in the subsequent peaks. For the defined flow conditions ($Q_0$ = 120.6 × $10^{-3}$ $Ls^{-1}$), when using the Joukwosky formula ($\Delta H = c\Delta U/g$, where $\Delta H$ is the change in the pressure head, $c$ is the fluid wave speed, $\Delta U$ is the change in the mean velocity, and $g$ is the gravity acceleration), the initial mean velocity errors of 1% and 10% correspond to a maximum head of 0.49 and 4.91 m, respectively. These values correspond to 2 and 20 times, respectively, that of the initial steady-state friction head losses. The use of more sophisticated numerical models, such as those based on Q2D or 3D CFD approaches, is not a worthwhile option, unless a high level of accuracy in the determination of the steady-state velocity profile can be achieved.

### *3.2. Unsteady-State Mesh Assessment*

#### 3.2.1. The Rational

The optimization of the $N_C$ and $C_R$ parameters of the GS mesh is investigated based on transient simulations for 80 different meshes. The total number of cylinders, $N_C$, varies between 60, to control the steady-state velocity profile error, and 120, to avoid very time-consuming simulation times. The $N_C$ increment is constant and equal to 20 (i.e., grids with 60, 80, 100, and 120 cylinders are used). The geometric rate, $C_R$, varies between 1% and 9%, with a step of 2%, as previously considered for steady-state evaluations (i.e., 1%, 3%, 5%, 7%, and 9%). For every combination of $N_C$ and $C_R$, four different axial discretizations of the pipe, $\Delta x$, are analyzed (i.e., 0.01, 0.02, 0.05 to 0.10 m).

Three parameters have been used to describe previous mesh characteristics (Table 1): $N_C$, the total number of concentric cylinders; $\Lambda_X$, the dimensionless mesh size in the pipe axial direction ($\Lambda_X = \Delta x/D$); and $\Lambda_R$, the ratio between the number of cylinders in the three outers region (laminar and buffer I and II), $N_{3R}$, and the total number of cylinders, $N_C$ ($\Lambda_R = N_{3R}/N_C$). The Courant–Friedrich–Lewy condition ($\Delta x/c\Delta t = 1$) is ensured when implementing the MOC along the pipe direction. Thus, by considering different $\Delta x$ values ($\Lambda_X$), the influence of $\Delta t$ is implicitly analyzed.

Since the energy dissipated decays exponentially with time, only the results for the first wave period are presented, but the results for five wave periods have also been calculated and analyzed, leading to similar conclusions. The reference mesh is $GS_{5\%}120$ with $\Delta x$ = 0.01 m, which corresponds to the highest $N_C$, $\Lambda_X$ (5.0) and $\Lambda_R$ (0.67) values without numerical instabilities. The results obtained for this mesh have been compared with those from Zielke's convolution model (i.e., lower $\Lambda_R$ or $N_C$ values do not ensure the same accuracy), and an exceptional accuracy in transient pressures, shear stresses, and mean velocities has been observed.

The influence of the mesh parameters on numerical accuracy is examined according to two main aspects: (i) the mesh capability to correctly calculate the velocity profile after the pressure surge arrival (i.e., properly describing the high-gradient velocity profile near the wall area), referred to as the short-term accuracy; and (ii) the grid capability to correctly describe the phenomenon when the created flow inversion ("vortex") moves into the pipe core, referred to as the long-term accuracy.

**Table 1.** Meshes' parameters considered in the 80 transient flow simulations.

| $N_C$ | $\Lambda_R$ | $\Lambda_X$ |
|---|---|---|
| 60 | 0.50, 1.00, 2.50, 5.00 | 0.18, 0.30, 0.43, 0.55, 0.63 |
| 80 | | 0.20, 0.36, 0.53, 0.63, 0.71 |
| 100 | | 0.22, 0.43, 0.60, 0.70, 0.75 |
| 120 | | 0.23, 0.49, 0.67, 0.77, 0.80 |

### 3.2.2. Short-Term Accuracy

Figure 4 shows the comparison of the results obtained for the reference mesh ($GS_{5\%}120$) with those from two grids with a lower $C_R$ value ($C_R$ = 1%): the first grid ($GS_{1\%}60$) with half of the grid points, corresponding to half of the computational effort; and the second mesh ($GS_{1\%}120$) with the same number of grid points as the benchmark. The results of the $GS_{5\%}120$ mesh show the highest initial peak and a considerable decay with time, with an almost 80% amplitude reduction during the three-wave periods (Figure 4a).

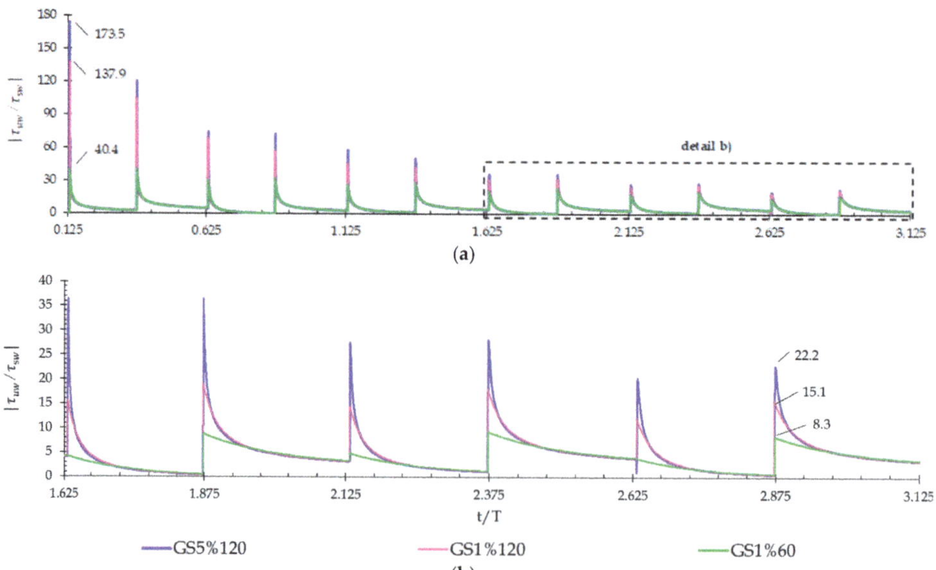


**Figure 4.** Wall shear stress time-history at pipe midsection for $GS_{1\%}60$, $GS_{1\%}120$, and $GS_5\%120$: (**a**) the complete series for $0.125 \leq t/T \leq 3.125$; (**b**) detail for $1.625 \leq t/T \leq 3.125$.

The $GS_{1\%}120$ mesh results only depict the reference grid behaviour after a short interval after each wall shear stress peak. The peaks tend to approximate to those of the reference mesh almost at the end of the simulation ($t/T$ = 3.125), but, in the first peak, this mesh leads to a variation in the amplitude of almost 15% (Figure 4a).The $GS_{1\%}60$ grid shows a lower accuracy in the description of the unsteady wall shear stress variation, with low and constant $\tau_{wu}$ peaks and a linear decay after each velocity profile inversion that does not describe the real exponential energy dissipation. Also, this grid does not correctly simulate the velocity profile changes, as shown by the inexistence of shear stress peaks after the wave reflection in the upstream reservoir (see $t/T$ = 2.125 and $t/T$ = 2.625 in Figure 4b).

Figure 5 illustrates the $C_R$ influence on the shear stress peaks for a constant $N_C$ value. The benchmark mesh results are compared with those from two meshes with half the $N_C$ value, but with higher or equal $C_R$ values ($GS_{5\%}60$ and $GS_{9\%}60$).

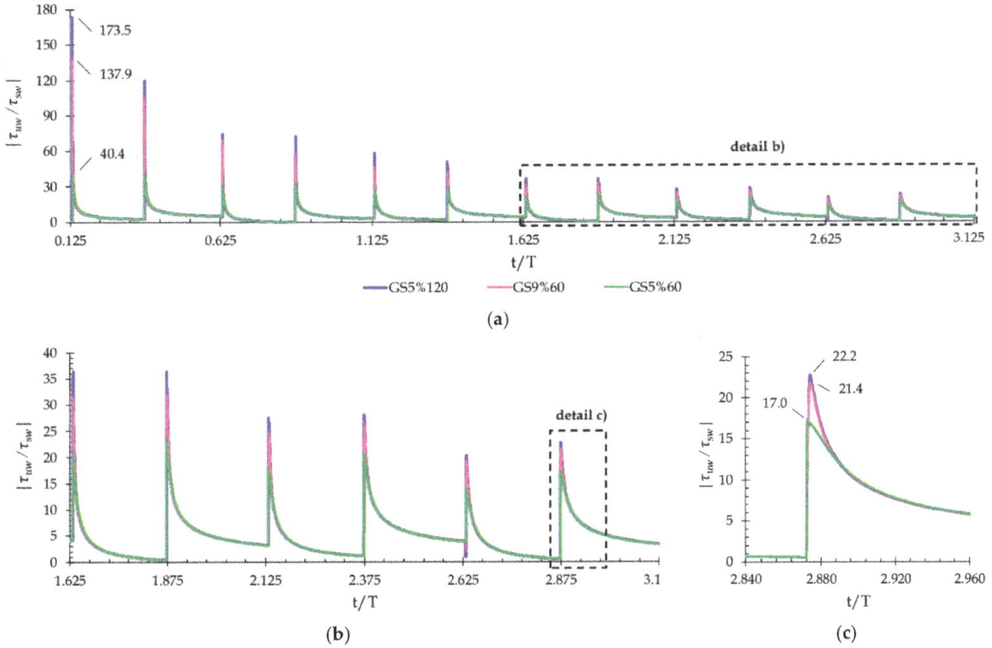


**Figure 5.** Wall shear stress time-history at pipe midsection for $GS_{7\%}60$ and $GS_{9\%}60$ (using $GS_{5\%}120$ as benchmark): (**a**) the complete series for $0.125 \leq t/T \leq 3.1$; (**b**) detail for $1.625 \leq t/T \leq 3.1$; and (**c**) detail for $2.840 \leq t/T \leq 2.960$.

The effect of the mesh refinement in the wall area strongly depends on the $C_R$ value. Higher $C_R$ values correspond to more refined meshes in the wall area and courser meshes in the pipe core (for the same $N_C$). A low resolution near the wall meshes leads to lower and constant $\tau_{wu}$ peaks and a linear decay after each velocity profile inversion. This decay does not describe the real exponential energy dissipation, with it being responsible for the obtained rectangular shape of the pressure wave, unlike the rounded shape observed in the benchmark mesh.

The $GS_{9\%}60$ mesh shows approximately the same performance as the benchmark mesh (Figure 5a). It allows a very precise energy dissipation in each pressure wave arrival and practically eliminates the discrepancy over time (Figure 5b). This excellent energy dissipation calculation is responsible for a pressure history identical to that of the benchmark mesh. The $GS_{5\%}60$ simulation results are less accurate than those for $GS_{9\%}60$ without the exponential $\tau_{wu}$ peak decay over time, and, for that reason, this grid does not describe the rounder $\tau_{wu}$ peak observed in the third wave period (Figure 5c). Nevertheless, $GS_{5\%}60$ mesh produces an almost perfect pressure time variation.

The previous examples highlight the importance of $C_R$ increasing to reproduce the real energy dissipation in unsteady flows (compare $GS_{1\%}60$ and $GS_{9\%}60$ simulations) and the higher effect of $C_R$ than of $N_C$ on the accuracy of the results (compare $GS_{1\%}120$ and $GS_{9\%}60$ results).

Figure 6 depicts the effect of pipe axial discretization for three values of $\Lambda_R$ (0.22, 0.60, and 0.75) with a fixed $C_R$ value (100), considering three spatial discretizations ($\Lambda_X$ = 5.0, 1.0, and 0.5). Results of the mesh with the lowest $\Lambda_R$ values ($\Lambda_R$ = 0.22) are independent of $\Lambda_X$ (no benefit from $\Delta x$ reduction). The mesh with an intermediate $\Lambda_R$ value ($\Lambda_R$ = 0.60) benefits considerably from the first reduction of $\Lambda_X$ (from 5.0 to 1.0) and presents similar results to the $\Lambda_R$= 0.75 mesh for $\Lambda_X$= 1.0. The highest resolution mesh at the wall ($\Lambda_R$ = 0.75) needs the maximum pipe discretization (the highest values for $\Lambda_X$= 0.5) to ensure the accurate calculation of the highest peaks. Therefore, low-resolution meshes (lower $\Lambda_R$ values) do not benefit from further pipe axial discretization. High-resolution meshes (higher $\Lambda_R$ values), without proper pipe axial discretization, present numerical instabilities and no improvement in terms of accuracy.

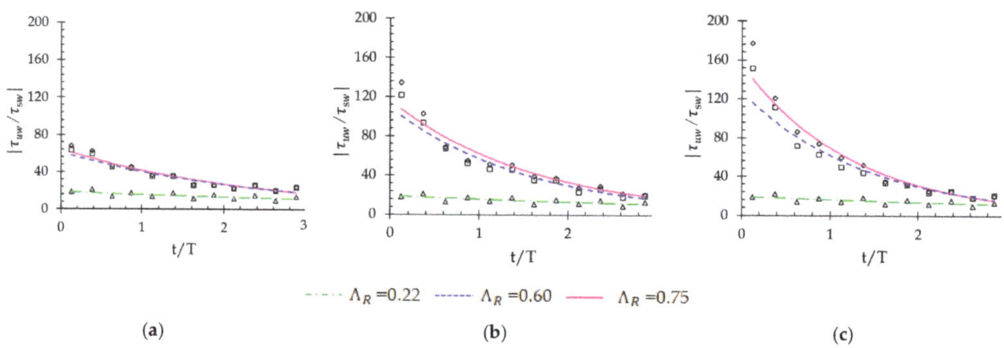


**Figure 6.** Wall shear stress maximum values at pipe midsection for three meshes with $N_C$ = 100, considering (**a**) $\Lambda_R$ = 5.0 ($\Delta x$ = 0.10 m); (**b**) $\Lambda_R$ = 1.0 ($\Delta x$ = 0.02 m); and (**c**) $\Lambda_R$ = 0.5 ($\Delta x$ = 0.01 m).

Table 2 presents the calculated $\tau_w$ peak in the first ($t/T$ = 0) wave period and compares these values with those from the benchmark mesh ($GS_{5\%}120$). The values presented between brackets represent the relative error between each simulation and the reference mesh simulation ("green" cell). The results of the simulations with numerical instabilities are not presented in this table (see "grey" cells).

For meshes with lower $\Lambda_R$ and $N_C$ values (i.e., low refinement in the wall area), there is no reason to decrease the $\Lambda_X$ parameter (i.e., increase the pipe discretization); this is shown in Table 2 marked by "orange" shade for $N_C$ = 60 with $\Lambda_R$ = 0.18 and 0.30; for $N_C$ = 80 meshes with $\Lambda_R$ = 0.20; and for $N_C$ = 100 meshes with $\Lambda_R$ = 0.22. A second set of meshes is flagged with a higher $\Lambda_R$ value, for which the $\Lambda_X$ reduction corresponds to a moderate accuracy improvement. These meshes are marked with a "red" shade. Finally, for higher $\Lambda_R$ meshes, the $\tau_w$ simulation considerably benefits from a $\Lambda_X$ decrease. All simulations benefit from a $\Lambda_R$ increase for the same $\Lambda_X$, independently of the $N_C$ value.

For the same accuracy, the computational effort can be optimized by combining $\Lambda_X$ and $\Lambda_R$. The $GS_{7\%}60$ mesh ($\Lambda_R$ = 0.55 and $N_C$ = 60) with $\Lambda_X$ = 5.0 ensures approximately six times a higher shear stress first peak at $t/T$ = 0 than a lower resolution mesh ($\Lambda_R$ = 0.18) with the same $N_C$. For $\Lambda_R$ = 0.55, the $\Lambda_X$ decrease significantly improves the accuracy, leading to a ten-fold higher peak for $\Lambda_X$ = 0.5 than $\Lambda_X$ = 5.0. To improve the accuracy of $GS_{7\%}60$ mesh, a careful optimization of the parameters must be carried out. For instance, for $N_C$ = 80, the $\Lambda_X$ value must be less than 1.0 and $\Lambda_R$ should be the highest value. For $N_C$ = 100 only, with the highest $\Lambda_R$ value (0.75), only the benchmark with $N_C$ = 120 has shown better results.

**Table 2.** Wall shear stress peak value at the pipe midsection for all simulations (80 meshes).

| $N_C$ | $\Lambda_R$ | $C_R$ | $\Lambda_X$ *5.0* | *2.5* | *1.0* | *0.5* |
|---|---|---|---|---|---|---|
| 60 | 0.18 | 1 | 5.22 (−95.1%) | 5.25 (−95.1%) | 5.27 (−95.1%) | 5.27 (−95.1%) |
| | 0.30 | 3 | 11.00 (−89.7%) | 11.20 (−89.5%) | 11.40 (−89.3%) | 11.40 (−89.3%) |
| | 0.43 | 5 | 21.40 (−79.9%) | 23.10 (−78.3%) | 24.20 (−77.3%) | 24.80 (−76.7%) |
| | 0.55 | 7 | 32.80 (−69.3%) | 39.90 (−62.5%) | 45.40 (−57.4%) | 51.20 (−52.0%) |
| | 0.63 | 9 | | 50.50 (−52.6%) | 63.00 (−40.9%) | 84.40 (−20.5%) |
| 80 | 0.20 | 1 | 7.90 (−92.6%) | 8.00 (−92.5%) | 8.10 (−92.4%) | 8.10 (−92.4%) |
| | 0.36 | 3 | 19.60 (−81.6%) | 20.80 (−80.4%) | 21.70 (−79.6%) | 22.10 (−79.3%) |
| | 0.53 | 5 | | 42.50 (−60.1%) | 51.90 (−51.3%) | 56.70 (−46.8%) |
| | 0.63 | 7 | | | | 94.80 (−11.0%) |
| | 0.71 | 9 | | | | |
| 100 | 0.22 | 1 | 11.40 (−89.3%) | 11.60 (−89.1%) | 11.80 (−88.9%) | 11.80 (−88.9%) |
| | 0.43 | 3 | 29.30 (−72.5%) | 33.90 (−68.2%) | 38.10 (−64.3%) | 39.80 (−62.6%) |
| | 0.60 | 5 | | | | 92.90 (−12.8%) |
| | 0.70 | 7 | | | | |
| | 0.75 | 9 | | | | |
| 120 | 0.23 | 1 | 15.10 (−85.8%) | 15.70 (−85.3%) | 16.10 (−84.9%) | 16.20 (−84.8%) |
| | 0.49 | 3 | | 45.40 (−57.4%) | 57.01 (−46.4%) | 63.70 (−40.2%) |
| | 0.67 | 5 | | | | 106.6 |
| | 0.77 | 7 | | | | |
| | 0.80 | 9 | | | | |

- (grey) Numerical instabilities did not allow to obtain a solution
- (yellow) Short-term accuracy was not influenced by a higher pipe discretization
- (orange) A higher conduit discretization had a limited influence on short term-accuracy
- (green) The benchmark (or reference) mesh

#### 3.2.3. Long-Term Accuracy

Long-term accuracy is assessed by comparing the value of the wall shear stress immediately before the arrival of the pressure wave reflected from the upstream reservoir for the 80 meshes. The objective is to determine whether the increase in $\Lambda_R$ threatens the long-term accuracy of the simulation due to the inability of the mesh to correctly calculate the shear stress as the flow vortex created by the valve closure moves towards the pipe core due to insufficient grid points.

The accuracy is generally high (compared to the benchmark), with relative errors lower than 1.5% in most cases. The worst long-term accuracy is observed for the lowest $N_C$ and $\Lambda_R$ meshes ($N_C$ = 60 with $\Lambda_R$ = 0.18), since the initial wall shear peak values are underestimated, though with relative errors being relatively low (6.2–6.9%). For $N_C$ = 60 meshes using higher $\Lambda_R$ values, the long-term accuracy is high. Overall, long-term accuracy tends to benefit from an increase in $\Lambda_R$ (especially for low $N_C$ meshes) and a decrease in $\Lambda_X$, the latter with less impact.

### *3.3. Numerical Stability*

For a specific $N_C$ value, the preceding simulations have demonstrated numerical instabilities as a consequence of an increase in $\Lambda_R$. A reduction in $\Lambda_X$ mitigates this effect. As the $N_C$ value increases, the numerical instabilities occur for lower $\Lambda_R$ values (see Table 2 where simulations with numerical instabilities are highlighted in "grey").

An analysis for the complete simulation set is carried out to identify the correlation between $N_C$, $\Lambda_R$, and $\Lambda_X$ and the occurrence of numerical instabilities. No correlation is observed between the numerical instability and a particular parameter, rather being a combination of parameters. In fact, numerical instabilities tend to appear for a higher number of cylinders in the outer layers, $N_{3R}$ (i.e., $N_C \times \Lambda_R$), because these correspond to a higher mesh resolution near the wall and, thus, lead to higher axial velocity gradients immediately after the pressure surge. On the other hand, to maintain the numerical stability for higher $\tau_w$ peak values, the time interval must decrease (lower $\Lambda_X$ values) since the method of characteristics used to calculate the transient in the axial direction is required to comply with the Courant–Friedrich–Lewy condition.

Figure 7 plots the complete set of simulations in terms of dimensionless parameters, identifying the results with or without numerical instabilities. A relation between $N_{3R}/N_X$ ($N_X$ being the number of sections in the axial direction) and $\Lambda_R$ and simulations with numerical instability is observed.

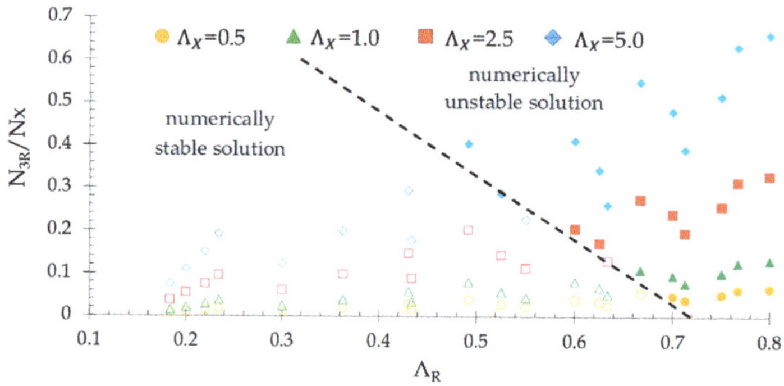


**Figure 7.** Numerical instability representation according to $N_{3R}/N_X$ and $\Lambda_X$. Filled marks represent numerically unstable solutions, and unfilled marks represent the numerically stable simulations.

For meshes with a $\Lambda_R$ ratio smaller than 0.5, no numerical instability is observed in the 80 simulations; for $\Lambda_X$ = 0.5 meshes ("yellow" marks), the $\Lambda_R$ can increase until 0.7 without any numerical instability; for $\Lambda_X$ = 1.0 meshes ("green" marks), the $\Lambda_R$ is limited to 0.6; for $\Lambda_X$ = 2.5 meshes ("red" marks), the $\Lambda_R$ is limited to 0.6; and for $\Lambda_X$ = 5.0 ("blue" marks), the $\Lambda_R$ is limited to 0.5. Figure 7 demonstrates that the numerical stability is a direct consequence of the mesh resolution in the wall area and is not solely dependent on $\Lambda_R$. Additionally, the stability dependence on $\Lambda_X$ is also recognized. Vardy and Hwang [8] found an optimal ratio $N_X/N_C$ = 1.5 for the accuracy and stability of the solutions, showing that a $N_X/N_C$ higher than 1.5 reduces the accuracy and a $N_X/N_C$ lower than 1.5 leads to numerical instabilities. This aspect is true for the equal area cylinder meshes used by Vardy and Hwang because there is a constant relation between $N_{3R}$ and $N_C$. Zhao and Ghidaoui [19] argued that, for a geometric sequence mesh (GS) with a fixed $N_C$ value, a correlation between $\Lambda_R$ and $\Lambda_X$ must be ensured. The presented stability analysis adds the importance of the $N_{3R}$ parameter and, consequently, the correlation between $C_R$ and $N_C$.

The observed limits for $\Lambda_R$ herein refer to this particular pipe system and the instantaneous valve closure. Should an S-shaped valve closure be simulated, the velocity gradient near the wall would be reduced, and the majority of the previous numerical instabilities would not be observed. In that case, higher $\Lambda_X$ values could be used for the same $\Lambda_R$.

### 3.4. The Most Efficient Mesh

The accuracy of transient simulation enhances significantly with a $\Lambda_R$ increase. Yet, increasing the $\Lambda_R$ is limited by the simulation numerical stability and, thus, must be followed by a $\Lambda_X$ decrease. To ensure an accurate simulation with a considerable reduction in the computer effort, the simultaneous optimization of both parameters, $\Lambda_R$ and $\Lambda_X$, is carried out for the 48 meshes without numerical instabilities (Table 2). The computational effort is measured by the ratio between the total number of grid points ($N_C \times N_X$) and the lowest number of grid points of the 48 meshes (i.e., 9000, corresponding to $N_C$ = 60 and $\Delta x$ = 0.10 m). Since the Zhao and Ghidaoui [19] model is used, the computational effort increases linearly with the total number of grid points. For example, a mesh with $N_C$ = 120 and $\Delta x$ = 0.01 m (total number of grid points equal to 180,000) has a computational effort 20 times that of the mesh with the lowest number of grid points.

The accuracy is assessed comparing the piezometric head results for each mesh with those of the highest resolution mesh ($GS_{5\%}120$ with $\Lambda_R = 0.49$ and $\Lambda_X = 0.5$) by means of the mean absolute percentage error (MAPE) in three-wave periods using the following equation:

$$\text{MAPE} = \frac{100}{n}\sum_{t=0}^{n}\left|\frac{X_i - \hat{X}_i}{X_i}\right| \tag{5}$$

where $X_i$ is the true value (given by benchmark mesh value) for time $t$, $\hat{X}_i$ is the estimated value (simulated value) for time $i$, and $n$ is the number of time steps in three-wave periods. The difference between $X_i$ and $\hat{X}_i$ is divided by the benchmark value $X_i$ again.

### 3.4.1. Transient Results Calculation for the Piezometric Head in the Dimensionless Form

The dimensionless form of the piezometric head, $(H{-}H_0)/(aU_0/g)$, is used herein. This approach only accounts for the unsteady-state error (i.e., it cancels the steady-state simulation error associated with $H_0$ and $U_0$). Figure 8a presents computational effort versus the MAPE for the 48 simulations, whereas Figure 8b only shows the solutions obtained with MAPE < 0.20%. The meshes are divided into four groups according to the $\Lambda_R$ parameter: (i) the "blue" group with $\Lambda_R \leq 0.2$; (ii) the "orange" group with $\Lambda_R$ between 0.2 and 0.3; (iii) the "grey" group with $\Lambda_R$ between 0.3 and 0.5; and (iv) the "yellow" group with the highest $\Lambda_R$ ($\geq$0.50) meshes.

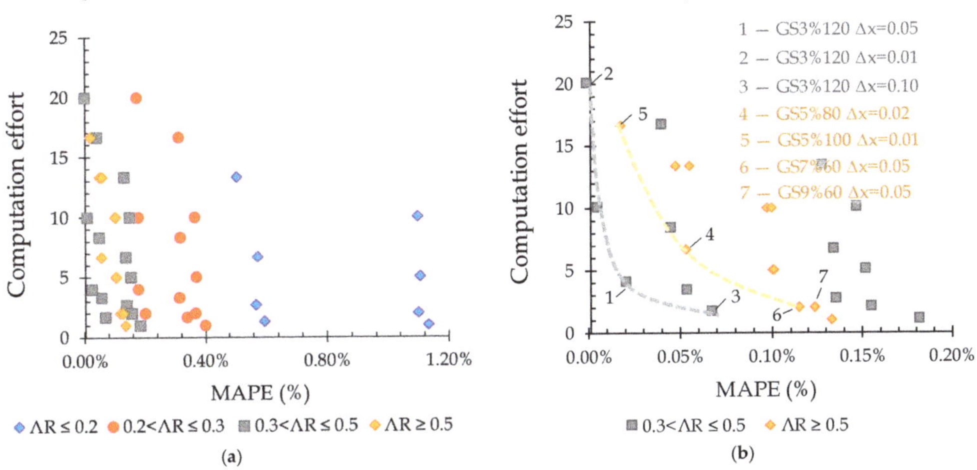


**Figure 8.** Computational effort versus transient simulation error for dimensionless form, considering (**a**) the four $\Lambda_X$ groups; (**b**) detail for MAPE < 0.20%.

The lowest $\Lambda_R$ meshes (the blue and orange groups) present the most unsatisfactory results, since they have a similar computational effort as the other groups but with lower accuracy (i.e., higher MAPE). The grey group contains meshes with the best results (Meshes 1, 2, and 3, see Figure 8b), though having a lower $\Lambda_R$ value than the yellow group. This result is explained by the fact that the meshes with higher $N_C$ values (yellow meshes) were removed from the analysis due to numerical instabilities. For instance, the yellow group does not have a valid mesh for $N_C = 120$ and has one single mesh with $C_R = 5\%$ (Mesh 5) for $N_C = 100$.

Mesh 1 (see Figure 8b) ensures the best relationship between computational effort and accuracy. If a higher accuracy is needed, the best option is to decrease the axial discretization interval to $\Delta x = 0.02$ m, as obtained by Mesh 2 (the same radial mesh, but a five-fold higher axial discretization). On the contrary, if a lower resolution is accepted, the best option is to decrease $N_C$ and to increase $C_R$, as represented by Mesh 3. This analysis highlights that the $\Delta x$ increase is a good solution, when no numerical instability is observed, and its decrease is only justified to control the occurrence of numerical instabilities. The dashed grey line, containing Meshes 1, 2, and 3, defines the Pareto front (i.e., the set of non-dominated solutions) of this two-objective optimization problem, and all the other meshes correspond to dominated solutions.

#### 3.4.2. Transient Results Calculation for the Piezometric Head in the Dimensional Form

A high-accuracy transient simulation is only justified if a high-accuracy steady-state velocity profile is ensured by the Q2D model, as previously stated (see Section 3.1). Figure 9 shows the computational effort versus MAPE for the 48 simulations using the piezometric head results to account for the steady-state velocity profile error.

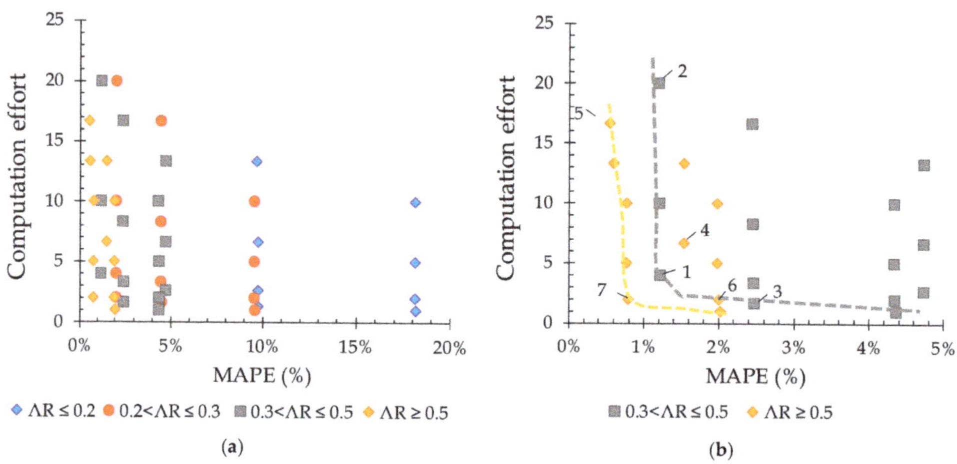


**Figure 9.** Computation effort versus transient simulation error for dimensionless form considering (**a**) the four $\Lambda_X$ groups; (**b**) detail for MAPE < 0.20%.

The MAPE shows a significant increase (compared with Figure 8), highlighting the importance of correctly calculating the steady-state velocity profile and the relatively low transient error provided by the Q2D models, as previously referred. The yellow group presents the lowest transient error since it has the best steady-state simulation, and, consequently, it presents better results than the grey meshes. The best result is obtained for Mesh 7 (corresponding to $GS_{9\%}60$ with $\Delta x = 0.05$ m mesh). The previous mesh has the highest $C_R$ value, and, consequently, the lowest steady-state error. The dashed yellow line corresponds to the Pareto front in this case.

In conclusion, the Q2D model with the proper selection of $\Lambda_R$ and $\Lambda_X$ correctly simulates the transient flow energy dissipation with the lowest $N_C$ value (i.e., $N_C = 60$). The steady-state error is commonly ignored, though it considerably influences the transient state results, which can be improved by using higher $C_R$ values.

Mesh 7 ($GS_{9\%}60$) is the best option when considering the piezometric head (i.e., considering steady-state mean velocity error), and Mesh 1 ($GS_{3\%}120$) is the best option when the dimensionless piezometric head is considered (i.e., when the steady-state error is neglect).

## 4. Performance of the Two Best Meshes

### *4.1. Aim and Meshes Characterization*

The results obtained for the two best meshes, $GS_{9\%}60$ and $GS_{3\%}120$, are compared for a five-period pressure wave simulation with an instantaneous valve closure. The comparison aims to identify differences in the results associated with the $C_R$ optimization.

The two best meshes, $GS_{9\%}60$ and $GS_{3\%}120$, have a $\Lambda_X$ value equal to 2.5 (i.e., $\Delta x = 0.05$ m). The $GS_{9\%}60$ mesh has a higher $\Lambda_X$ value (0.63 compared with the 0.49 value of the $GS_{3\%}120$ mesh) to compensate the smaller $N_C$ value. The $GS_{9\%}60$ mesh reduces to half the first cylinder thickness. Comparing the two meshes number of grid points, the $GS_{9\%}60$ mesh has approximately the same number of grid points in the viscous layer, half the number of points in the buffer I and II layers, and one third of the points in the central regions (logarithmic and core regions).

### 4.2. The Steady-State Flow Simulation

Figure 10 presents the dimensionless velocity profile for the steady-state flow for the two meshes ($u^+$ as a function of $y^+$), with both velocity profiles showing the same increase with $y+$. The turbulent steady-state dimensionless velocity profile is particularly sensitive to the radial mesh interval between the viscous and log-layer due to the sharp increase in velocity observed. The $GS_{9\%}60$ mesh shows a correct transition between the viscous and log-layer with a courser grid. Both meshes show an excellent performance, demonstrated by (i) the velocity profile matching the law of the wall $\left(u^+ = \frac{1}{\kappa} lny^+ + C\right)$ for values of $y^+ > 30$; (ii) the velocity profile variation close to the wall (viscous sublayer) being linear with $y^+$ ($u^+ = y^+$); and (iii) the gradual velocity profile variation observed between the viscous and log-layer. Yet, $GS_{9\%}60$ has half the grid points of $GS_{3\%}120$ for the same accuracy (performance).

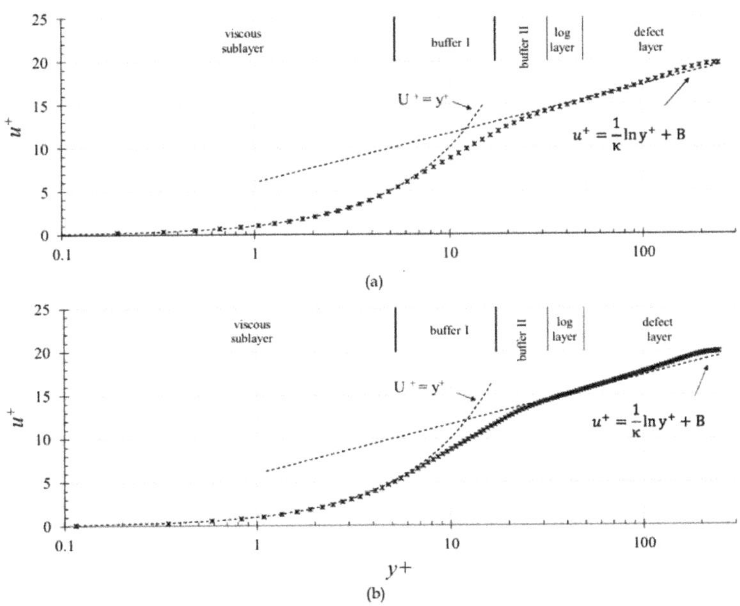


**Figure 10.** Steady velocity profile for a turbulent boundary layer for the best two meshes used in the Q2D model: (**a**) $GS_{9\%}60$; (**b**) $GS_{3\%}120$.

### 4.3. The Unsteady Flow Simulation

The time-history of piezometric head at the pipe mid and end sections in the first five periods of the transient are simulated by the Q2D model and are presented in Figure 11 for both meshes ($GS_{9\%}60$ and $GS_{3\%}120$). Results are presented in the dimensionless form. Hardly any differences are observed: the two grids lead to an identical pressure wave damping, associated with the correct simulation of energy dissipation after the velocity shift, and the development of a similar S-shape wave form associated with the energy dissipation immediately after the pressure surge.

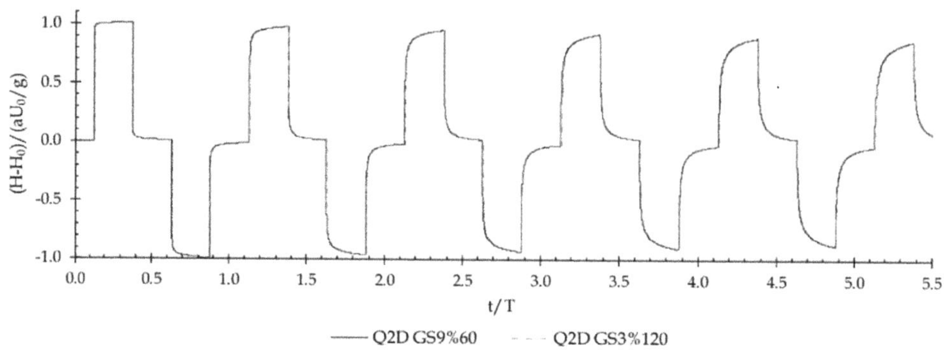


**Figure 11.** Piezometric head time-history simulation using $GS_{3\%}120$ and $GS_{9\%}60$ at the pipe midsection.

The wall shear stress results for both meshes are presented in Figure 12. Once again, differences between the results of the two meshes are minor and restricted to the first pressure peaks and during a short time interval after the arrival of each wave front, after which the results converge very quickly. The $GS_{9\%}60$ mesh results have slightly higher wall shear stress peaks.

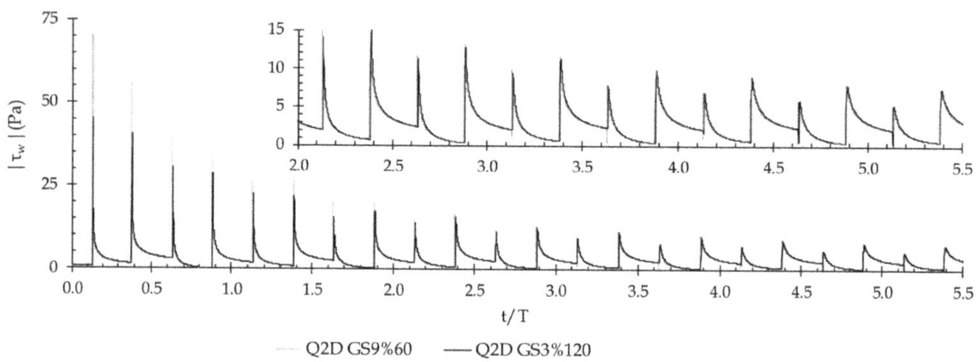


**Figure 12.** Absolute wall shear stress time-history using $GS_{3\%}120$ and $GS_{9\%}60$ with the Q2D model at pipe midsection.

The correlation between the wall shear stress and the mean velocity at the pipe midsection for both meshes ($GS_{3\%}120$ and $GS_{9\%}60$) is shown in Figure 13. Results are presented for the first, the second, and the third wave periods. In the first pressure surge ($t/T$=0.125, see Figure 13a), the mean velocity reduction is abrupt (with a rectangular pressure wave shape, as observed in Figure 11). The two meshes show different peak values, with the $GS_{9\%}60$ mesh generating a higher $\tau_W$ value, which reduces the maximum mean velocity variation compared with $GS_{3\%}120$. Higher $\tau_W$ values lead to a smoother pressure wave. The $GS_{9\%}60$ mesh increases the pressure wave smoothing, and doing so tends to attenuate the difference between the two meshes results. Consequently, the results in the third period are almost identical for both meshes (Figure 13c). As the simulation proceeds, there is a rounding of the pressure wave and a progressive decay of the maximum mean velocity variation between time intervals, as observed by the elliptical shape in Figure 13c. The two meshes present similar performances, as previously verified for the wall shear stress analysis, because the mesh accuracy requirement in the wall region decreases.

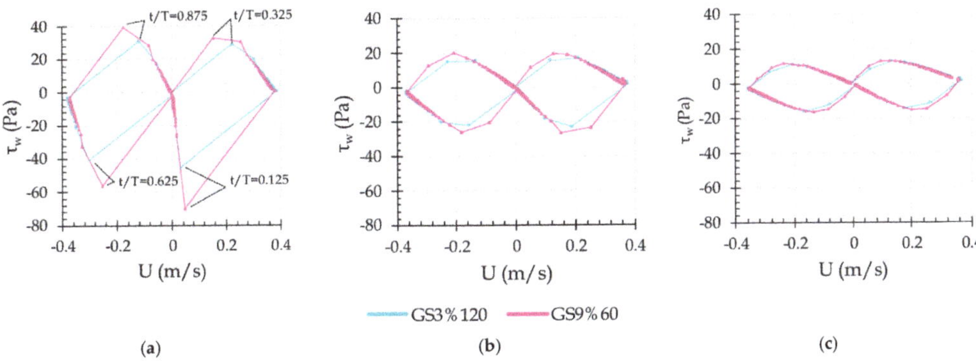


**Figure 13.** Unsteady wall shear stress and mean velocity at the conduit midsection for $GS_{9\%}60$ and $GS_{3\%}120$ meshes with (**a**) $0.125 \leq t/T \leq 1.125$ (1T); (**b**) $1.125 \leq t/T \leq 2.125$ (2T), (**c**) $2.125 \leq t/T \leq 3.125$ (3T).

The velocity profiles are also evaluated, and minimal differences are observed between the two meshes, with the exception of regions in close proximity to the wall and immediately following the arrival of the initial pressure wave. These observations are coherent with the previous results.

The two meshes yielded comparable results. However, the use of a low-resolution mesh on the wall (low $N_C$ combined with low $C_R$) will result in an inability to accurately calculate the peak value of wall shear stress, and the transformation from a rectangular shape into an S-type shape pressure wave will not occur. The time-history of the piezometric head will continue to exhibit abrupt variations without the associated dissipation of transient state energy.

## 5. Transient Dynamic Characterization

The dynamic characterization of the transient phenomenon is further developed using the $GS_{9\%}60$ mesh. The dimensionless mean velocity time-history is presented in Figure 14. The profiles $t_0$ to $t_8$ describe the instants immediately before and after each pressure variation in the first period ($0.125 \leq t/T \leq 1.125$) at the pipe midsection. Instants $t_9$ to $t_{17}$ describe the homologous instants, but for the fifth wave period ($4.125 \leq t/T \leq 5.125$).

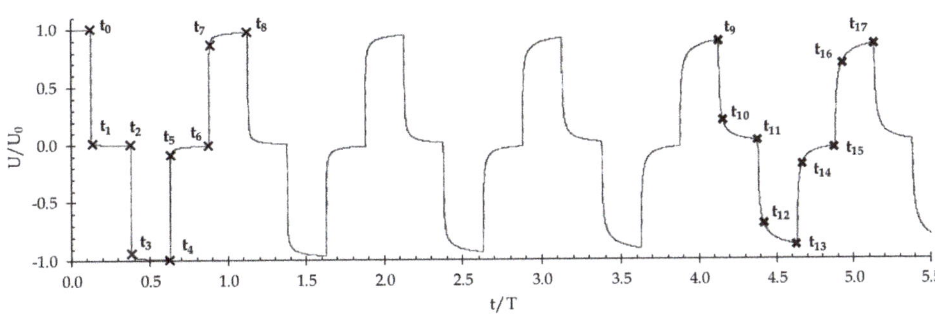


**Figure 14.** Mean velocity time-history at pipe midsection with the $GS_{9\%}60$ mesh.

Figure 15 shows the local acceleration and the wall shear stress time-history, both in absolute values. The critical features include the following: (i) the wall shear stress peak values directly correspond to the local acceleration peaks; and (ii) the switch from an instantaneous to gradual variation, as previously observed concerning the piezometric head time-history. However, the wall shear stress presents a slower decay and tends to a non-zero value, while the local acceleration presents an initial peak, followed by a fast decay and almost immediately tending to zero. This difference justifies the instantaneous acceleration-based model's main shortcoming.

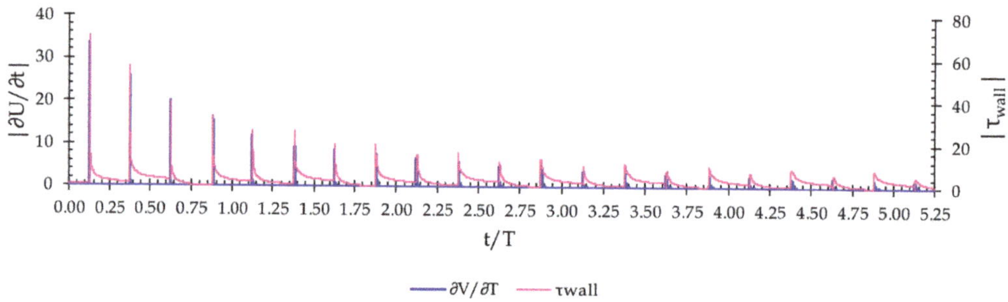


**Figure 15.** Mean velocity time-history at the pipe midsection with the $GS_{9\%}60$ mesh.

For the analyzed instants ($t_0$ to $t_{16}$), the shear stress profile and the dimensionless velocity profile are presented in Figures 16 and 17, respectively, both plotted with *y*+ in a logarithmic scale to better distinguish the five turbulent regions.

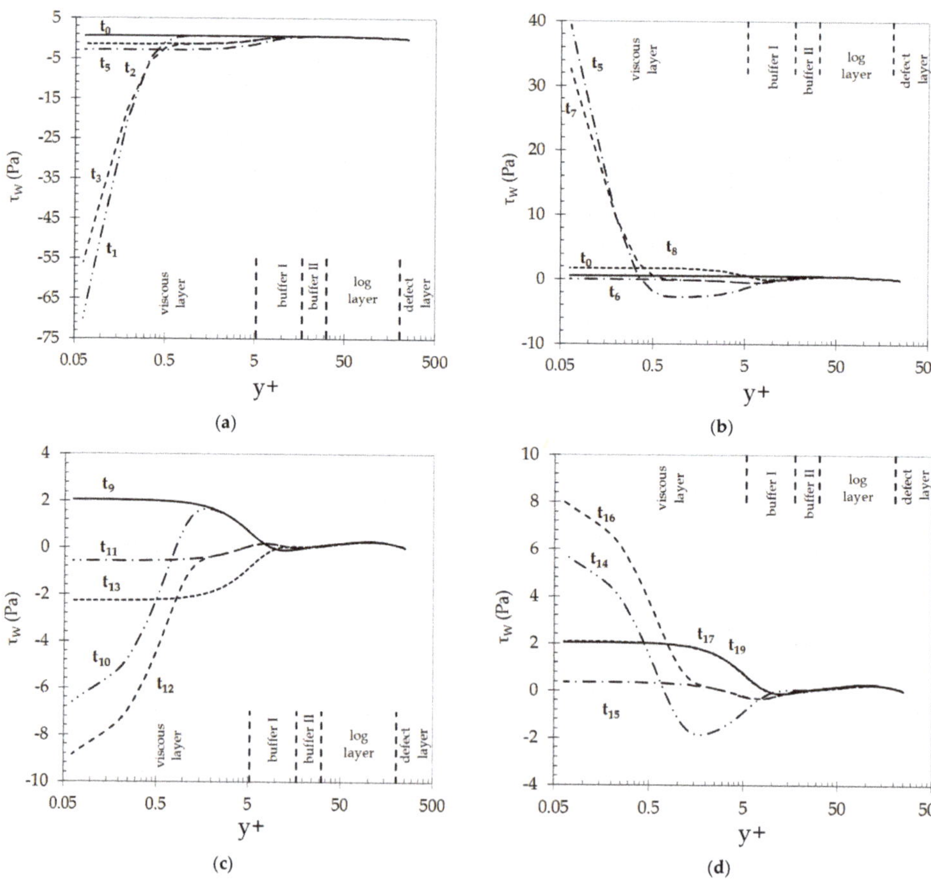


**Figure 16.** Dimensionless shear stress profiles close to the pipe wall for GS9%60. (**a**) $0 \leq t/T \leq 0.5$; (**b**) $0.5 \leq t/T \leq 1.0$; (**c**) $4.0 \leq t/T \leq 4.5$; (**d**) $4.5 \leq t/T \leq 5.0$.

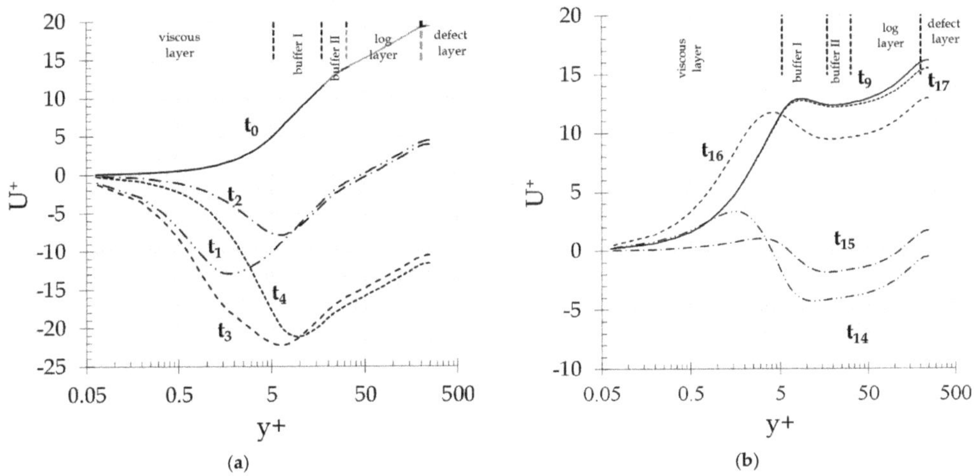


**Figure 17.** Dimensionless velocity profiles close to the pipe wall for $GS_{9\%}60$: (**a**) $0 \leq t/T \leq 0.5$; (**b**) $4.5 \leq t/T \leq 5.0$.

The shear stress profile allows to analyze the mesh influence on the simulation accuracy because the numerical scheme truncation error at each grid point corresponds to the local shear stress gradient multiplied by the grid dimension. Thus, a high shear stress gradient area should correspond to a higher mesh refinement, and, conversely, a linear profile allows for the use of courser meshes. The shear stress profile reacts to the pressure surge and the no-slip condition with an immediate change in the wall region (for an illustration of this, see $t_1$ and $t_5$ in Figure 16). In the equilibrium phase, the peak reduces as the "vortex" sheet moves into the pipe core. The wall shear stress peaks diminish with time due to the pressure surge smoother variation, observed by the rounder wall shear stress profile (the previous S-shape effect observed in Figure 14). Nevertheless, the surge influence is limited to the viscous and buffer I layers and is not considerably changed along the simulation. Inside the buffer I layer, the wall shear stress presents a lower and linear gradient that is accurately described by a coarser mesh.

The velocity profile is the result of the equilibrium process up to that moment. As the transient propagates, successive vortices are created that move toward the pipe core, and their effect is strongly dissipated along time. In Figure 17, the first peak ($t_1$) effect during the initial quarter-wave period spreads until the buffer I layer ($t_2$). At the end of the first half-wave period ($t_4$), the steady profile ($t_0$) is mainly affected from the wall to the buffer II layer, whereas, at the end the fifth period ($t_{16}$), the effect reaches the log layer. Nevertheless, during the complete analysis, the high shear stress gradients are limited to the viscous and buffer I layers. This explains why the two meshes present similar results ($GS_{9\%}60$ has 31 grid points and $GS_{3\%}120$ has 43 grids points in the two outer layers).

Figure 18 illustrates the axial velocity time-history at different distances, r, from the pipe centerline described as a percentage of the pipe radius, R, from r = 50%R to 92.5%R. Figure 18a presents four grid points that are situated within the core, log, and buffer II layer (the log and buffer II layers' upper limits correspond to r = 85%R and r = 92.5%R). Figure 18b provides a detailed description of the axial velocity at several points in the viscous layer (99%R to 99.9%R). All points within the pipe core (Figure 18a) follow the mean velocity trace and are not affected by the high-velocity gradient in the pipe wall (with no spikes and quite smooth changes). Figure 18b illustrates a contrasting behaviour amongst the core points, which can be readily distinguished from the wall region points with a fast response to the pressure variations. This change is evident, with the viscous layer points exhibiting a performance analogous to that of the wall shear stress. The closer the points are to the pipe wall, the similar the velocity traces to the wall shear stress are, as observed in Figure 18b.

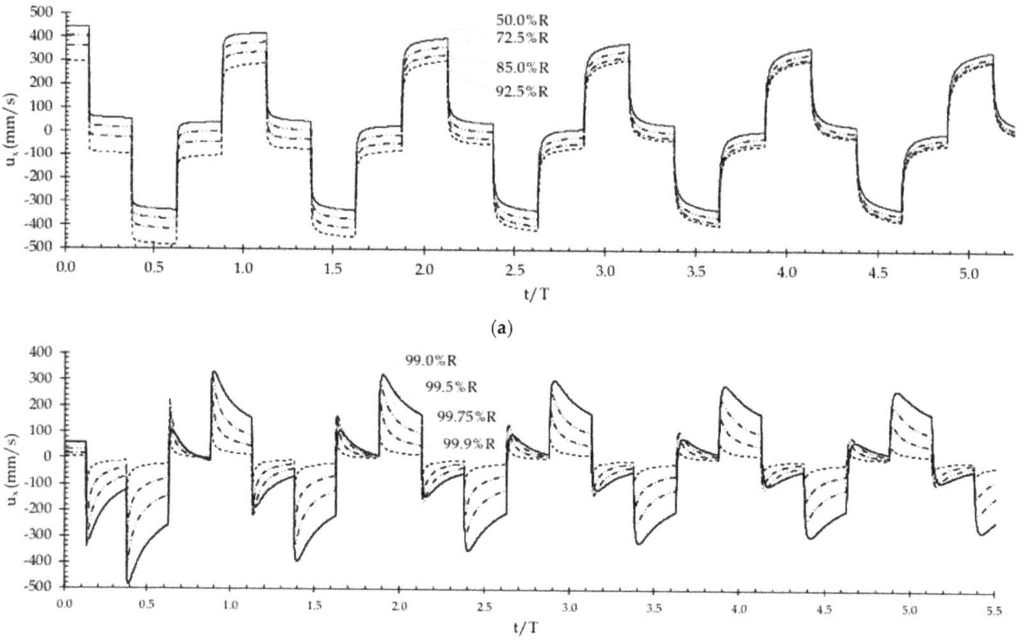


**Figure 18.** Axial velocity time-histories at different distances from the wall: (**a**) from 50% R to 92.5% R; (**b**) from 99.0% R to 99.9% R.

The preceding evaluations have highlighted the need to increase the mesh refinement in the outer layers (predominantly in the viscous and buffer I layers), given that this is the region where the flow dynamics undergo the most significant and rapid changes. This provides further evidence to support the hypothesis that an increase in $C_R$ values may be beneficial. Although the results show some variation in the core over time, the fluctuations are markedly less pronounced when this occurs. The necessity for controlling the finite difference truncation error indicates that a reduction in the mesh may be a viable approach in instances where higher shear stress variations are observed.

## 6. Conclusions

The utilization of a geometric sequence grid for the simulation of transient turbulent flow is supported by the high concentration of mesh points distributed between the pipe wall and the log layer. The accuracy of the unsteady simulation is enhanced by higher $C_R$ values, which allows the use of a high-resolution mesh in the wall region, while maintaining a low $N_C$ value and a lower computation time. The value of the conduit axial discretization should be sufficiently high to prevent numerical instabilities.

The comparison between the two best performance meshes—$GS_{9\%}60$ and $GS_{3\%}120$—with a total number of 60 and 120 cylinders and $C_R$ of 9% and 3%, respectively, has demonstrated the benefit of the $C_R$ optimization. The two meshes show small differences for the instantaneous valve closure, which are limited to a reduced number of time steps after the pressure surge arrival and solely during the first pressure waves. The $GS_{9\%}60$ mesh has the advantage of requiring half of the computational time. The transient dynamic characterization reveals the in-depth consistency between the model results and the hydraulic transient phenomenon. The high shear stress gradient is confined to the viscous and buffer I layer (the inner part of the flow is in phase with the mean velocity), thereby justifying the refinement of the radial mesh in this specific area.

**Author Contributions:** Conceptualization, P.L.F. and D.I.C.C.; methodology, P.L.F. and D.I.C.C.; software, P.L.F.; validation, P.L.F.; formal analysis, P.L.F.; investigation, P.L.F. and D.I.C.C.; resources, P.L.F. and D.I.C.C.; data curation, D.I.C.C.; writing—original draft preparation, P.L.F.; writing—review and editing, P.L.F. and D.I.C.C.; visualization, P.L.F.; supervision, D.I.C.C.; project administration, D.I.C.C.; funding acquisition, D.I.C.C. All authors have read and agreed to the published version of the manuscript.

**Funding:** The authors would like to thank the Fundação para a Ciência e Tecnologia (FCT) through funding UIDB/04625/2020 from the research unit CERIS.

**Data Availability Statement:** No available data.

**Acknowledgments:** The authors would like to acknowledge João Paulo Ferreira for assisting the data collection in the experimental facility and for providing additional data.

**Conflicts of Interest:** The authors declare no conflicts of interest.

## Nomenclature

| | |
|---|---|
| $C_R$ | geometric sequence common ratio (-) |
| $c$ | pressure wave speed (m/s) |
| $H$ | piezometric head (m) |
| $g$ | acceleration due to gravity (m/s$^2$) |
| $L$ | pipe length (m) |
| $Q$ | flow rate or discharge (m$^3$/s) |
| $N_C$ | total number of cylinders of the radial mesh (-) |
| $N_{HR}$ | total number of cylinders in the three outer regions (turbulence model) |
| $N_X$ | number of axial mesh points in the pipe direction (-) |
| $R$ | pipe inner radius (m) |
| $r$ | distance from the axis in the radial direction (m) |
| $t$ | time (s) |
| $U$ | mean velocity of the fluid in the pipe cross section (m/s) |
| $\Delta t$ | numerical time step (s) |
| $\Delta x$ | numerical spatial step (m) |
| $\Delta r_j$ | wall thickness of the $j$th cylinder $\left(\Delta r_j = r_j - r_{j-1}\right)$ |
| $\varepsilon$ | pipe wall roughness (m) |
| $u$ | axial velocity (m/s) |
| $u'$ | turbulence velocity fluctuation on the axial direction (m/s) |
| $v$ | radial velocity (m/s) |
| $v'$ | turbulence velocity fluctuation on the radial direction (m/s) |
| $x$ | distance along the pipe (m) |
| $\nu$ | kinematic viscosity of liquid (m$^2$/s) |
| $\varrho$ | liquid density (kg/m$^3$) |
| $\tau_w$ | wall shear stress (Pa) |
| $\tau_{sw}$ | steady wall shear stress (Pa) |
| $\tau_{uw}$ | unsteady wall shear stress (Pa) |

**Abbreviations**

| | |
|---|---|
| CFD | computation fluid dynamics |
| MAPE | mean absolute percentage error |
| Q2D | axisymmetric or quasi-2D model |

## References

1. Vardy, A.E.; Brown, J.M.B. Transient Turbulent Friction in Smooth Pipe Flows. *J. Sound Vib.* **2003**, *259*, 1011–1036. [CrossRef]
2. Ferras, F.; Manso, A.; Schleiss, A.; Covas, I.C.D. Experimental distinction of damping mechanisms during hydraulic transients in pipe flow. *J. Fluids Struct.* **2016**, *66*, 424–446. [CrossRef]

3. Ghidaoui, M.; Zhao, M.; McInnis, D.; Axworthy, D. A review of water hammer theory and practice. *Appl. Machanics Rev.* **2005**, *58*, 49–76. [CrossRef]
4. Vitkosky, J.; Bergant, A.; Simpson, A.; Lambert, M. Systematic evaluation of one-dimensional unsteady friction models in simple pipelines. *J. Hydraul. Eng.* **2006**, *132*, 696–708. [CrossRef]
5. Brunone, B.; Golia, U.; Greco, M.; IAHR-Group. Modelling of fast transient by numerical methods. In Proceedings of the International Conference on Hydraulic Transients with Water Column Separation, Madrid, Spain, 4–6 September 1991; pp. 273–280.
6. Brunone, B.; Golia, U.; Greco, M.; IAHR-Group. Some remarques on the momentum equation for fast transients. In Proceedings of the International Conference on Hydraulic Transients with Water Column Separation, Madrid, Spain, 4–6 September 1991; pp. 201–209.
7. Vitkosky, J.; Lambert, M.; Simpson, A.; Bergant, A.; BHr Group. Advances in unsteady friction modelling in transient pipe flow. In Proceedings of the 8th International Conference on 'Pressures Surges', Hague, The Netherlands, 12–14 April 2000; pp. 471–481.
8. Vardy, A.E.; BHR Group. Acceleration-dependent unsteady friction revisited. In Proceedings of the 13th International Conference on Pressure Surge, Bordeaux, France, 14–16 November 2018; pp. 265–282.
9. Vardy, A.; Brown, J.; He, S.; Ariyaratne, C.; Gorji, S. Applicability of frozen-viscosity models of unsteady wall shear stress. *J. Hydraul. Eng.* **2015**, *141*, 04014064. [CrossRef]
10. Zielke, W. Frequency-dependent friction in transient in pipe flow. *J. Basic Eng.* **1968**, *90*, 109–115. [CrossRef]
11. Vardy, A.E.; Hwang, K.-L.; Brown, J.M. A weighting function model of transient turbulent pipe friction. *J. Hydraul. Res.* **1993**, *31*, 533–548. [CrossRef]
12. Trikha, A. An efficient method for simulating frequency-dependent friction in transient liquid flow. *J. Fluids Eng.* **1975**, *97*, 97–105. [CrossRef]
13. Martins, N.; Brunone, B.; Meniconi, S.; Ramos, H.; Covas, D. Efficient Computational Fluid Dynamics Model for Transient Laminar Flow Modelling: Pressure Wave propagation and Velocity Profile Changes. *J. Fluids Eng.* **2018**, *140*, 011102. [CrossRef]
14. Vardy, A.E.; Hwang, K.-L. A characteristic model of transient in pipes. *J. Hydraul. Res.* **1991**, *29*, 669–684. [CrossRef]
15. Cebecci, T. *Analysis of Turbulent Flows with Computer Programs*, 3rd ed.; Elsevier: Oxford, UK, 2013.
16. Abreu, J.; Betâmio de Almeida, A. Timescale behavior of the wall shear stress in unsteady laminar pipe flows. *J. Hydraul. Eng.* **2009**, *135*, 415–424. [CrossRef]
17. Zhao, M.; Ghidaoui, M. Investigation of turbulence behavior in pipe transient using a k-e model. *J. Hydraul. Res.* **2010**, *44*, 682–692. [CrossRef]
18. Riasi, A.; Nourbakhsh, A.; Raisee, M. Unsteady turbulent pipe flow due to water hammer using k-ω turbulence model. *J. Hydraul. Res.* **2009**, *47*, 429–437. [CrossRef]
19. Zhao, M.; Ghidaoui, M. Efficient quasi-two-dimensional model for water hammer problems. *J. Hydraul. Eng.* **2003**, *129*, 1007–1013. [CrossRef]
20. Korbar, R.V. Efficient solution method for quasi two-dimensional model of water hammer. *J. Hydraul. Res.* **2014**, *52*, 575–579. [CrossRef]
21. Korbar, R.V. Truncated method of characteristics for quasi-two dimensional water hammer model. *J. Hydraul. Eng.* **2014**, *140*, 04014013. [CrossRef]
22. Ferreira, P.L.; Covas, D.I.C. New Optimized Equal-Area Mesh Used in Axisymmetric Models for Laminar Transient Flows. *Water* **2023**, *15*, 1402. [CrossRef]
23. Hirsch, C. *Numerical Computation of Internal & External Flows*, 2nd ed.; Elsevier: Oxford, UK, 2007.
24. Ferreira, J.P.; Martins, N.M.C.; Covas, D.I.C. Ball Valve Behavior under Steady and Unsteady Conditions. *J. Hydraul. Eng.* **2018**, *144*, 04018005. [CrossRef]
25. Ferras, D.; Manso, P.A.; Schleiss, A.J.; Covas, D.I.C. Fluid-structure interaction in straight pipelines with different anchoring conditions. *J. Sound Vib.* **2017**, *394*, 348–365. [CrossRef]
26. Wilcox, D.C. *Turbulence Modeling for CFD*, 2nd ed.; DCW Industries: La Canada, CA, USA, 1998.

fluids MDPI

Article

# Features of Motion and Heat Transfer of Swirling Flows in Channels of Complex Geometry

Sergey Dmitriev [1], Alexey Sobornov [1] and Andrey Kurkin [2,*]

[1] Department of Atomic and Thermal Stations, Nizhny Novgorod State Technical University n.a. R.E. Alekseev, Nizhny Novgorod 603155, Russia; dmitriev@nntu.ru (S.D.); wisestjedi@gmail.com (A.S.)
[2] Department of Applied Mathematics, Nizhny Novgorod State Technical University n.a. R.E. Alekseev, Nizhny Novgorod 603155, Russia
* Correspondence: aakurkin@nntu.ru

**Abstract:** The computational and experimental study results of swirling single-phase coolant motion and heat transfer for the standard operation parameters of a nuclear power plant are presented. The experimental model is a vertical heat exchanger of a "pipe in a pipe" type with the countercurrent movement of coolants. Six different swirlers (three with a constant twist pitch and three with a variable pitch) were considered. The heat exchanger temperature field was measured at various combinations of coolant flow rates, and a channel pressure drop for each swirl was determined. Computational studies were performed using the Omega-based Reynolds stress model and SST model with a correction for curvature streamlines. A good agreement between numerical and experimental data was obtained. Based on the velocity and temperature fields, swirling flow motion features in channels with a variable swirl pitch were discovered. For each intensifier, the effectiveness criterion in comparison with a pipe channel was determined.

**Keywords:** swirling flow; heat transfer intensification; hydrodynamics; PWR heat exchangers; computational fluid dynamics; Omega Reynolds stress model

## 1. Introduction

To reach the highest level of modern nuclear power engineering, the energy efficiency of heat exchanger equipment needs to be increased. The most rational way to solve this problem, together with avoiding a significant increase in the heat exchange equipment's weight and size, is the intensification of heat and mass transfer processes.

The intensification methods of convective heat exchange can be of two types: active and passive [1]. The active methods require additional energy costs (mechanical, electromagnetic, etc.) to affect the heat exchange mechanism. The energy transfer to a fluid flow during active intensification can be carried out by various means: rotation and vibration of the heat exchange surface, vibration of the coolant, injection, or suction of the coolant through a porous surface, and imposition of electrostatic or electromagnetic fields. However, the active methods listed above are either technically difficult to implement or energetically inefficient, and as a result, they do not find practical application in heat exchangers of power plants. Passive methods of heat exchange intensification are widely used in NPU units [2]. Their action is aimed at turbulization or destruction of the boundary layer, turbulization of the entire fluid flow, as well as an increase in the heat exchange surface, especially from the side where the heat exchange intensity is too low.

One of the effective passive ways to increase energy efficiency is to create curved heat exchange surfaces, leading to a swirl of the coolant flow [3–5]. The flow's swirling can also be carried out in straight-tube bundles with their inherent high compactness, using various types of screw inserts. These inserts can be belt and blade swirlers or screw inserts made by cutting or winding wire onto a rod. A special place among the passive methods of heat exchange intensification is occupied by tape flow swirlers [6]. They, as well as auger swirlers, make it possible to increase heat transfer intensity due to secondary flows arising under the action of centrifugal forces, which intensify heat exchange between the core of the flow and the boundary layer. Tape swirlers are the simplest ones and are not expensive to manufacture compared to other methods of intensification [7–9]. The channels with tape swirlers have less pressure drop compared to auger turbulators with the same heat transfer coefficient. This intensification method is characterized by lower metal consumption. This intensification method can be used to modernize heat exchangers [10]. In this case, there is no need to design a new heat exchanger; it is enough to insert a twisted tape into the heat exchange channel. The most acceptable intensifier geometry is selected by varying the winding step or using a variable step. An important drawback of any heat transfer intensification method is the additional pressure drop that arises. Regardless of intensification aims and conditions, the pressure drop of the improved channel must be evaluated. The efficiency assessment result is based on a combination of coefficients that take into account changes in the heat exchange surface area and heat flow and the increase in channel pressure drop.

In nuclear reactors, like the PWR or WWER, the heat exchange equipment functions within unique operation parameters. The most important of these is the high-temperature field gradient. The temperature field change along the length of the channel leads to significant changes in the physical properties of the coolant. For example, at 10 MPa of pressure, if the temperature changes from 35 to 200 degrees, the dynamic viscosity decreases by 5.2 times, the kinematic viscosity by 6.3 times, and the Prandtl number by 5.3 times [11]. The generalized dependencies $Nu = f\ (Re, Pr)$, which are experimentally obtained within a narrow range of parameter changes, can be used in the heat exchange surface design very carefully.

It will be more accurate to use modern CFD codes, which will make it easy to take into account all factors. At the moment, there is a wide variety of CFD codes. Most of them are based on the Navier–Stokes equation's solution for coolant turbulent flow. Since the direct solution (DNS) of a system of equations requires high-performance computer systems, turbulence models have become the most widespread. These are additional equations to the system, the main aim of which is to simplify and speed up equation solving. The most popular among them are RANS (k-e, k-ω, and SST), DES and LES models. All turbulent models are based on additional experimental studies. They need to be carried out for various coefficient adjustments, making it possible to improve the quality of the resulting solution despite the introduced simplifications.

In this study, the authors try to evaluate the application of Omega-based Reynolds stress (ORS) [12] and SST-CC [13,14] with a correction for curvature streamline models of turbulence for coolant flow modeling, moving in the swirling channel with large temperature changes. Numerical modeling was carried out by Ansys CFX. Numerical approaches' validation was carried out by an experimental study of heat transfer within PWR operation parameters [15–17]. The high level of numerical details made it possible to identify some swirling flow features that occur when the twist pitch changes.

## 2. Swirl Flow Study Overview

In the context of the physical processes considered in this article, the theoretical and experimental studies of single-phase coolant swirling flows are of main interest. Extensive studies in this field were conducted by Koch [18], Shchukin [19], R.M. Manglik and A.E. Bergles [20], and others [21,22]. The main geometric characteristics of a channel with an inserted twisted tape are the diameter ($d$), tape thickness ($\delta$), and swirl pitch ($s$), corresponding to a 180° turn of the flow around the pipe axis (Figure 1).

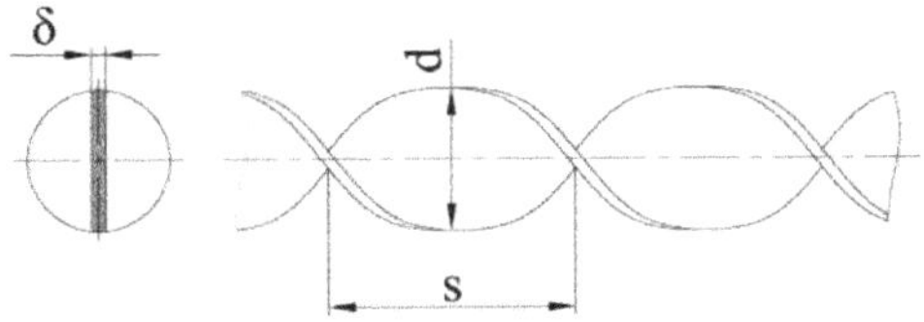


**Figure 1.** Twisted tape configuration.

The flow swirl intensity can be assessed in several ways as follows:

- The relative pitch of the helical line, where *s/d* is a simple and less effective parameter;
- The max swirl angle ($\theta$), which is the local parameter of the limiting streamline deviation from the axial direction [19]:

$$\tan(\theta) = \frac{\tau_\varphi}{\tau_z}, \quad (1)$$

where $\tau_\varphi$, $\tau_z$ is the tangential stress projections onto the angular and axial directions;

- The integral twist parameter ($Sn$) takes into account the velocity uneven distribution in the radial ($r$) direction as follows:

$$Sn = \frac{\int_0^R \rho u_\phi u_z r^2 dr}{R\int_0^R \rho u_z^2 r dr}, \quad (2)$$

where $u_\varphi$ and $u_z$ are the velocity projections onto the angular and axial directions, $\rho$ is the density, and $R$ is the tube channel cross-section radius. This parameter characterizes the relative magnitude of momentum flows transferred in the angular and axial directions [23]. $Sn$ accurate calculation is difficult. In article [24], the authors show that the $Sn$ value has high sensitivity due to the method and limits of integration.

Researchers from [19,23] have found a strong connection between the integral and local swirl parameters, for example, for axial-vane swirlers:

$$\tan(\theta) = 1.13 \cdot Sn^{0.82}, \quad (3)$$

Starting from a threshold value of integral twist parameter $Sn^*$, ($0.23 < Sn^* < 0.3$) [23], the tangential velocity distribution takes the form as seen in Figure 2.

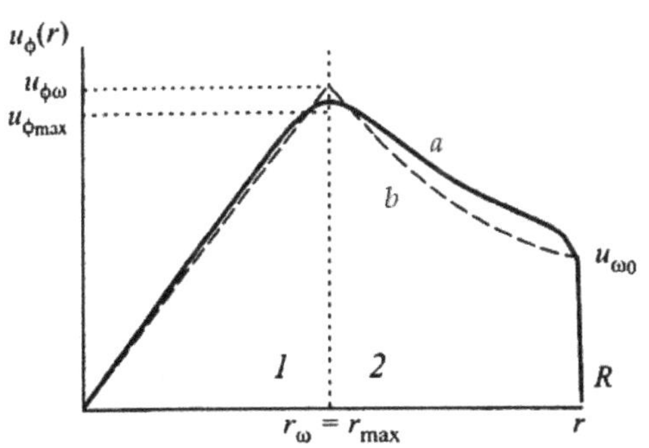


**Figure 2.** Tangential velocity distribution when $Sn > Sn^*$: (a)—real profile; (b)—theoretical; 1—zone of quasi-rigid rotation; 2—zone of quasi-potential flow.

The theoretical profile calculated by Formula (4):

$$u_{\Phi}(r) = \begin{cases} \omega \cdot r \text{ if } 0 < r < r_{\omega}, \\ \frac{u_{\Phi\omega} r_{\omega}}{r} \text{ if } r_{\omega} < r < R, \end{cases} \tag{4}$$

where $\omega$ is the angular velocity in a quasi-rigid rotation region, $r_{\omega}$ is the radius corresponding to the tangential velocity maximum $u_{\varphi\omega}$. At $Sn < Sn^*$, the tangential velocity maximum is achieved near the wall region but outside the viscous sublayer. Regions with positive and negative gauge pressure appear at $Sn > Sn^*$. Swirl attenuation during the rotation degeneration process leads to pressure profile equalization, i.e., to the pressure increasing in the central region. The negative axial pressure gradient leads to reverse flow emergence.

During the coolant movement along a swirling channel, inertial force excess action leads to secondary flows' formation. The laser Doppler anemometry (LDA) method application in the experimental studies and the direct numerical modeling (DNS) method for numerical ones allowed for more accurate localization of stable vortex structures [25]. Two spiral rotating vortices arise in the water flow at room temperature after passing a short swirl insert. Computational and experimental velocity fields refuted the hypothesis of vortices' counter-rotation. Figure 3 shows the evolution of velocity vector distribution at $s/d = 2.36$ and $Re = 77{,}000$. The authors showed that single-rotating vortices were formed on each side of the insert as a result of pressure imbalance that occurs near the inlet region of the twisted tape.

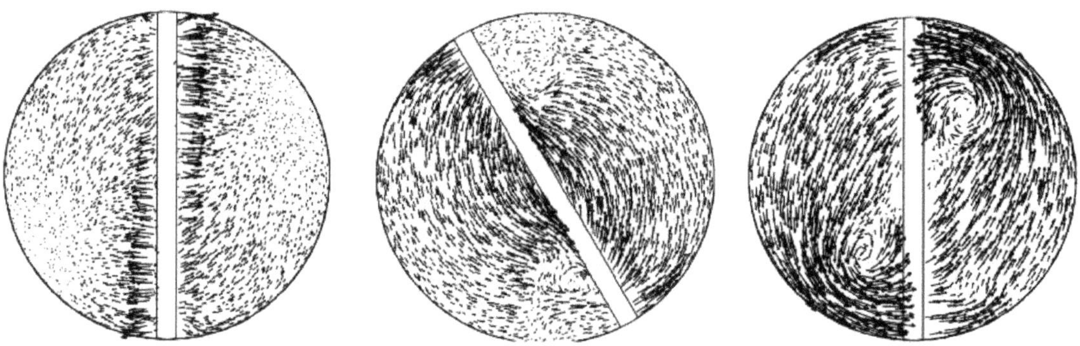

**(a)** **(b)** **(c)**

**Figure 3.** Velocity distribution across the channel cross-section with a tape insert at the inlet (**a**), with flow turning by 75° (**b**) and 180° (**c**).

Some experimental generalized dependencies for heat transfer coefficients and hydraulic friction are presented below. According to Koch R. [18]:

$$\lambda = \frac{6.34}{Re^{0.474}} \left(0.5 + \frac{8}{\pi^2}\left(\frac{s}{d}\right)^2\right)^{-0.263} + \frac{25.6}{Re}, \tag{5}$$

$$Nu = 0.3 Re^{0.6} Pr_f^{0.43} \left(0.5 + \frac{8}{\pi^2}\left(\frac{s}{d}\right)^2\right)^{-0.135}, \text{ if} \tag{6}$$

$$Re_I^{crit} = 11.6\sqrt{0.5 + \frac{8}{\pi^2}\left(\frac{s}{d}\right)^2} < Re < Re_{II}^{crit} = 38900\left(\frac{s}{d}\right)^{-1.16} + 2300. \tag{7}$$

According to R.M. Manglik and A.E. Bergles [20], for laminar flow:

$$\lambda = \frac{15.767}{Re}\left[\frac{\pi d + 2d - 2\delta}{\pi d - 4\delta}\right]^2 \left(1 + 10^{-6} Sw^{2.55}\right)^{\frac{1}{6}} \left[\frac{\pi d}{\pi d - 4\delta}\right]\left[1 + \left(\frac{\pi d}{2s}\right)^2\right], \tag{8}$$

$$Nu = 4.612\left(\frac{\mu_f}{\mu_w}\right)^{0.14}\left(\left[\left(1+0.0951Gz^{0.894}\right)^{2.5}+6.413\cdot10^{-9}\left(Sw\cdot Pr^{0.391}\right)^{3.835}\right]^2 +2.132\cdot10^{-14}(Re\cdot Ra)^{2.23}\right)^{0.1}, \quad (9)$$

$$Sw = \frac{Re}{\sqrt{s/d}}\left[\frac{\pi d}{\pi d-4\delta}\right]\sqrt{1+\left(\frac{\pi d}{2s}\right)^2}; \quad (10)$$

and for liquid turbulent flow in a heated channel:

$$\lambda = \frac{0.0791}{Re^{0.25}}\left[1+\frac{2.752}{(s/d)^{1.29}}\right]\left[\frac{\pi d}{\pi d-4\delta}\right]^{1.75}\left[\frac{\pi d+2d-2\delta}{\pi d-4\delta}\right]^{1.25}, \quad (11)$$

$$Nu = 0.023Re^{0.8}Pr^{0.4}\left[1+\frac{0.769}{s/d}\right]\left[\frac{\pi d}{\pi d-4\delta}\right]^{0.8}\left[\frac{\pi d+2d-2\delta}{\pi d-4\delta}\right]^{0.2}\left(\frac{\mu_f}{\mu_w}\right)^{0.18}, \quad (12)$$

where *Re*, *Gz*, *Pr*, *Nu*, and *Ra* are the Reynolds, Graetz, Prandtl, Nusselt, and Rayleigh numbers, $\lambda$ is the friction coefficient, and $\mu_f$ and $\mu_w$ are the dynamic viscosity of coolant at the fluid flow center and wall temperature.

## 3. Experimental Study

This big study was conducted in two stages: experimental and numerical. The experimental study of heat transfer intensification was carried out on the "FT-80" testing branch, which was created to study hydrodynamic and heat transfer processes in steam-generating systems. The test branch includes three hydraulic circuits. The first and the second circuits, filled with bi-distilled water, are hydraulically closed and used under excess pressure, and the third circuit is hydraulically open and made for cooling main equipment. The ranges of operating parameters are as follows:

- Pressure of the first and the second circuits: $P_1 = 8 \div 11$ MPa, $P_2 = 3 \div 5$ MPa;
- Inlet temperature of the first and the second circuits: $T_1 = 245 \div 260$ °C, $T_2 = 30 \div 50$ °C;
- Mass flow rate of the first and the second circuits: $G_1 = 11 \div 277$ kg/h, $G_2 = 4 \div 114$ kg/h;
- Reynolds numbers for the input sections: $Re_1 = 1250 \div 31{,}600$, $Re_2 = 140 \div 3900$.

The main aim of the experimental study was to obtain the temperature field and pressure drop of the swirling coolant flow in the heat exchanger channel. The general view of the experimental model is shown in Figure 4a, and the schematic view of the interior space is shown in Figure 4b. The model is a heat exchanger channel made according to the "pipe in pipe" principle. The case of the model consists of two parts, which are after welding, and the longitudinal welds form a circular cross-section channel with a 17 mm inner diameter. A heat exchange tube with a 13 mm outer diameter and 1.5 mm wall thickness, made of titanium alloy, is placed along the channel axis (Figure 4c).

The intensifier is a strip made of 12Cr18Ni10Ti stainless steel 1 mm thick and 9.8 mm wide, twisted around the central axis (Figure 5).

The length of the heat exchanger intensifier is equal to the active part length of the experimental model and is equal to 2280 mm. Six metallic bands were manufactured and studied: three with constant twist pitches ($s$ = 20, 30, and 40 mm) and three with a variable twist pitch created on the basis of different combinations of these fragments.

The countercurrent flow of the coolants is organized in the model. The second circuit coolant moves upwards along the inner tube and is heated by the first circuit coolant that moves from the top down in the annular gap. Such a coolants' movement scheme is the most optimal because it allows them to achieve the maximum temperature gradient value with minimum impact of thermal driving head forces on the heat transfer process and coolants' movement. The heat exchange surface's active part length is 2440 mm. The channel total length in which thermocouples are installed is 2280 mm.

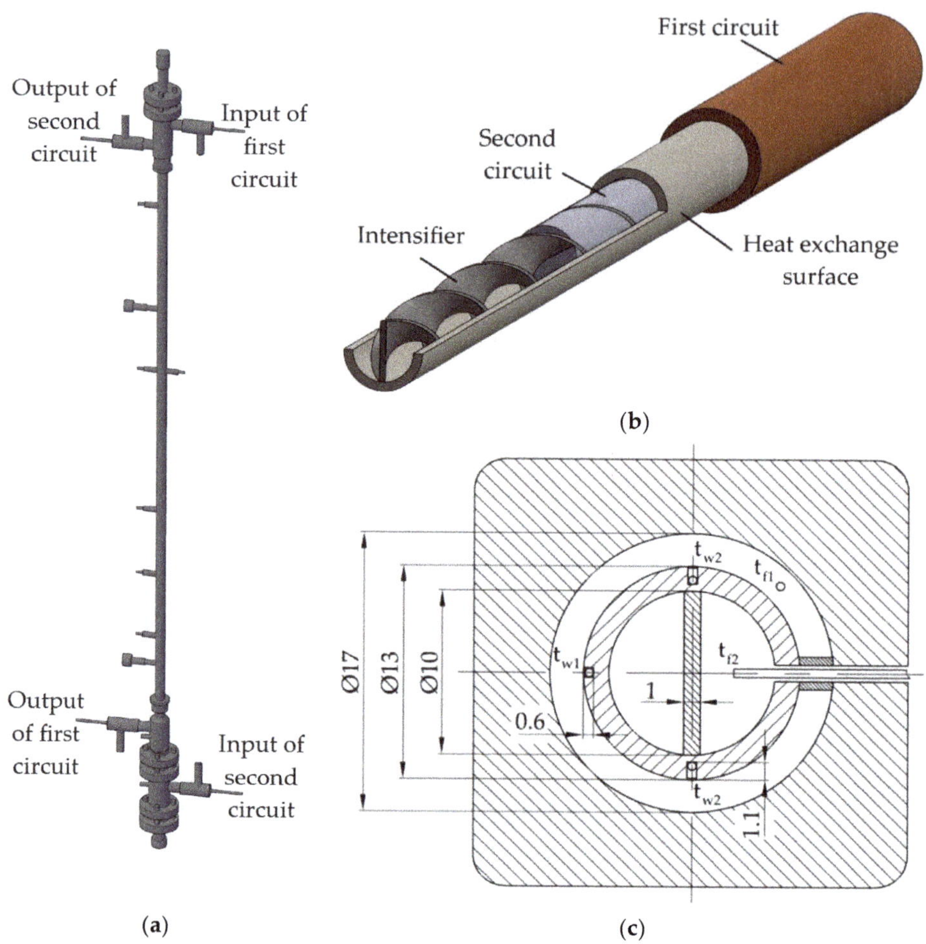


**Figure 4.** The experimental model: (**a**) general view; (**b**) schematic view of the interior space; (**c**) temperature measurement control points in the section of the model.

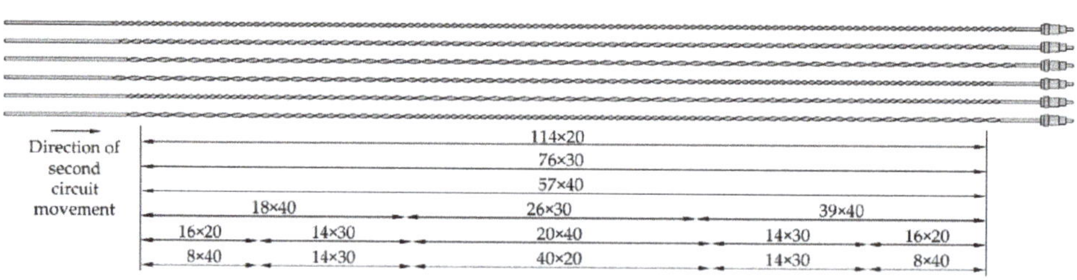


**Figure 5.** Schematic view of the intensifiers.

Operating parameters and geometric dimensions of the heat exchange surface are selected in such a way as to ensure heat transfer, improving the possibility for swirling heated water: $\alpha_I > \alpha_{II}$. For the specified range of coolants parameters, the heat transfer coefficients' average values, calculated using formulas applicable to annular (13) and round (14), (15) [26] channels, are: $\alpha_I = 3400 \div 45{,}000$ W/(m$^2$·K), $\alpha_{II} = 330 \div 5300$ W/(m$^2$·K).

With a significant difference in flow rates, which may occur during the actual operation of heat exchange equipment, a high level of intensification was not achieved along the entire channel length (discussed below).

$$Nu = 0.017Re^{0.8}Pr^{0.4}\left(\frac{Pr_f}{Pr_w}\right)^{0.25}\left(\frac{d_{out}}{d_{in}}\right)^{0.18} \text{ if } Re > 2300, \quad (13)$$

$$Nu = 0.015Re^{0.8}Pr_f^{0.43}\left(\frac{\mu_f}{\mu_w}\right)^{0.14} \text{ if } Re > 2300, \quad (14)$$

$$Nu \cong 3.66 \text{ if } Re < 2300. \quad (15)$$

The temperature of the coolants and the heat exchange surface is measured in 11 sections along the height of the model. These sections include a different set of thermocouples. In the general case, there are:

- Two sets (2 × 11 units) of thermocouples ($T_{W2a}$, $T_{W2b}$) embedded to 1.1 mm depth inside the heat exchange surface, designed to measure the wall temperature from the heated coolant side;
- Four at a 0.6 mm depth to measure the wall temperature from the heating coolant side ($T_{W1}$);
- Eight to measure the heating coolant temperature ($T_{F1}$);
- Five temperature sensors for the heated coolant ($T_{F2}$).

The temperature sensors' active part is located in the central part of the flow. The thermocouples $T_{W1}$ and $T_{W2}$ metallization were performed by plasma spraying in an inert gas (argon) flow. The applied coatings are treated flush with the main surface. All thermocouples (type K) have individual calibration, so the maximum temperature measurement error was 0.5 °C. The maximum value of the heat loss does not exceed 0.5%. Such small heat loss values allow us to exclude them from further numerical analysis.

The channel pressure drop was measured at the heat exchange process when present and absent. At states with $T \neq$ const, the pressure drop is measured by the differential pressure sensor, Yokogawa EJX110A. The maximum relative measurement error value was 0.5%. A significant contribution to the pressure drop measurement error was made by the back pressure of the water column in the measuring tubes connected to the measuring device located at the bottom of the model. For this reason, additional studies were conducted on channel pressure drop with a horizontal model arrangement (Figure 6).

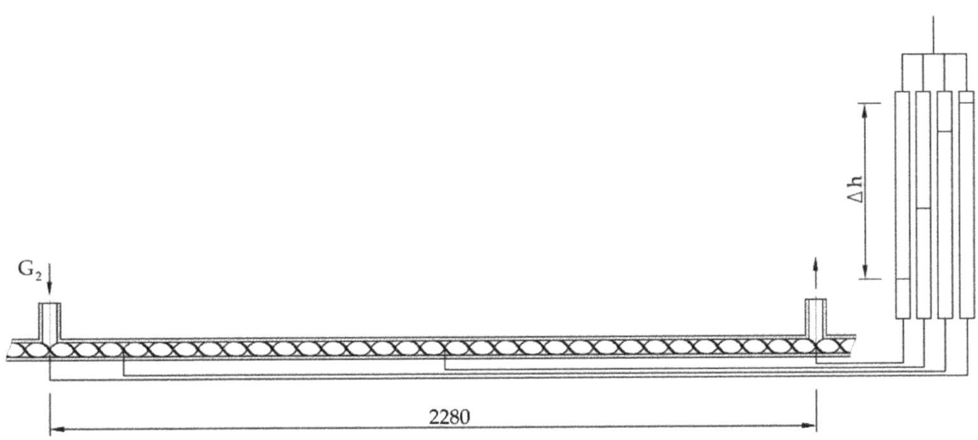


**Figure 6.** The isothermal flow pressure drop measuring.

Using an isothermal formulation ($T$ = 20 °C), it was possible to significantly increase the measurement accuracy (0.5 mm of water column or ΔP = 5 Pa) by connecting piezometers to the fittings of temperature sensors. More information about the experimental studies can be obtained from [15–17].

## 4. Experimental Data Processing

From the point of view of thermal process modeling in heated channels, the heat transfer coefficient distribution along the model length has the greatest representation. A computational and experimental distribution comparative analysis makes it possible to evaluate the effectiveness of using a variable twist step. The characteristic temperature distribution in Figure 7 shows that the initial set of thermocouples does not allow for control of the heated coolant temperature in each section of the experimental model. In the first step, the $T_{F2}$ measured values were approximated by different functions in order to determine the missing values.

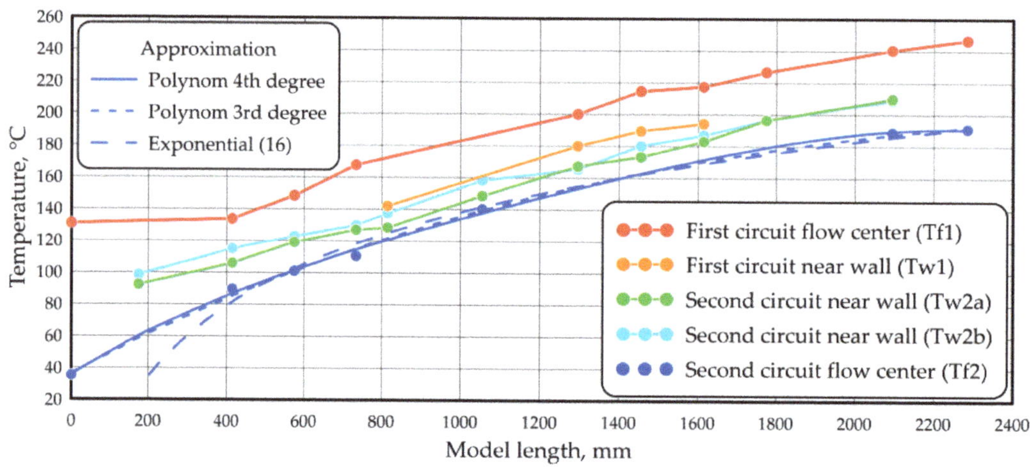


**Figure 7.** An example of temperature distribution.

The classical analytical solution for the heat exchanger temperature field with the coolants' counter-movement gives a function for heated coolant temperature $T_2(x)$ [27]:

$$T_2(x) = Ae^{Bx} + C, A, B, C - \text{const}. \quad (16)$$

This type of approximation has several disadvantages: firstly, it is obtained with the film coefficient constant value along the length, which is not observed in the experiment; secondly, it does not have significant flexibility and provides rough approximations to the true values. Polynomial functions are simpler and are also suitable for modes with significant differences in flow rates. The optimal polynomial degree is determined from the condition of the minimum value of the sum of deviations from the experimental values (least squares). The application of higher-order polynomials provides non-physical results. Table 1 contains values of least squares sums for different orders of polynomials, shown in Figure 7. As a result, a fourth-degree polynomial was chosen for all experimental modes.

**Table 1.** Residual sum of squares for approximation polynomials.

| Type of Swirlers | Residual Sum of Squares | Type of Swirlers | Residual Sum of Squares |
|---|---|---|---|
| 1st degree | 992.41 | 3rd degree | 47.02 |
| 2nd degree | 50.63 | 4th degree | 33.12 |

To obtain heat transfer coefficient distribution, the heat exchange surface was divided into 10 fragments based on the vertical coordinate of 11 measuring sections. There are two sets of $T_{W2}$ thermocouples as well as a significant uncertainty of their position relative to the installed tape, and their values were averaged over the cross-section. In the next step, the following parameters were determined for each fragment:

- Average ($t_f$) and boundary values of the flow temperature ($t_{in}$ and $t_{out}$) and average wall temperature ($t_w$);
- The heat flow to heated coolant ($Q$) was determined based on the mass flow rate ($G_2$) and the enthalpy difference ($h_{in}$ and $h_{out}$):

$$Q = G_2(h_{out} - h_{in}); \tag{17}$$

- Using the fragment length ($l$), channel diameter ($d$), and intensifier thickness ($\delta$), the heat transfer area ($S_T$) and heat transfer coefficient ($\alpha_e$) were determined as follows:

$$S_T = (\pi d - 2\delta)l; \tag{18}$$

$$\alpha_e = \frac{Q}{S_T(t_F - t_W)}. \tag{19}$$

## 5. Numerical Study

The following two-stage numerical approach was used:

- Isothermal swirling-flow modeling was conducted on a lightweight mesh model to obtain velocity and pressure fields using the SST and ORS turbulence models. Also, at this stage, the selection of RANS turbulence models was carried out, which showed the highest accuracy of the solution. These results are not included in this article;
- Heat transfer process simulation was conducted in cases of oncoming movement of coolants using selected turbulence models. For the heating coolant, the SST turbulence model was used in all calculations.

The computational models presented in Figure 8 are geometrical copies of the experimental model. Two mesh models for each intensifier were made as follows:

- Full model—for heat transfer modeling with coolant counterflows—50 million elements (Figure 8a);
- Lightweight model—for isothermal modeling of only one heated flow—12 million elements (Figure 8b).

When creating the analysis area, the following assumptions were made:

- The manifolds necessary for sealing flows' input and output have been discarded;
- Flow rotation presence at 90°, the complex manifold's interior geometry, and a small distance between the entry point and the first measuring section made it possible to consider the heated flow structure on the inlet as uniform: *V(x,y,z) = const;*
- The model created is more streamlined: the faces and other geometrical shape inaccuracies are not taken into account because it would significantly increase the dimensionality of the task and require more computational resources and time for modeling;
- Thermal losses are not taken into account: when specifying boundary conditions, we assume that there is no thermal flow on the outer wall.

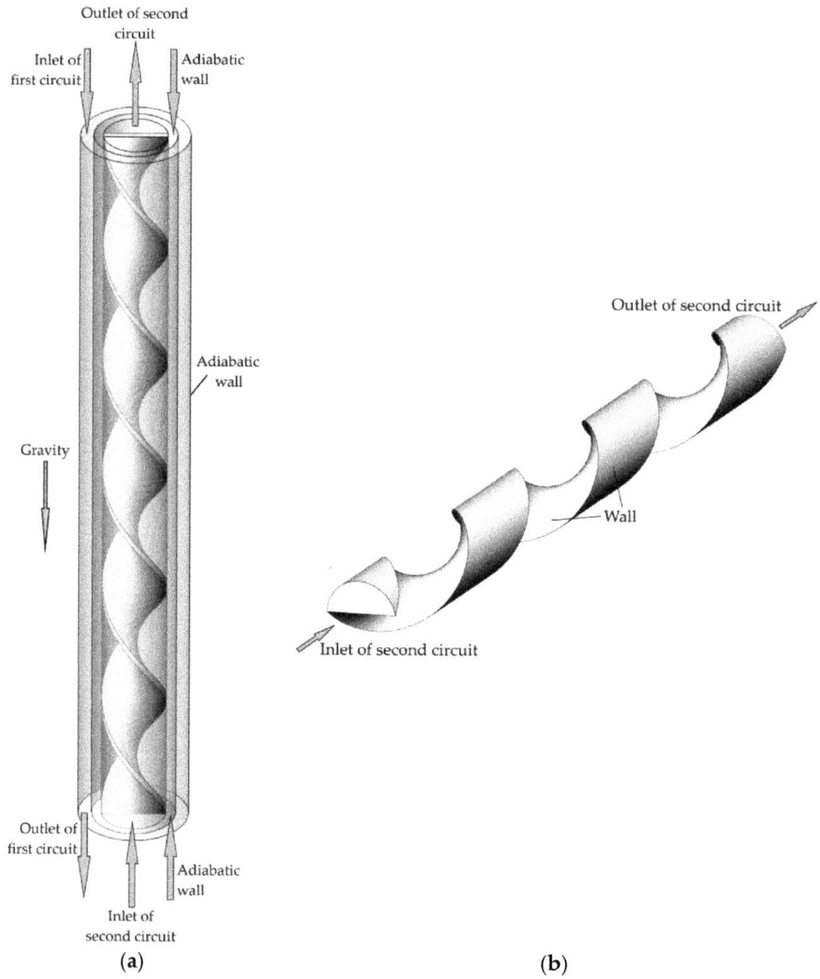


**Figure 8.** Numerical models: (**a**) schematic view of a full-scale model; (**b**) schematic view of a lightweight model.

Mathematically, the problem's solution is based on solving the basic system of continuity (20), momentum (21) and (22), energy transfer for fluids (23) and (24), and for solids (25) equations:

$$\frac{\partial \rho}{\partial t} + \nabla \cdot \left( \rho \vec{U} \right) = 0, \tag{20}$$

$$\frac{\partial}{\partial t}\left( \rho \vec{U} \right) + \nabla \cdot \left( \rho \vec{U} \otimes \vec{U} \right) = -\nabla P + \nabla \cdot \tau + \left( \rho - \rho_{ref} \right) \vec{g}, \tag{21}$$

$$\tau = \mu \left( \nabla \vec{U} + \left( \nabla \vec{U} \right)^T - \frac{2}{3} \delta \nabla \cdot \vec{U} \right), \tag{22}$$

$$\frac{\partial}{\partial t}\left( \rho \left( E + \frac{U^2}{2} \right) \right) + \nabla \cdot \left( \rho \vec{U} \left( h + \frac{U^2}{2} \right) \right) = \nabla \cdot \left( \left( \lambda + \frac{\mu_t}{\Pr_t} C_P \right) \nabla T \right), \tag{23}$$

$$h = \int_{298.15}^{T} C_P dT + \frac{P}{\rho}, E = h - \frac{P_{OP} + P}{\rho}, \quad (24)$$

$$\frac{\partial}{\partial t}\left(\rho \int_{298.15}^{T} C_p dT\right) = \nabla \cdot (\lambda \nabla T), \quad (25)$$

where $\mu$, $\lambda$, and $C_P$ are the dynamic viscosity, thermal conductivity, and heat capacity of the material, respectively, $U$ is the fluid velocity, $\rho_{ref}$ is the density in reference point, $\tau$ is the stress tensor, $\delta$ (or $\delta_{ij}$) is the Kronecker delta, $\mu_t$ is the turbulent viscosity, $Pr_t = 1.1$ is the turbulent Prandtl number for water [28], $E$ and $h$ are the fluid internal energy and enthalpy, and $P$ and $P_{OP}$ are the gauge and operating pressure.

The use of the SST-CC [12–14] turbulence model implies the solution of two additional transport equations: turbulence kinetic energy ($k$) (26) and the rate of dissipation of eddies ($\omega$) (27):

$$\frac{\partial(\rho k)}{\partial t} + \frac{\partial(\rho U_j k)}{\partial x_j} = P_k - 0.09\rho\omega k + \frac{\partial}{\partial x_j}\left(\left(\mu + \frac{\mu_t}{\sigma_k}\right)\frac{\partial k}{\partial x_j}\right), \quad (26)$$

$$\frac{\partial(\rho\omega)}{\partial t} + \frac{\partial(\rho U_j \omega)}{\partial x_j} = \alpha\frac{\rho\omega}{k}P_k - \beta\rho\omega^2 + \frac{\partial}{\partial x_j}\left(\left(\mu + \frac{\mu_t}{\sigma_\omega}\right)\frac{\partial \omega}{\partial x_j}\right) + 1.712(1 - F_1)\frac{\rho}{\omega}\frac{\partial k}{\partial x_j}\frac{\partial \omega}{\partial x_j}, \quad (27)$$

$$P_k = f_r\left(\mu_t\left(\frac{\partial U_i}{\partial x_j} + \frac{\partial U_j}{\partial x_i}\right)\frac{\partial U_i}{\partial x_j} - \frac{2}{3}\frac{\partial U_k}{\partial x_k}\left(3\mu_t\frac{\partial U_k}{\partial x_k} + \rho k\right)\right), \quad (28)$$

$$\mu_t = \frac{0.31\rho k}{\max(0.31\omega, SF_2)}, \quad (29)$$

$$S = \sqrt{2S_{ij}S_{ij}}, S_{ij} = \frac{1}{2}\left(\frac{\partial U_i}{\partial x_j} + \frac{\partial U_j}{\partial x_i}\right), \quad (30)$$

$$\sigma_k = 0.176F_1 + 1, \alpha = \frac{26F_1}{225} + 0.44, \beta = 0.0828 - 0.0078F_1, \sigma_\omega = \frac{89F_1}{107} + \frac{125}{107}, \quad (31)$$

$$f_r = \max\left[\min\left(\frac{4 \cdot r'}{1 + r'}\left(1 - tan^{-1}(2r^*)\right) - 1, 1.25\right), 0\right], \quad (32)$$

where $U_i$ is the averaged velocity component, $S$ is the magnitude of the strain rate tensor ($S_{ij}$), functions $r'$ and $r^*$ depend on combinations of strain rate and vorticity tensors, and $F_1$ and $F_2$ are the blending functions. Their formulation is based on the distance to the nearest surface and the flow variables. Various coefficients in Formula (31) are empirical and change their value depending on the value of the blending functions. It is necessary to take into account the empirical function fr increasing turbulence of the flow due to rotation, proposed by the authors from [13,14]. It is a multiplier of the production term $P_k$ in Formula (28). The specific limiter 1.25 provided a good compromise for different test cases that have been considered with the SST model (for example, flow through a U-turn, flow in a hydro cyclone, and flow over a NACA 0012 wing tip vortex [14]).

When studying heat transfer using the ORS turbulence model, the mathematical formulation becomes more complicated [12] since, firstly, the momentum (33) and $\omega$ (34)–(36) transfer equations are changed, and secondly, six additional turbulent stress transfer equations are added to the system (37)–(40):

$$\frac{\partial(\rho U_i)}{\partial t} + \frac{\partial(\rho U_i U_j)}{\partial x_j} - \frac{\partial}{\partial x_j}\left(\mu\left(\frac{\partial U_i}{\partial x_j} + \frac{\partial U_j}{\partial x_i}\right)\right) = -\frac{\partial}{\partial x_i}\left(P + \frac{2}{3}\mu\frac{\partial U_k}{\partial x_k}\right) - \frac{\partial(\rho\overline{u_i u_j})}{\partial x_j}, \quad (33)$$

$$\frac{\partial(\rho\omega)}{\partial t} + \frac{\partial(\rho U_k \omega)}{\partial x_k} = \frac{5}{9}\frac{\rho\omega}{k}P_k - 0.075\rho\omega^2 + \frac{\partial}{\partial x_k}\left(\left(\mu + \frac{\mu_t}{2}\right)\frac{\partial \omega}{\partial x_k}\right), \quad (34)$$

$$P_k = \mu_t\left(\frac{\partial U_i}{\partial x_j} + \frac{\partial U_j}{\partial x_i}\right)\frac{\partial U_i}{\partial x_j} - \frac{2}{3}\frac{\partial U_k}{\partial x_k}\left(3\mu_t\frac{\partial U_k}{\partial x_k} + \rho k\right), \quad (35)$$

$$\mu_t = \rho\frac{k}{\omega} = \frac{\rho}{2\omega}(\overline{u_l u_l}), \quad (36)$$

$$\frac{\partial(\rho\overline{u_i u_j})}{\partial t} + \frac{\partial(U_k\rho\overline{u_i u_j})}{\partial x_k} = P_{ij} - 0.06\rho\omega k\delta_{ij} + \Phi_{ij} + \frac{\partial}{\partial x_k}\left(\left(\mu + \frac{\mu_t}{2}\right)\frac{\partial\overline{u_i u_j}}{\partial x_k}\right), \quad (37)$$

$$P_{ij} = -\rho\overline{u_i u_k}\frac{\partial U_j}{\partial x_k} - \rho\overline{u_j u_k}\frac{\partial U_i}{\partial x_k}, P = 0.5P_{kk}, \quad (38)$$

$$\Phi_{ij} = 0.162\rho\omega\left(\frac{2}{3}k\delta_{ij} - \overline{u_i u_j}\right) - \frac{213}{275}\left(P_{ij} - \frac{2}{3}P\delta_{ij}\right) - \frac{54}{275}\left(D_{ij} - \frac{2}{3}P\delta_{ij}\right) - \frac{136}{275}\rho k\left(S_{ij} - \frac{1}{3}S_{kk}\delta_{ij}\right), \quad (39)$$

$$D_{ij} = -\rho\overline{u_i u_k}\frac{\partial U_k}{\partial x_j} - \rho\overline{u_j u_k}\frac{\partial U_k}{\partial x_i}, \quad (40)$$

where $u_i$ is the fluctuating component of the velocity, $P_{ij}$ is the production tensor of Reynolds stresses, $\phi_{ij}$ is the pressure–strain correlation, and $D_{ij}$ is the tensor from the pressure–strain model. The buoyancy effects on turbulence were not taken into account according to the recommendation from [12] because the ratio of the Grashof and Reynolds numbers at the lowest flow rate is $Gr/Re^2 < 1$. At the wall region, demonstrated in the motion and heat transfer modeling, the unified wall law according to B.A. Kader was used [29]. The wall temperature at the liquid–wall boundary was determined from the condition of heat fluxes' equality.

The calculation was carried out in a stationary formulation with a timescale (time step) equivalent to the Courant number within the 30–50 range. The simulation stopped when the following conditions were met simultaneously: the RMS of the residuals was less than $10^{-4}$, and imbalances of velocity components, mass flow, and heat flux for the mesh regions were less than $10^{-3}$. CFX's solver uses an element-based finite volume method. The mesh is used to construct finite volumes, which are used to conserve relevant quantities such as mass, momentum, and energy. The equations for mass, momentum, and a passive scalar are integrated over each control volume. The advection scheme is high resolution (HRS) [12]. It is based on a first-order upwind difference scheme (UDS), but HRS uses a special nonlinear recipe for the specified blend factor ($\beta$) at each node, computed to be as close to 1 as possible. The equality $\beta$ to 1 formally leads to second-order accuracy in space, and the resulting discretization will more accurately reproduce steep spatial gradients than the first-order UDS. The tri-linear shape functions are used to evaluate spatial derivatives for all the diffusion terms of the finite-element model part [12]. The summation is carried out over all the shape functions for the element. The Cartesian derivatives of the shape functions are expressed in terms of their local derivatives. The shape function gradients are evaluated at the actual location of each integration point.

Boundary conditions for the full grid model of each intensifier were various combinations of the first and second circuits' mass flow rates and input temperatures (Table 2). The fluids' movement and gravity action directions are shown in Figure 8a. Outlet pressure for each circuit was set equal to 0. The channels' sidewalls had no slipping. The operating water pressure was in accordance with the experimental values. There were no heat losses in any direction. Temperature dependence of water's physical properties was taken into account.

**Table 2.** The regime's parameters for heat transfer study.

| Circuit | Parameter | Regimes | | | | |
|---|---|---|---|---|---|---|
| I | $G_1$ (kg/h)/$T_{in1}$ (°C) | 11/245 | 46/250 | 71/255 | 184/255 | 277/260 |
| II | $G_2$ (kg/h)/$T_{in2}$ (°C) | 4/35 | 19/36 | 38/38 | 76/42 | 114/50 |

The boundary condition for the isothermal swirling-flow study was the mass flow rate in the range of 8–80 kg/h. Gravity force was not taken into account (Figure 8b). The pressure at the outlet was 0. The operating pressure of water was 1 atm. The channel sidewall had no slipping. Water's physical properties were taken at 20 °C. For all numerical studies in the inlet boundaries, the ratio of the root mean square of velocity fluctuations ($u$) and the mean flow velocity ($U$), named turbulence intensity, is equal to 0.05.

The grid independence study was based on a comparison of heat transfer coefficient averaged values and the tangential velocity profiles determined. The elements' number and thickness growth coefficient in the wall region varied in the channel cross-section (Figure 9a,c,e). For each cross-section version, three grid linear density variants were considered (Figure 9b,d,f). Nine grid model variants were studied in total.

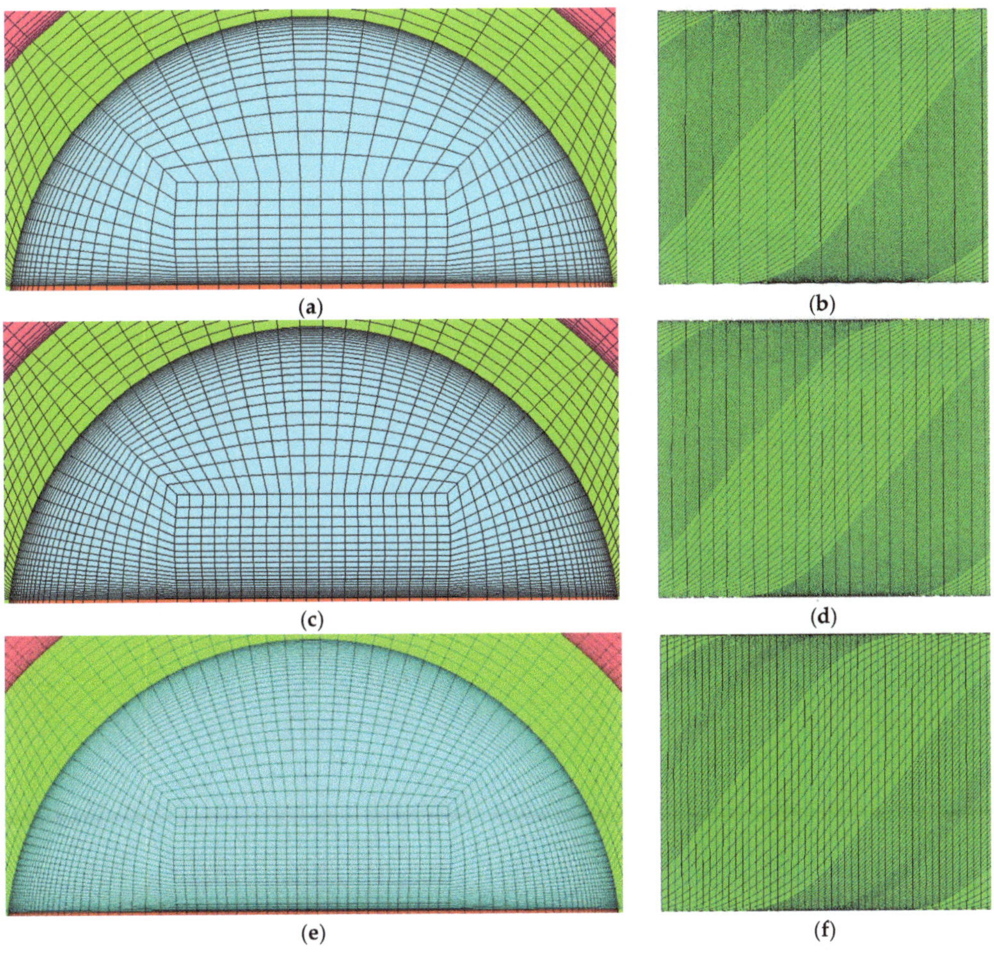

**Figure 9.** General view of grids for one twist step: (**a**) section type 1 (S1); (**b**) length type 1 (L1); (**c**) section type 2 (S2); (**d**) length type 2 (L2); (**e**) section type 3 (S3); (**f**) length type 3 (L3).

All meshes consist of hexahedral elements adapted to the swirling streamline orientations. The finite element and volumetric mesh models were connected "node to node". The total elements' numbers for each type of grid are given in Table 3. The most heat-stressed heat exchanger state ($G_1/G_2$ = 277/144) with the $s/d$ = 2 intensifier was modeled. The simulations using SST and ORS turbulence models were performed.

**Table 3.** Elements' numbers for different grid types.

| Grid Section Type | L1 | L2 | L3 |
|---|---|---|---|
| S1 | $17 \times 10^6$ | $33 \times 10^6$ | $66 \times 10^6$ |
| S2 | $26 \times 10^6$ | $50 \times 10^6$ | $99 \times 10^6$ |
| S3 | $33 \times 10^6$ | $66 \times 10^6$ | $123 \times 10^6$ |

Figure 10 shows the dependence of the ratio's average numerical Nu value on its experimental value for grids of different types and dimensions.

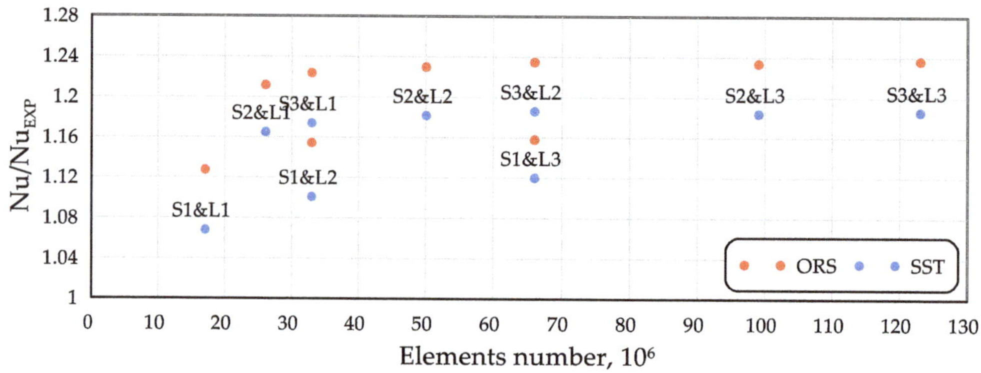


**Figure 10.** Heat transfer coefficient averaged values.

The grid model's low density in the section did not allow correct modeling of the velocity field (Figure 11). Comparing the tangential velocity component radial distributions, it can be noted that a thickness growth coefficient higher than 1.2 should be avoided, as well as not allowing a significant excess of the longitudinal elements' size of the transverse ones.

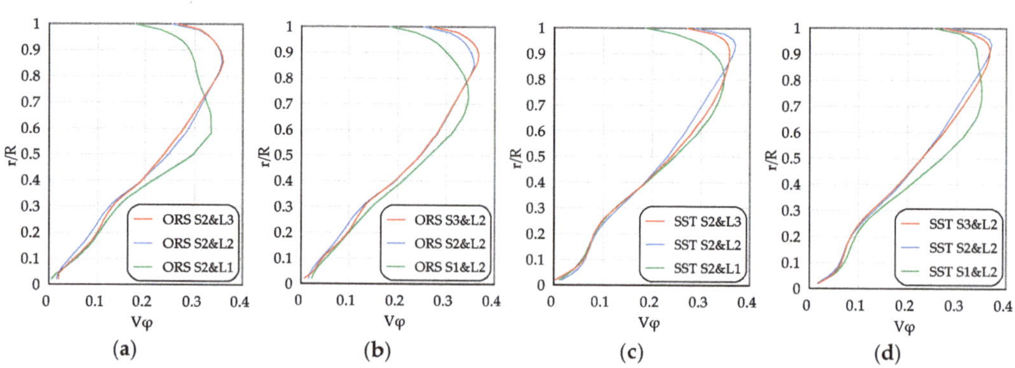


**Figure 11.** Tangential velocity profile from grid independence study: grid linear density variations with ORS (**a**) and SST (**b**) models; grid surface density variations with ORS (**c**) and SST (**d**) models.

As a result, the mesh type S2&L2 was selected (Figure 12). Other twisting steps models had similar settings. Grids with tetrahedral elements were not considered because these elements' normal vectors are not oriented along the swirling streamlines. This may lead to a numerical diffusion significant influence on the results in a long channel.

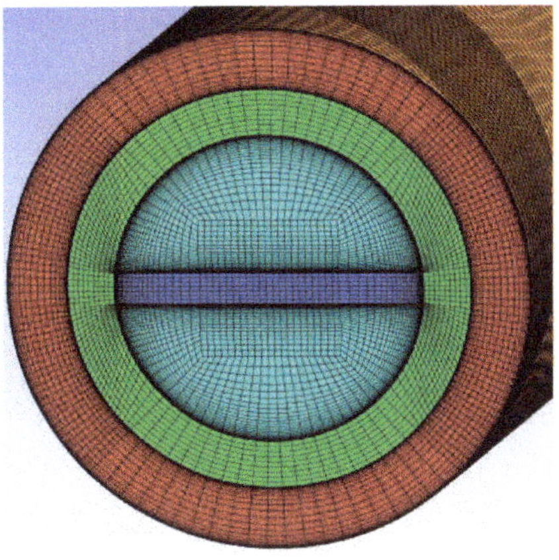

Figure 12. General view of selected type of mesh (S2&L2).

## 6. Numerical Data Processing

By analogy with experimental modes, the main aim of the numerical temperature field analysis was to determine the heat transfer coefficient distributions. Chaotically occurring zones of temperature gradients in the wall region were discovered (Figure 13a). It was necessary to determine the local heat transfer coefficient values in several directions. To solve this problem, a set of cross-sections was built according to the model height—one-by-one flow step rotation through 90°. Three main directions (Roman numerals) of heat transfer were identified (Figure 13b).

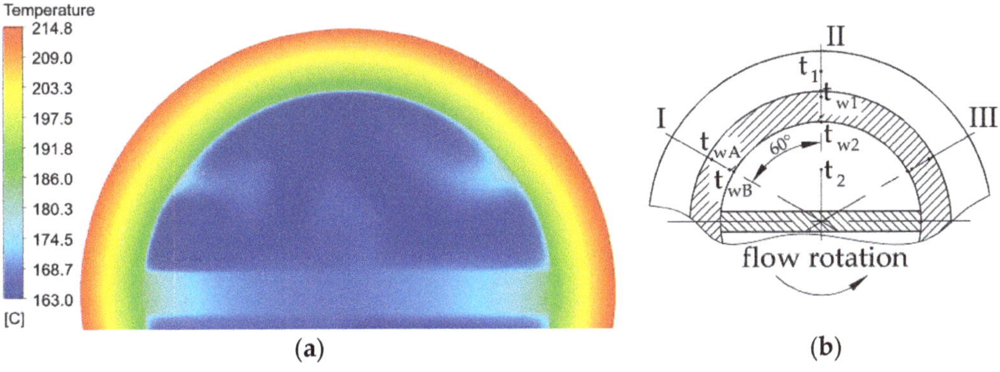


Figure 13. Processing of numerical data: (a) instantaneous temperature field in a random cross-section; (b) schematic view of temperature control points.

Each section has the following temperature control points:

- Heating coolant temperature: one ($T_1$) for the center and one ($T_{W1}$) for inside the wall region (center of first wall element);
- Three sets of $T_{WA}$, $T_{WB}$ (1 mm distance) for metal temperature control and the linear thermal flux density ($q_l$) calculation:

$$q_l = \frac{Q}{l} = \frac{2\pi\lambda_{metal}}{\ln(r_{TwA}/r_{TwB})}(t_{wA} - t_{wB}) = \frac{2\pi\lambda_{metal}}{0.175}(t_{wA} - t_{wB}); \quad (41)$$

- Three $T_{W2}$ in the wall region of heated flow (center of first wall element) for the heat transfer coefficient calculation ($\alpha$) for each direction:

$$\alpha = \frac{q_l}{(\pi d - 2\delta)(t_{w2} - t_2)} = \frac{q_l \cdot 10^3}{2.94(t_{w2} - t_2)}; \quad (42)$$

- Heated coolant temperature ($T_2$).

An example of the local heat transfer coefficients' distribution along the model length is shown in Figure 14. The outer (I), middle (II), and inner (III) flow zones are characterized by a different velocity profile and flow turbulence intensity. The distribution analysis confirmed the absence of spatial regularity of temperature gradient regions' occurrence in the wall region. A more detailed spectral distribution analysis was not carried out since this requires additional correct velocity and temperature field modeling: at the least, nonstationary LES formulation with the Courant number < 1. At the stage of final comparison with the experiment, the value of $\alpha$ was averaged ($\alpha_c$):

$$\alpha_c = \frac{\alpha_I + \alpha_{II} + \alpha_{III}}{3}. \quad (43)$$

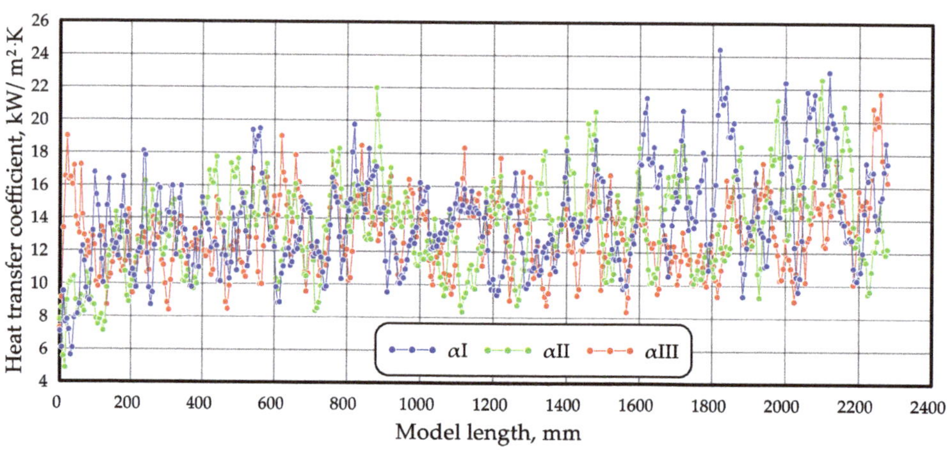


**Figure 14.** Local heat transfer coefficients' distribution along the model length for intensifier $s/d = 2$ at $G_1/G_2 = 277/114$.

## 7. Results

The friction factor at $T$ = const with swirlers of constant and variable twist pitch is shown in Figure 15.

The experimental values of R.M. Manglik and A.E. Bergles were calculated by dependencies (8). The experimental values of Koch R. were calculated by dependencies (5), which are valid under conditions of laminar motion mode with macro vortices, and the Reynolds critical numbers by (7) (Table 4).

**Table 4.** The Re critical numbers for intensifiers with const step.

| Type of Swirlers | $s/d = 2$ | $s/d = 3$ | $s/d = 4$ |
|---|---|---|---|
| $Re_{cr1}$ | 22.4 | 32.4 | 42.6 |
| $Re_{cr2}$ | 19,710 | 13,170 | 10,090 |

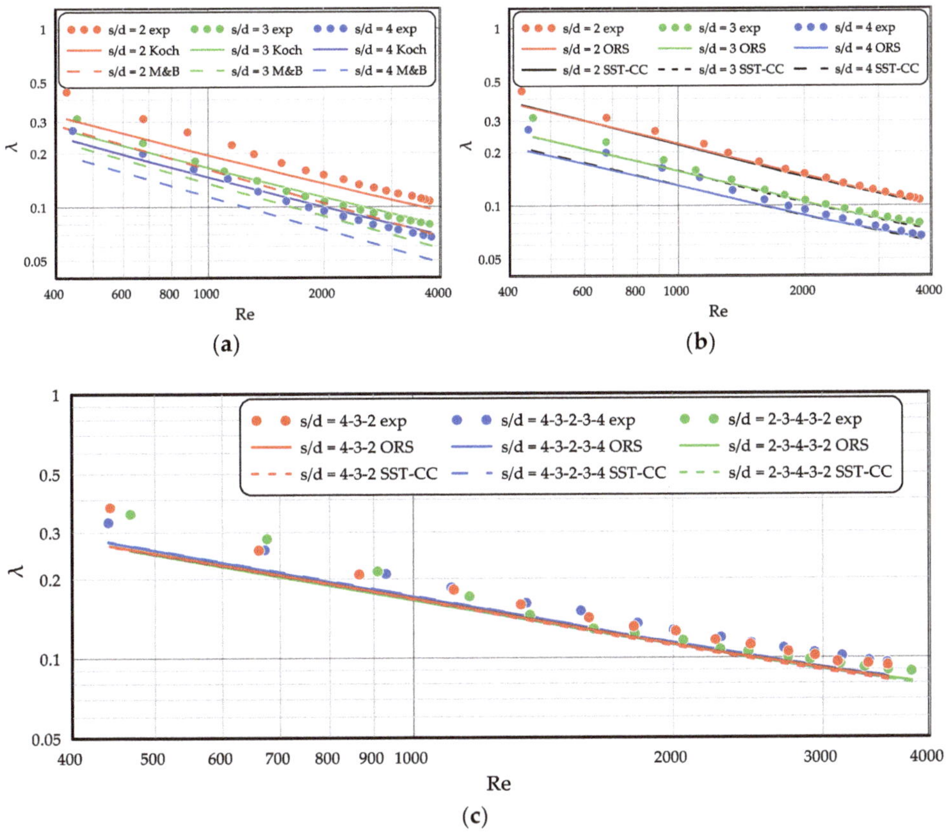


**Figure 15.** Channels' friction factor at $T$ = const: (**a**) constant twist pitch (only exp); (**b**) constant twist pitch (exp with num); (**c**) variable twist pitch.

The RMS deviation of the experimental values from numerical and generalized dependences, calculated by Formula (44), are shown in Table 5.

$$\sigma_{\lambda} = \sqrt{\sum_{i=1}^{N} \frac{\left(\lambda_i - \lambda_{\exp}\right)^2}{N}}. \tag{44}$$

**Table 5.** RMS experimental values deviation from numerical and generalized dependences.

| Type of Swirlers | Koch R. | Manglik and Bergles | ORS | SST-CC |
|---|---|---|---|---|
| s/d = 2 | 0.0416 | 0.1162 | 0.0265 | 0.0261 |
| s/d = 3 | 0.0153 | 0.0671 | 0.0234 | 0.0235 |
| s/d = 4 | 0.0211 | 0.0551 | 0.0361 | 0.0351 |
| s/d = 4-3-2 | - | - | 0.0398 | 0.0398 |
| s/d = 4-3-2-3-4 | - | - | 0.0243 | 0.0237 |
| s/d = 2-3-4-3-2 | - | - | 0.0349 | 0.0349 |

The calculated values are slightly lower than the experimental ones because of the high idealization degree of the geometric model. Swirlers consist of separate fragments connected by spot welding, which cannot ensure complete side surfaces' docking of the connected parts. Variable pitch swirlers have additional such connections. The numerical values (ORS and SST-CC) of pressure drops in the presence of heat exchange differ by less than 1%. More complex experimental study conditions, associated first of all with a large number of inaccuracies in the inner channel geometric shape, cause significant differences from the generalized dependencies obtained at an atmosphere or low gauge pressure.

The heat exchanger temperature conditions can be divided into three groups:

- The mode of fast swirled-flow warm-up at $G_1/G_2 > 2$ (Figure 16a);
- The mode with an optimal flow ratio at $2 > G_1/G_2 > 0.5$ (Figure 16b);
- The mode of slow swirled-flow warm-up at $0.5 > G_1/G_2$ (Figure 16c).

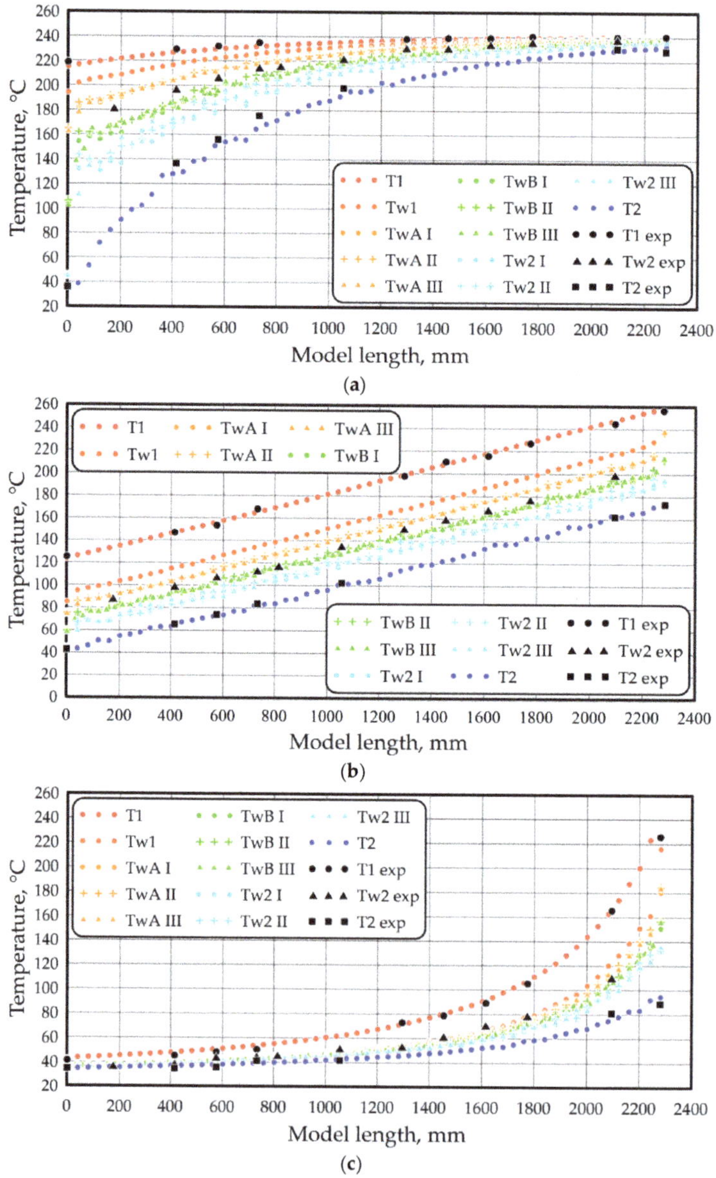


**Figure 16.** Temperature fields (ORS and exp): (**a**) $G_1/G_2 > 2$; (**b**) $2 > G_1/G_2 > 0.5$; and (**c**) $0.5 > G_1/G_2$.

The small deviations in the calculated values from the general trend are explained by the local heat transfer peculiarities due to the vortex structure generation, the thermogravitation effect, and the intensifier heat conduction influence.

The RMS deviations between numerical and experimental temperature data's overall flow rates range, calculated using a formula similar to 44, are given in Table 6.

**Table 6.** RMS deviations between the numerical and experimental temperature data (°C).

| Type of Swirlers | ORS ($T_{F2}$) | SST-CC ($T_{F2}$) | ORS ($T_{W2}$) | SST-CC ($T_{W2}$) |
|---|---|---|---|---|
| $s/d$ = 2 | 6.4 | 6.4 | 11.3 | 11.5 |
| $s/d$ = 3 | 6.4 | 6.5 | 10.9 | 11.0 |
| $s/d$ = 4 | 6.3 | 6.4 | 11.4 | 11.2 |
| $s/d$ = 4-3-2 | 5.9 | 6.0 | 10.8 | 10.7 |
| $s/d$ = 4-3-2-3-4 | 5.9 | 5.9 | 10.7 | 10.8 |
| $s/d$ = 2-3-4-3-2 | 5.8 | 5.9 | 10.9 | 10.9 |

It is possible to note a worse coincidence of temperature fields with a significant difference between coolant flow rates. Significant axial temperature gradients are more difficult to localize. Using a larger number of thermocouples for the swirling flow temperature would reduce the error. Significant errors in the wall temperature can be associated with a large logarithmic mean temperature difference between the flows and a greater inaccuracy in the spatial arrangement of the experimental thermocouples. Obtaining a complete match of temperature fields is impossible due to the high level of numerical model idealization.

A more correct approach is to compare the average values of the heat transfer coefficients. The experimental *Nu* (*Re*) obtained for the intensifiers with a constant swirl pitch with an almost uniform temperature difference along the model length are shown in Figure 17. The heat flux constancy condition along the channel length allows us to make a comparison with the results of semi-empirical formulas. During the swirling flow heating, a significant change in the Prandtl number occurs. This leads to several values of Nusselt number presence at the same Reynolds number. The RMS deviations between the experimental and semi-empirical Nu for modes with an optimal flow ratio, calculated using a formula similar to 44, are given in Table 7. There is good agreement between all experimental data on heat transfer.

**Table 7.** RMS deviations between experimental and semi-empirical Nu.

| Type of Swirlers | Koch R. | Manglik & Bergles |
|---|---|---|
| $s/d$ = 2 | 4.70 | 5.91 |
| $s/d$ = 3 | 5.28 | 4.96 |
| $s/d$ = 4 | 4.64 | 6.97 |

The thermal efficiency of variable pitch intensifiers application can be estimated based on total thermal load (45) comparison for $G_1/G_2 = 277/114$. The values are given in Table 8.

$$Q_{tot} = G_2(H_{out} - H_{in}), \tag{45}$$

where $H_{in}$ and $H_{out}$ are the secondary circuit enthalpies at the inlet and outlet of the model.

Analyzing this table, it is important to take into account the non-complete coincidence of the operating parameters during experimental studies. A small discrepancy between the inlet temperatures and coolant flow rates can cause up to 5% changes in the values presented in the table. Larger values of heat loads for the numerical calculations can be explained by the heat loss' complete absence in all directions. For the overwhelming quantity of modes, it can be noted that both numerical approaches lead to almost identical values of $Q$. The thermal load difference between the tapes is explained by the fragment $s/d$ = 2 different lengths since it provides the highest local heat transfer coefficients.

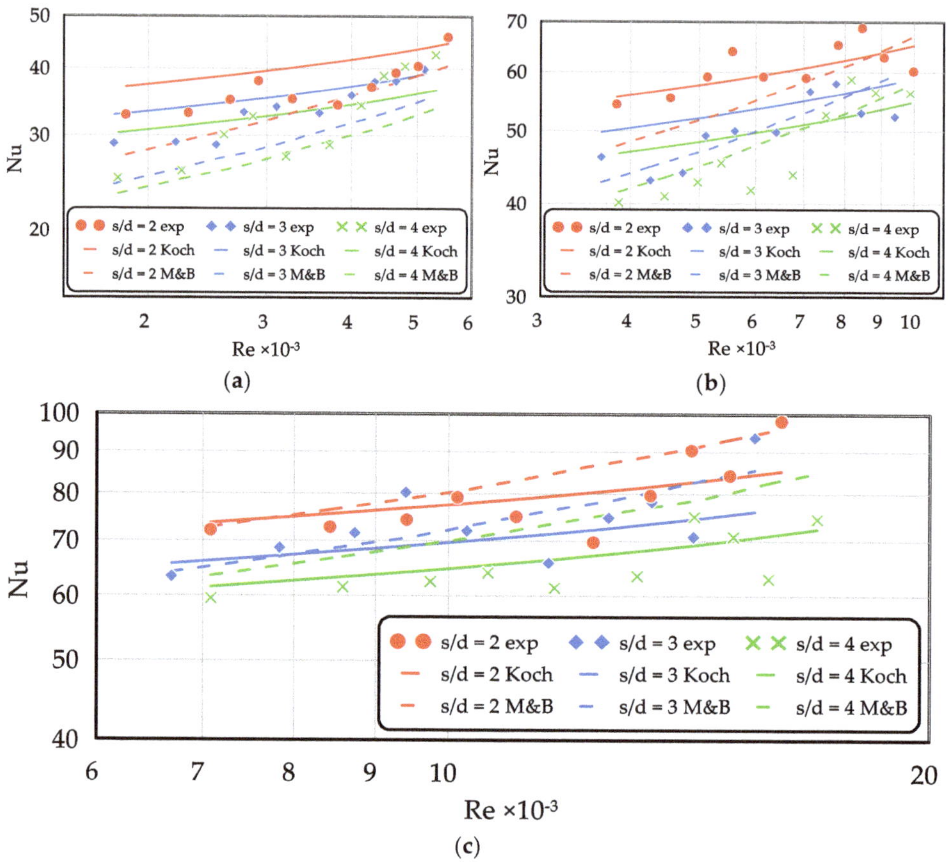


**Figure 17.** The $Nu = f(Re)$: (**a**) $G_1/G_2 = 46/38$; (**b**) $G_1/G_2 = 71/76$; and (**c**) $G_1/G_2 = 184/114$.

**Table 8.** Total thermal load (W) at maximum coolant flow rates.

| Type of Swirlers | Exp | ORS | SST-CC |
|---|---|---|---|
| $s/d = 2$ | 21,100 | 21,520 | 21,519 |
| $s/d = 3$ | 20,370 | 20,872 | 20,870 |
| $s/d = 4$ | 19,760 | 20,145 | 20,143 |
| $s/d$ = 4-3-2 | 20,060 | 20,391 | 20,389 |
| $s/d$ = 4-3-2-3-4 | 20,490 | 20,920 | 20,919 |
| $s/d$ = 2-3-4-3-2 | 19,420 | 19,823 | 19,821 |

Numerical models allow us to conduct a more detailed swirling flow structure analysis. Some flow features, such as local hydrodynamic or temperature effects, can be identified. A velocity fields comparative analysis for thermal and isothermal states revealed several features: at high Reynolds values in the presence of heating, due to a decrease in viscosity, the flow detachment zone in the central part of the channel occurs much earlier (Figure 18a); at low Reynolds numbers, the temperature gradient in the wall region leads to the zone's appearance, mentioned earlier (Figure 18b). These effects are more clearly visible in the fields obtained using the ORS turbulence model. It is important to note the high qualitative coincidence level of the axial velocity component fields, calculated by both models and the experimental values obtained in air [19,21] (Figure 19).

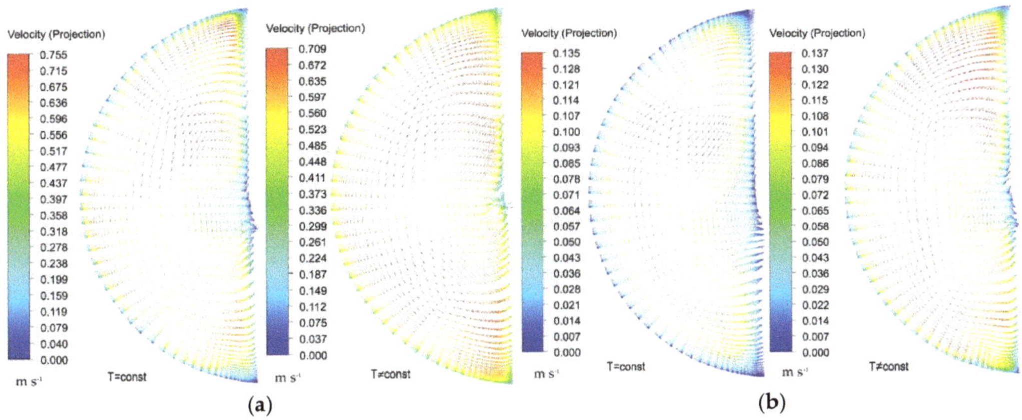


**Figure 18.** Velocity fields (ORS model): (**a**) $Re$ = 3900 and (**b**) $Re$ = 650.

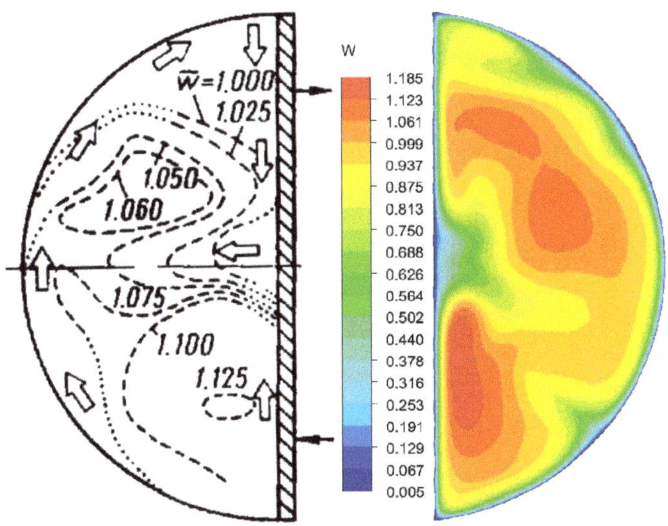


**Figure 19.** Axial velocity component fields: experimental (**left**); calculated (**right**).

Vector fields of the velocity tangential component for both models confirm the presence of two vortex regions rotating in the same direction (Figure 20a,b). Such vortex structures arise not only in the model's initial section but in other sections and at different flow rates. The tangential velocity field in model outlet section is also similar for both models (Figure 20c,d). Only a small difference can be seen in the near-wall region. It is associated with the local velocity gradients' occurrence due to the temperature gradient. The spatial zone distribution depends on the time step (the iteration number for quasi-statical calculations) and radial velocity distribution. Figure 21 shows the process of tangential velocity field evolution. It can be noted that the radius corresponding to the maximum tangential velocity for the SST model is always larger, i.e., most of the flow has rotation according to the rigid body law.

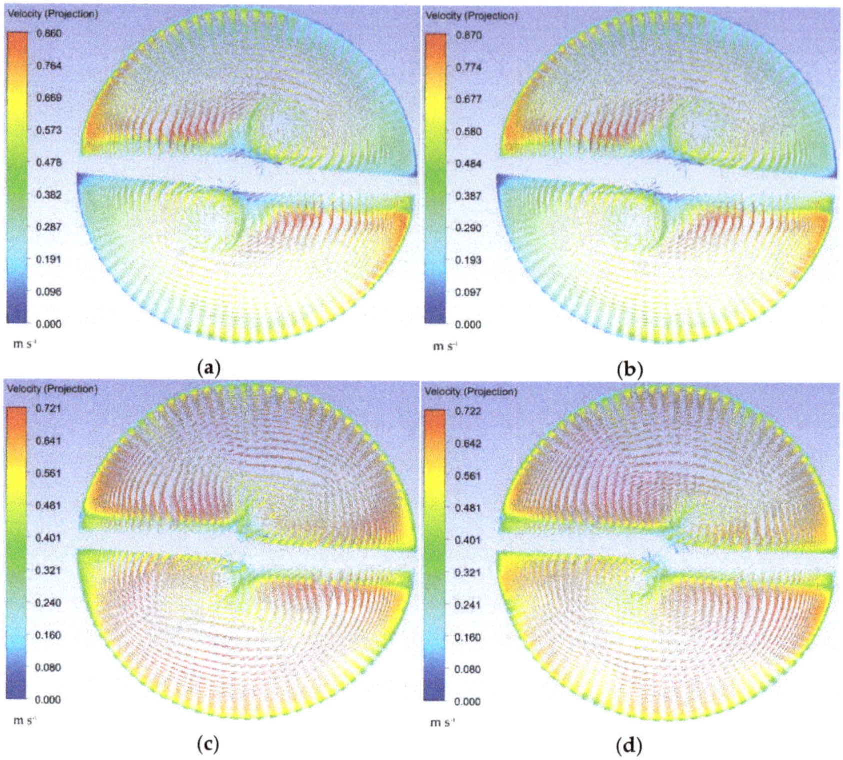

Figure 20. Tangential velocity component fields at $G_1/G_2$ = 277/144 after 180° flow turn: (a) ORS; (b) SST-CC; and in model outlet section: (c) ORS; (d) SST-CC.

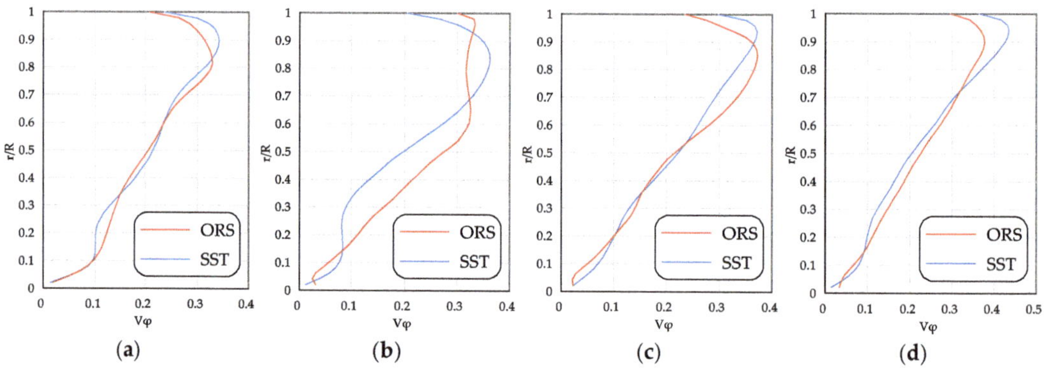

Figure 21. Tangential velocity profile evolution for intensifier $s/d$ = 2 at $G_1/G_2$ = 184/144: (a) l = 4s; (b) l = 44s; (c) l = 72s; and (d) l = 96s.

For variable pitch intensifiers, the most interesting point is how the flow responds to twist pitch changes. The example of a step transition from $s/d$ = 3 to 2 (Figure 22) shows that after two full turns (2 × 360°) of the flow, the velocity profile on the ORS model has been restored, while on the SST model, it has not.

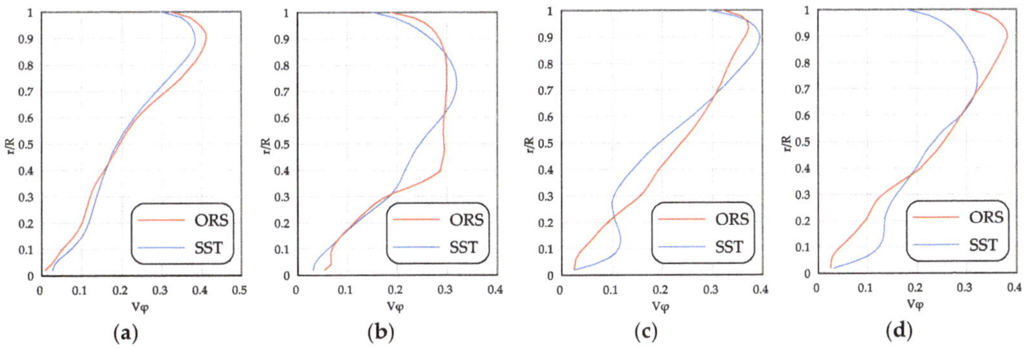

**Figure 22.** Tangential velocity profile evolution for intensifier *s/d* = 4-3-2 at $G_1/G_2$ = 184/144: (**a**) one step before change *s/d* = 3 to 2; (**b**) change section; (**c**) 360° after; and (**d**) 720° after.

The study of local heat transfer effects includes a comparative analysis of heat transfer coefficient distributions along the model length. When the flow rate of the heating coolant exceeds the flow rate of the heated coolant by more than two times, a heat exchange 'acceleration' on the initial section occurs, followed by a low-intensity process of heat transfer. With the opposite ratio, the heat flow from the first circuit is too low to obtain high values for the heat transfer coefficient.

The states with ratio $G_1/G_2$ = 184/114 = 1.61 are of greatest interest (Figure 23).

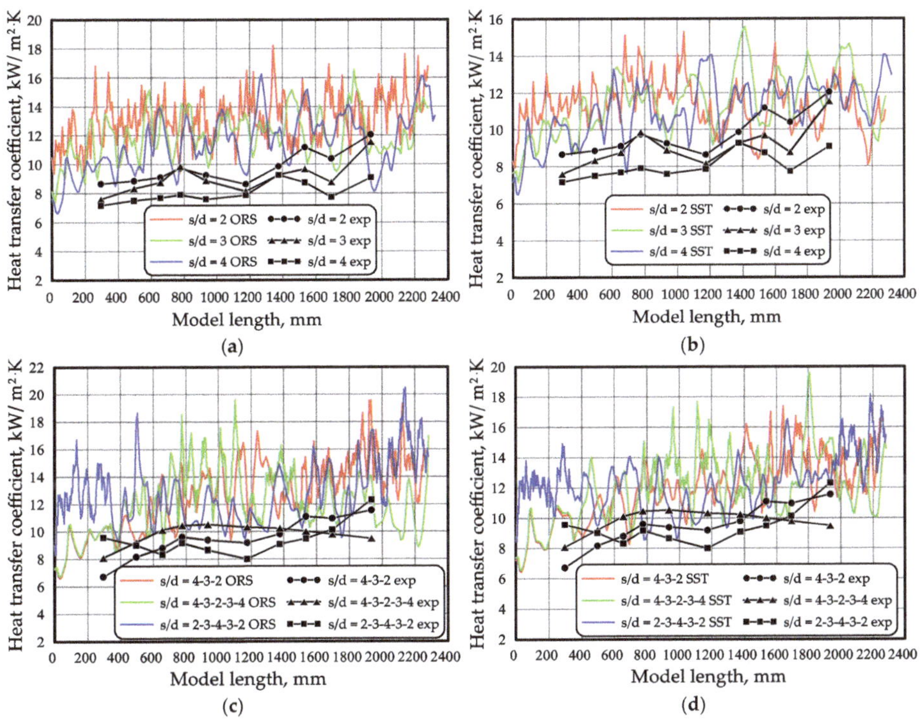

**Figure 23.** Heat transfer coefficient distributions at $G_1/G_2$ = 184/114: (**a**,**b**) constant twist pitch; (**c**,**d**) variable twist pitch.

A higher level of temperature field details in the calculation allows us to estimate the contribution of the changing twist pitch to the heat transfer process. Each step pitch corresponds to a specific level of heat transfer coefficient value. This observation is also true for variable pitch intensifiers. In sections with a 20 mm step, there is an increase in the heat transfer coefficient by 1.2–1.4 times compared to a step equal to 40 mm. Periodic spikes of the heat transfer coefficient value are explained by the control point hit in the zones of high-temperature gradients.

Variable twist pitch intensifiers are also more interesting from a numerical modeling point of view. A heat transfer coefficient numerical distributions comparison (SST-CC and ORS) is shown in Figure 24:

- The qualitative and quantitative coincidence of distributions in the initial fragment (flow stabilization);
- The heat transfer coefficient value (averaged by length) for the SST model is 5% less;
- Smaller values of heat transfer coefficient gradients arise when the twist pitch changes in the SST model.

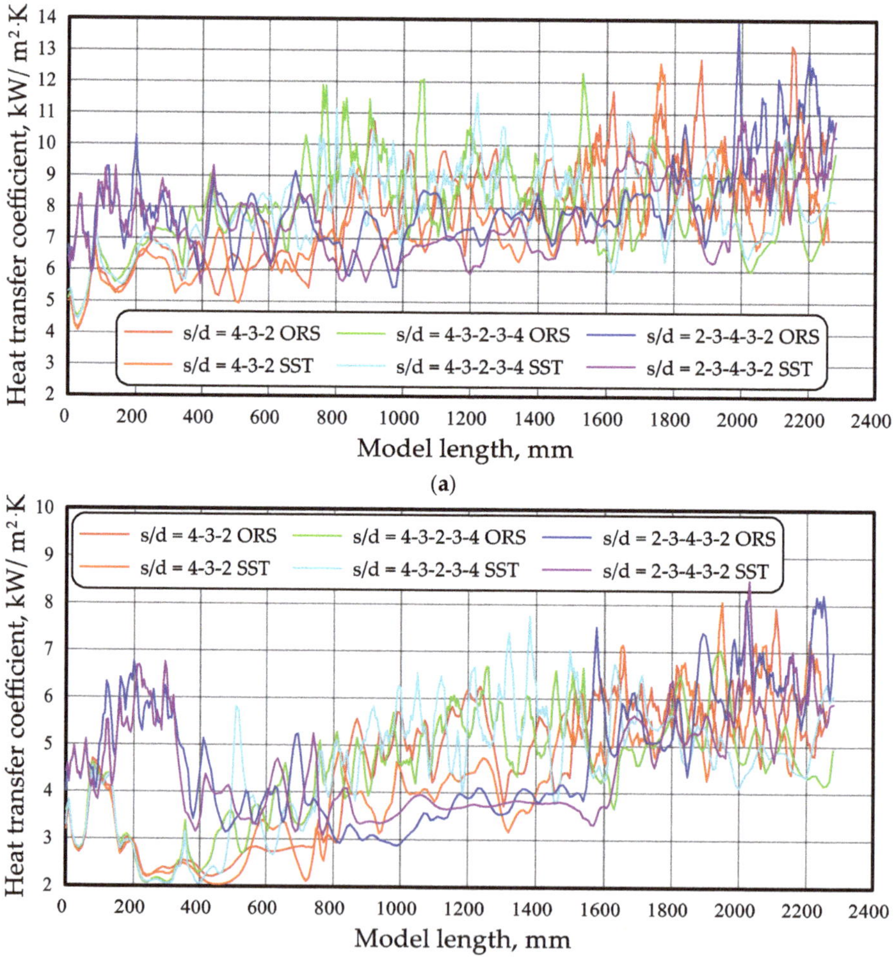


**Figure 24.** Heat transfer coefficient distributions: (**a**) $G_1/G_2$ = 71/76 and (**b**) $G_1/G_2$ = 184/38.

A high value of average heat transfer coefficient is maintained by the presence of twist pitch of 20 mm fragments. Fragments with a larger pitch have a lower friction coefficient. In the zones of a twist pitch change, a local heat transfer increase is observed. It is associated with the restructuring of the boundary layer. All these swirling flow movement features lead to a high efficiency of intensifiers with variable twist pitch.

The intensification effectiveness coefficients by tape swirlers were estimated using Formula (46) [30]. The states with the optimal flow rates ratio were processed.

$$k_{eff} = \frac{\alpha_{int}}{\alpha_{tube}} \frac{F_{int}}{F_{tube}} \text{ if } \frac{\Delta P_{int} G_{int}}{\Delta P_{tube} G_{tube}} \approx 1. \quad (46)$$

where $F$ is the heat exchange surface area, $\Delta P$ is the channel pressure drop, $G$ is the mass flow rate, $\alpha$ is the heat transfer coefficient, and the *int* and *tube* indexes indicate that the value belongs to a channel with and without an intensifier. The calculation results, based on experimental data, are presented in Table 9. The experimental temperature and velocity fields for the channel without an intensifier are not included in this article. With a non-optimal flow rates ratio, intensifiers with a variable twist pitch also performed best [15].

**Table 9.** Intensification effectiveness coefficients.

| Type of Swirler | *k* | Type of Swirler | *k* |
|---|---|---|---|
| s/d = 4 | 1.07 | s/d = 2-3-4-3-2 | 1.11 |
| s/d = 3 | 1.11 | s/d = 4-3-2 | 1.21 |
| s/d = 2 | 1.14 | s/d = 4-3-2-3-4 | 1.31 |

## 8. Conclusions

Variable twist pitch intensifiers provide greater efficiency because they enable high heat transfer coefficients to be achieved at lower hydraulic power consumption for coolant motion. This becomes possible despite a pressure drop increase due to a decrease in the coolant average density (at the same mass flow rate) due to the swirling flow's improved heating compared to a whole pipe. With a significant difference in coolant flow rates in the heat exchanger (not nominal or transition mode), its efficient operation is also ensured.

Using the Omega Reynolds stress turbulence to model the motion and heat transfer computations of swirling flows in complex geometry channels allows us to study the structure of thermal and hydrodynamic boundary layers with sufficient accuracy. The SST model, with a correction for curvature streamlines, allows us to obtain a slightly worse result but in a shorter time: the calculation time is 15–20% lower since many fewer transfer equations are solved. The best results can be achieved by the LES approach to turbulence modeling, but this will significantly increase the calculation time. Comparative analysis of the velocity fields showed a small advantage with the ORS model. This is due to the solution of additional equations for the transfer of turbulent stresses. Experimental values of the heat transfer coefficient are lower than the numerical ones. In many ways, the discrepancies between the calculated and experimental data are due to the complexity of the experiment and the high level of idealization of the calculated model.

**Author Contributions:** Conceptualization, S.D. and A.K.; data curation, A.S.; formal analysis, A.S.; investigation, S.D., A.K., and A.S.; methodology, S.D. and A.K.; software, A.S.; supervision, S.D.; validation, A.S.; visualization, A.S.; writing—original draft, A.K. and A.S.; writing—review and editing, A.K. and A.S. All authors have read and agreed to the published version of the manuscript.

**Funding:** This research received no external funding.

**Data Availability Statement:** The original contributions presented in the study are included in the article, further inquiries can be directed to the corresponding author.

**Conflicts of Interest:** The authors declare no conflicts of interest.

## Nomenclature

| | |
|---|---|
| ORS | Omega-Based Reynolds Stress Model |
| SST-CC | Shear Stress Transport Model with Curvative Correction |
| PWR | Power Water Reactor |
| NPU | Nuclear Power Unit |
| WWER | Water–Water Energetic Reactor |
| CFD | Computational Fluid Dynamics |
| DNS | Direct Numerical Simulation |
| RANS | Reynolds-averaged Navier–Stokes |
| DES | Detached Eddy Simulation |
| LES | Large Eddy Simulation |
| RMS | Root Mean Square |

## References


1. Kalinin, E.K.; Dreitser, G.A.; Kopp, I.Z.; Myakochin, A.S. *Effective Heat Exchange Surfaces*; Energoatomizdat: Moscow, Russia, 1998.
2. Chunchula, R.B.; Santhosh, K.G. CFD Analysis of Heat Transfer Enhancement by Using Passive Technique in Heat Exchanger. *Int. J. Recent Adv. Mech. Eng.* **2015**, *4*, 99–111.
3. Khafaji, H.Q.A.; Abdul Wahhab, H.A.; Alsaedi, S.S.; Al-Maliki, W.A.K.; Alobaid, F.; Epple, B. Thermal Performance Evaluation of a Tubular Heat Exchanger Fitted with Combined Basket–Twisted Tape Inserts. *Appl. Sci.* **2022**, *12*, 4807. [CrossRef]
4. Bhuiya, M.; Roshid, M.; Talukder, M.; Rasul, M.; Das, P. Influence of perforated triple twisted tape on thermal performance characteristics of a tube heat exchanger. *Appl. Therm. Eng.* **2020**, *167*, 114769. [CrossRef]
5. Bhuiya, M.; Azad, A.; Chowdhury, M.; Saha, M. Heat transfer augmentation in a circular tube with perforated double counter-twisted tape inserts. *Int. Commun. Heat Mass Transf.* **2016**, *74*, 18–26. [CrossRef]
6. Bucak, H.; Yılmaz, F. The current state on the thermal performance of twisted tapes: A geometrical categorization approach. *Chem. Eng. Process. Process Intensif.* **2020**, *153*, 107929. [CrossRef]
7. Piriyarungrod, N.; Kumar, M.; Thianpong, C.; Pimsarn, M.; Chuwattanakul, V.; Eiamsaard, S. Intensification of thermo-hydraulic performance in heat exchanger tube inserted with multiple twisted-tapes. *Appl. Therm. Eng.* **2018**, *136*, 516–530. [CrossRef]
8. Khoshvaght-Aliabadi, M.; Feizabadi, A. Performance intensification of tubular heat exchangers using compound twisted-tape and twisted-tube. *Chem. Eng. Process. Process Intensif.* **2020**, *148*, 107799. [CrossRef]
9. Salam, B.; Biswas, S.; Saha, S.; Bhuiya, M.M.K. Heat Transfer Enhancement in a Tube using Rectangular-cut Twisted Tape Insert. *Procedia Eng.* **2013**, *56*, 96–103. [CrossRef]
10. Man, C.; Lv, X.; Hu, J.; Sun, P.; Tang, Y. Experimental study on effect of heat transfer enhancement for single-phase forced convective flow with twisted tape inserts. *Int. J. Heat Mass Transf.* **2017**, *106*, 877–883. [CrossRef]
11. Wagner, W.; Pruß, A. The IAPWS Formulation 1995 for the Thermodynamic Properties of Ordinary Water Substance for General and Scientific Use. *J. Phys. Chem. Ref. Data* **2002**, *31*, 387–535. [CrossRef]
12. Ansys. *Ansys 19.2 Help*; ANSYS Inc.: Canonsburg, PA, USA, 2019.
13. Spalart, P.R.; Shur, M.L. On the Sensitization of Turbulence Models to Rotation and Curvature. *Aerosp. Sci. Technol.* **1997**, *1*, 297–302. [CrossRef]
14. Smirnov, P.E.; Menter, F.R. Sensitization of the SST turbulence model to rotation and curvature by applying the Spalart-Shur correction term. *J. Turbomach.* **2009**, *131*, 041010. [CrossRef]
15. Sobornov, A.E.; Dmitriev, S.M.; Ryazapov, R.R.; Kotin, A.V. Experimental efficiency evaluation of the of various geometry twisted bands for heat transfer intensification in the heat-exchange channels of a nuclear power unit equipment. *J. Phys. Conf. Ser.* **2021**, *2088*, 012046. [CrossRef]
16. Dmitriev, S.M.; Ryazapov, R.R.; Sobornov, A.E.; Dmitriy, V.S. Numerical and Experimental Research of Swirling Flow Heat Transfer in a Pipe with Equally Twisted Metallic Band. In Proceedings of the 9th International Topical Meeting on Nuclear Thermal-Hydraulics, Operation and Safety (NUTHOS-9), Kaohsiung, Taiwan, 9–13 September 2012; 13p.
17. Sobornov, A.E.; Dmitriev, S.M.; Ryazapov, R.R.; Kotin, A.V. Experimental research of heat transfer in a viscous flow in channels with twisted metallic bands different geometry. *Proc. Nizhny Novgorod State Tech. Univ. R.E. Alekseev №* **2013**, *2*, 134–141.
18. Koch, R. *Pressure Loss and Heat Transfer for Turbulent Flow*; No. AEC-TR-3875; Atomic Energy Commission: Washington, DC, USA, 1960.
19. Shchukin, V.K. *Heat Transfer and Hydrodynamics of Internal Flows in the Mass Forces Fields*; Mashinostroenie: Moscow, Russia, 1980; 240p.
20. Manglik, R.M.; Bergles, A.E. Bergles, Swirl flow heat transfer and pressure drop with twisted-tape inserts. *Adv. Heat Transf.* **2003**, *36*, 183–266.
21. Smithberg, E.; Landis, F. Friction and forced convection heat transfer characteristics in tubes with twisted tape swirl generators. *J. Heat Transf.* **1964**, *86*, 39–49. [CrossRef]
22. Ibragimov, M.M.; Nomofelov, E.V.; Subbotin, V.I. Heat transfer and hydraulic resistance with swirl-type motion of liquid in pipes. *Teploencrgetika* **1961**, *8*, 57–60.

23. Senoo, Y.; Nagata, T. Swirl Flow in Long Pipes with Different Roughness. *Bull. JSME* **1972**, *15*, 1514–1521. [CrossRef]
24. Vignat, G.; Durox, D.; Candel, S. The suitability of different swirl number definitions for describing swirl flows: Accurate, common and (over-) simplified formulations. *Prog. Energy Combust. Sci.* **2022**, *89*, 100969. [CrossRef]
25. Cazan, R.; Aidun, C. Analysis of the secondary vortices in rotating Flow induced by a twisted tape inside a pipe. *Sci. Iran.* **2018**, *25*, 3161–3172. [CrossRef]
26. Mills, A.F. *Heat Transfer*, 2nd ed.; Prentice-Hall: Englewood Cliffs, NJ, USA, 1999.
27. Hepbasli, A.; Biyik, E.; Ekren, O.; Gunerhan, H.; Araz, M. A key review of wastewater source heat pump (WWSHP) systems. *Energy Convers. Manag.* **2014**, *88*, 700–722. [CrossRef]
28. Lushchik, V.G.; Makarova, M.S. Turbulent Prandtl number in the boundary layer on a plate: Effect of the molecular Prandtl number, injection (suction) and longitudinal pressure gradient. *Thermophys. Aeromech.* **2018**, *25*, 169–182. [CrossRef]
29. Kader, B.A. Temperature and concentration profiles in fully turbulent boundary layers. *Int. J. Heat Mass Transf.* **1981**, *24*, 1541–1544. [CrossRef]
30. Gukhman, A.A. Intensification of convective heat transfer and the comparative evaluation problem for heat exchange surfaces. *Therm. Eng.* **1977**, *4*, 5–8.

fluids

MDPI

*Article*

# A Validation of the Junction Riemann Problem in Networks of Channels Under Transitory Flow Conditions

**Juan Mairal *, Javier Murillo and Pilar Garcia-Navarro**

Institute of Engineering Research of Aragon (I3A), University of Zaragoza, C. de Mariano Esquillor Gomez, 50018 Zaragoza, Spain; Javier.Murillo@unizar.es (J.M.); pigar@unizar.es (P.G.-N.)
* Correspondence: mairal@unizar.es

**Abstract:** Transitory states and supercritical regimes in channel junctions have been challenging to model using 1D models, and it has been common to resort to 2D models to locally solve networks of channels. In this paper, we present a methodology based on the Junction Riemann Problem that manages to include supercritical solutions by making use of suitable limiting coefficients. The method is compared against a pure 2D model in a series of test cases with challenging geometries that include transitory flow conditions. The results show that the method hereby presented is robust across all regimes and is able to capture the main features of wave propagation along a network of channels.

**Keywords:** shallow water; junction; network; transitory; supercritical; Riemann problem

## 1. Introduction

The flow of water in channels has often been modeled using one-dimensional (1D) formulations, where only one spatial coordinate is deemed important. Water channels [1,2], gas pipes [3–5], and blood vessels [6–10] are a few examples where these models have been successfully applied. The usefulness of this type of approximation lies in the reduction in the complexity of the models and methods and, more importantly, in the number of computational cells. Thus, 1D models are an excellent choice when the domain of interest is mostly laid in one direction, as the loss of information caused by dimension reduction is compensated by the enormous increase in efficiency, which in turn allows for faster simulations, higher resolutions, or larger domains [11]. Thus, robust methods to solve wave dynamics and propagation in networks are necessary to include other relevant phenomena, such as pollutant or solute transport [12].

The subject of study in this paper are networks of open channels, where free-surface flow is modeled by the shallow water equations, with a source term to account for changes in bed level, channel width, and friction. While there are many well-established methods, this system of hyperbolic partial differential equations models flow in the channels themselves, and their coupling at the junctions requires special consideration. Traditionally, in the 1D approximation, basic conditions including the equality of water levels have been used [13]. More complicated models have been devised based on momentum loss and recirculation due to the geometry of the junction [14,15].

In other cases, junctions are included as elements of higher dimension (2D) [1,16], but this approach results in a loss of efficiency and is mainly proposed to overcome the shortcoming of pure 1D coupling. Indeed, the conditions used up to this point are typically limited to the subcritical regime and even to relatively slow flows ($Fr < 0.35$) [17]. Some strategies have been devised to solve junction problems in the supercritical regime, but only for some specific geometries and a general formulation was missing [18–20].

Particularly, reference [1] argued in favor of local 2D elements to represents junctions in 1D networks, showing that a 1D method based on Sherwin's method for blood vessels [21] could not produce accurate results, and would not run altogether in supercritical conditions. In this paper, we showcase a methodology also first used to solve blood vessels [9,10] that is able to solve the junction problem in the supercritical regime.

The proposed strategy in this paper makes use of the Junction Riemann Problem [22,23], a generalization of the Riemann problem to a junction of channels, each initially in a uniform state. The Riemann problem is a well-known [24] initial value problem (IVP) in the theory of hyperbolic systems of PDEs that is the foundation of popular methods [25] to solve them. While these methods often make use of approximate solutions, the coupling of the terminal cells of the channel seeks an exact solution, whose existence has been proven for some configurations [23,26] in the subcritical regime.

The method presented here is coupled to an energy-balanced finite volume (FV) scheme of first-order accuracy that makes use of the Augmented Riemann method to solve for the inter-cell fluxes [27]. It is worth noting that, since solving the Junction Riemann Problem yields the exact solution at the edge of the terminal cell, it can be used in combination with other schemes, even of higher order. The properties of these types of methods are explained in detail in [28], including the complete derivation of an approximate Riemann solver for general hyperbolic systems of equations with non-conservative terms and its application to the shallow water equations in both 1D and 2D.

This paper does not delve into the mathematical theory of the problem, but rather shows its practical application, including supercritical flow conditions. A more detailed explanation is presented in [29], where the different possible combinations of flow regimes and flow orientations are analyzed following the groundwork laid out in [9] for blood vessels. The novelty of this paper is the validation against reference solutions produced in 2D domains through a well-established numerical method [30] that is currently used in practical cases [31]. The emphasis in the test cases is put on the transitory flow conditions by looking at how shock waves propagate through junctions. Thus, we are interested in the accuracy of the method in capturing the long-term behavior of the flow in the network, and not in the details at the junction. It is clear that a 1D method will never be able to capture the complex flow patterns that arise in 2D domains, but it is expected that the method is able to reproduce the main features of the flow to the accuracy of the numerical method.

This paper is organized as follows: The mathematical models and their main equations are given in the next section, followed by a description of the numerical methods used to solved them. Both the 1D model of interest and the 2D model used to validate it are presented, with a special emphasis on the former. The Junction Riemann Problem is also introduced in Section 3. Section 4 contains a description of the test cases and an analysis of the results obtained. Finally, the conclusions are drawn in Section 5.

## 2. Mathematical Models

### 2.1. One-Dimensional Shallow Water Equations

The one-dimensional shallow water equations are given by a system of partial differential equations derived from mass and momentum conservation laws. Assuming a rectangular channel of constant width $B$, the equations in conservative form are given by

$$\frac{\partial \mathbf{U}(x,t)}{\partial t} + \frac{\partial \mathbf{F}(x,t)}{\partial x} = \mathbf{S}(x,\mathbf{U}). \tag{1}$$

The vector of conserved variables **U** and the vector of fluxes both depend on space and time, and are given by

$$\mathbf{U} = \begin{pmatrix} A \\ Q \end{pmatrix}, \qquad \mathbf{F} = \begin{pmatrix} Q \\ \frac{Q^2}{A} + \frac{1}{2}gh^2 \end{pmatrix}, \tag{2}$$

where $A$ is the cross-sectional wetted area, $Q$ is the total discharge, $h$ is the water depth, computed as $h = A/B$, and $g$ is the acceleration due to gravity. The source term accounts for mass and momentum sources and sinks, of which we will only consider the latter, and is given by

$$\mathbf{S} = \begin{pmatrix} 0 \\ \mathcal{T}(\mathbf{U}) \end{pmatrix} = \begin{pmatrix} 0 \\ gA\left(S_o - S_f\right) \end{pmatrix}, \tag{3}$$

where $S_o = -\partial z_b/\partial x$ is the bed slope term and $S_f$ is the friction slope term. In this case, we will assume frictionless channels to obtain a better understanding of the behavior of the method hereby presented.

### 2.2. Two-Dimensional Shallow Water Equations

The two-dimensional shallow water equations are given by a similar system of partial differential equations:

$$\frac{\partial \mathbf{U}(x,y,t)}{\partial t} + \frac{\partial \mathbf{F}(x,y,t)}{\partial x} + \frac{\partial \mathbf{G}(x,y,t)}{\partial y} = \mathbf{S}(x,y,\mathbf{U}). \tag{4}$$

The vector of conserved variables **U** and the vector of fluxes depend on space and time, and are given by

$$\mathbf{U} = \begin{pmatrix} h \\ hu \\ hv \end{pmatrix}, \qquad \mathbf{F} = \begin{pmatrix} hu \\ hu^2 + \frac{1}{2}gh^2 \\ huv \end{pmatrix}, \qquad \mathbf{G} = \begin{pmatrix} hv \\ huv \\ hv^2 + \frac{1}{2}gh^2 \end{pmatrix}, \tag{5}$$

where the velocity vector is split in two components: $u$ along the $x$ direction and $v$ along the $y$ direction. The source term accounts for sinks and sources of mass and momentum, of which only the latter are present:

$$\mathbf{S} = \begin{pmatrix} 0 \\ \mathcal{T}_x(\mathbf{U}) \\ \mathcal{T}_y(\mathbf{U}) \end{pmatrix} = \begin{pmatrix} 0 \\ gh\left(S_{ox} - S_{fx}\right) \\ gh\left(S_{oy} - S_{fy}\right) \end{pmatrix}, \tag{6}$$

where the bed slope term is calculated in each direction

$$S_{ox} = -\frac{\partial z_b}{\partial x}, \qquad S_{oy} = -\frac{\partial z_b}{\partial y}. \tag{7}$$

Again, no friction is considered in this case.

## 3. Numerical Methods

The numerical model used in both domains is a first-order finite volume scheme. In general, the system of differential equations can be written in integral form over an arbitrary control volume $\Omega$

$$\frac{\partial}{\partial t}\int_\Omega \mathbf{U} d\Omega + \int_\Omega \left(\frac{\partial \mathbf{F}}{\partial x} + \frac{\partial \mathbf{G}}{\partial y}\right) d\Omega = \int_\Omega \mathbf{S} d\Omega, \tag{8}$$

and by using the Gauss–Ostrogrdsky theorem

$$\frac{\partial}{\partial t}\int_{\Omega}\mathbf{U}d\Omega + \oint_{\Omega}(\mathbf{F}n_x + \mathbf{G}n_y)d\Omega = \int_{\Omega}\mathbf{S}d\Omega, \tag{9}$$

where $n_x$ and $n_y$ are Cartesian components of the unitary outward pointing vector $\mathbf{n}_{\Omega}$ of the surface surrounding cell $\Omega$.

## 3.1. One-Dimensional Numerical Method

Equation (9) can be particularized in the case of a one-dimensional domain where a single cell $i$ is defined by its left and right walls $\Omega = [x_L, x_R]$. By also choosing an arbitrary discretization of time in time-steps $\Delta t = t^{n+1} - t^n$, an average value of vector $\mathbf{U}$ in the cell can be found:

$$\mathbf{U}_i^{n+1} = (x_R\mathbf{U}(x_R, t^n) - x_L\mathbf{U}(x_L, t^n)) - (\mathbf{F}(x_R, t) - \mathbf{F}(x_L, t))\Delta t + \int_{t^n}^{t^{n+1}}\int_{x_L}^{x_R}\mathbf{S}dxdt. \tag{10}$$

It is important to note that integration, or linearization, of the source term can be carried out in different manners. In this case, in accordance with the previous literature [27] the bed level source term is computed at the inter-cell walls and introduced into the numerical flux.

$$\hat{\mathbf{U}}_i^{n+1} = \mathbf{U}_i^n - \frac{\Delta t}{\Delta x}\left(\mathbf{F}_{i+\frac{1}{2}}^{-} - \mathbf{F}_{i-\frac{1}{2}}^{+}\right) \tag{11}$$

where $\mathbf{F}_{i+\frac{1}{2}}^{\pm}$ represents the numerical flux at the inter-cell wall labeled $i + \frac{1}{2}$ between cells $i$ and $i + 1$, which includes the contribution of the source term.

Thus, to compute the conserved variables at the next step, the source term needs to be discretized, and the fluxes need to be found at the cell walls. First, ignoring the linearization of the source term, Equation (1) can be written in non-conservative form

$$\frac{\partial\mathbf{U}(x,t)}{\partial t} + \mathbf{J}(\mathbf{U})\frac{d\mathbf{U}(x,t)}{dx} = \mathbf{S}(x, \mathbf{U}), \tag{12}$$

through the Jacobian matrix $\mathbf{J}(\mathbf{U})$:

$$\mathbf{J}(\mathbf{U}) = \begin{pmatrix} 0 & 1 \\ c^2 - u^2 & 2u \end{pmatrix}, \tag{13}$$

where $u = Q/A$ is the flow velocity in the $x$ direction and $c$ is the celerity of the surface waves, defined as $c = \sqrt{gh}$. The hyperbolic property of the convective part is confirmed by the two real eigenvalues:

$$\lambda_1 = u - c, \qquad \lambda_2 = u + c. \tag{14}$$

If $\lambda_1$ and $\lambda_2$ have different signs, the regime is said to be subcritical, as the Froude number fulfills $\mathrm{Fr} = u/c < 1$. If $\lambda_1$ and $\lambda_2$ are both positive or negative, then the regime is said to be supercritical and $\mathrm{Fr} > 1$. The point at which $u = c$ and $\mathrm{Fr} = 1$ is the critical point and is very relevant to define flow transitions and hydraulic jumps.

Additionally, eigenvectors $\mathbf{e}^k$ can be constructed from the eigenvalues:

$$\mathbf{e}^1 = \begin{pmatrix} 1 \\ \lambda_1 \end{pmatrix}, \qquad \mathbf{e}^2 = \begin{pmatrix} 1 \\ \lambda_2 \end{pmatrix}, \tag{15}$$

which can then be organized into the matrix $\mathbf{P} = (\mathbf{e}^1, \mathbf{e}^2)$ to diagonalize the Jacobian. In this way, the Roe scheme is used in order to compute the fluxes at the walls in spite of the non-linearity of the Jacobian matrix by expressing the difference of the conserved variables across the cell walls as the sum of waves associated with the averaged eigenvectors:

$$\delta\mathbf{U}_{i+\frac{1}{2}} = \sum_{m=1}^{2} (\tilde{\alpha}_m \mathbf{e}_m)_{i+\frac{1}{2}} \tag{16}$$

while the inter-cell numerical fluxes can be written as

$$\mathbf{F}^-_{i+\frac{1}{2}} = \mathbf{F}_{i+1} - \sum_{m\,|\,\tilde{\lambda}_m<0} (\tilde{\gamma}_m \mathbf{e}_m)_{i+\frac{1}{2}}, \qquad \mathbf{F}^+_{i-\frac{1}{2}} = \mathbf{F}_i + \sum_{m\,|\,\tilde{\lambda}_m>0} (\tilde{\gamma}_m \mathbf{e}_m)_{i-\frac{1}{2}}, \tag{17}$$

where $\gamma$ summarizes the contribution of each wave to the fluxes and the tilde signifies the averaged value at the cell wall:

$$\tilde{\gamma}_{m,i+\frac{1}{2}} = (\tilde{\lambda}\tilde{\alpha} - \tilde{\beta})_{m,i+\frac{1}{2}}. \tag{18}$$

According to the literature [27], these averaged values are chosen to be

$$\tilde{\lambda} = \tilde{u} \pm \tilde{c}, \qquad \tilde{c}_{i+\frac{1}{2}} = \sqrt{g\left(\frac{A_i}{B_i} + \frac{A_{i+1}}{B_{i+1}}\right)}, \qquad \tilde{u}_{i+\frac{1}{2}} = \frac{\sqrt{A_i}u_i + \sqrt{A_{i+1}}u_{i+1}}{\sqrt{A_i} + \sqrt{A_{i+1}}} \tag{19}$$

to compute

$$\tilde{\alpha}_{1,i+\frac{1}{2}} = \frac{\tilde{\lambda}_2 \delta A - \delta Q}{2\tilde{c}}, \qquad \tilde{\alpha}_{2,i+\frac{1}{2}} = \frac{-\tilde{\lambda}_1 \delta A + \delta Q}{2\tilde{c}} \tag{20}$$

and the source term contribution:

$$\tilde{\beta}_{1,i+\frac{1}{2}} = -\tilde{\beta}_{2,i+\frac{1}{2}} = \left(\frac{\tilde{S}}{2\tilde{c}}\right)_{i+\frac{1}{2}}. \tag{21}$$

It is convenient to express the effect of inter-cell fluxes using fluctuations $\mathbf{\Delta M}^{\pm}_{i+\frac{1}{2}}$:

$$\mathbf{\Delta M}^-_{i+\frac{1}{2}} = \left(\sum_{m\,|\,\tilde{\lambda}_m<0} \tilde{\gamma}_m \mathbf{e}_m\right)_{i+\frac{1}{2}}, \qquad \mathbf{\Delta M}^+_{i-\frac{1}{2}} = \left(\sum_{m\,|\,\tilde{\lambda}_m>0} \tilde{\gamma}_m \mathbf{e}_m\right)_{i-\frac{1}{2}} \tag{22}$$

$$\hat{\mathbf{U}}^{n+1}_i - \mathbf{U}^n_i = \frac{\Delta t}{\Delta x}\left(\mathbf{\Delta M}^-_{i+\frac{1}{2}} + \mathbf{\Delta M}^+_{i-\frac{1}{2}}\right). \tag{23}$$

Integration of the bed slope term is carried out ensuring the conservation of hydraulic energy in most situations by averaging two different formulations of the discretization [27]. If smooth variation of the variables is assumed, the thrust exerted by changes in the bed can be written in terms of hydrostatic pressure:

$$(S^a_o)_{i+\frac{1}{2}} = -g(\tilde{A}\delta(h+z))_{i+\frac{1}{2}} + \left(\tilde{c}^2 \delta A\right)_{i+\frac{1}{2}}, \tag{24}$$

where $\tilde{A}_{i+\frac{1}{2}}$ and $\tilde{c}_{i+\frac{1}{2}}$ are the Roe-averaged wetted area and wave celerity of the two neighboring cells and $\delta d = (z+h)_{i+1} - (z+h)_i$ is the difference in level of each cell. A suitable alternative if the bed level function includes discontinuities is the integral formulation, obtained by integrating the thrust along the $z$ axis taking into account the relative height of water level and bed at each side of the discontinuity.

$$\left(S^b_o\right)_{i+\frac{1}{2}} = -gA_j\delta(h+z) + \tilde{c}^2\delta A, \tag{25}$$

where index $j$ depends upon the orientation of the discontinuity:

$$A_j = \begin{cases} A_i & \text{if } A_i \leq A_{i+1} \\ A_{i+1} & \text{if } A_i > A_{i+1} \end{cases}. \tag{26}$$

Hence, the energy-conserving source term discretization is formulated as a weighted average:

$$S_o^c = (1 - \mathcal{A})S_o^a + \mathcal{A}S_o^b, \tag{27}$$

where the parameter $\mathcal{A}$ is found assuming conservation of the hydraulic energy

$$H = h + z + \frac{u^2}{2g} \tag{28}$$

and can be written as

$$\mathcal{A}_{i+\frac{1}{2}} = \left( \frac{\delta(Q^2/A) - \frac{1}{2}\tilde{A}\delta(Q^2/A^2)}{S_o^b - S_o^a} \right)_{i+\frac{1}{2}}, \tag{29}$$

and is typically bounded by $0 \leq \mathcal{A} \leq 1$. If friction is not considered, it is a well-known result [24] that energy is conserved in most stationary cases, except when there is a hydraulic jump. This structure, comprised of static discontinuity in water surface elevation, appears when a supercritical flow is forcibly decelerated by a downstream condition. Hence, $\mathcal{A}_{i+\frac{1}{2}} = 1$ is locally imposed when $\text{Fr}_i > 1$ and $\text{Fr}_{i+1} < 1$, assuming positive flow velocity.

The time-step $\Delta t$ in Equations (11) and (23) is variable at each iteration, limited by cell size and the approximate velocity of the waves $\tilde{\lambda}$:

$$\Delta t = \text{CFL} \min_{i,m} \left\{ \frac{\Delta x}{|\tilde{\lambda}_m|_i} \right\} \tag{30}$$

where CFL is the Courant–Friedrich–Lewy number that for this scheme must be less than 1 in order to ensure stability.

### 3.2. Junction Riemann Problem

The point where 1D channels meet is a junction or confluence, and the cells adjacent to it, the terminal cells, must be solved in a special way. The method proposed in this paper involves treating the junction as an internal boundary condition between the terminal cells, finding the inter-cell fluxes in an alternative way. If an arbitrary junction connects $N_c$ channels, there will be $N_c$ terminal cells, which can be separated into *parents* and *children* according to their relative position to the junction with respect the absolute $x$-axis. Hence, the orientation of channel $k$, comprising $N_k$ total cells, will be given by its parameter $\eta_k$, defined as follows:

$$\eta_k = \hat{\mathbf{n}}_k \cdot \hat{\mathbf{x}} = \begin{cases} 1 & \text{if} \quad i_k = N_k \\ -1 & \text{if} \quad i_k = 1 \end{cases}, \tag{31}$$

where $\hat{\mathbf{n}}_k$ is the unit vector pointing from the channel towards the junction and $\hat{\mathbf{x}}_k$ is the unit vector in the $x$-axis. Thus, it is possible to distinguish when flow is entering ($\eta_k u_k^n < 0$) or leaving ($\eta_k u_k^n > 0$) the channel regardless of whether the terminal cell is located at the beginning or end of the channel. Additionally, the channel Speed Index (SI) is defined as a signed extension to the Froude number

$$\text{SI}_{gk}^n = \eta_k \text{Fr}_k^n, \tag{32}$$

so it contains information on both the flow regime and its direction. The actualization of terminal cell $i_k$ of channel $k$ is a modification of Equation (51):

$$\mathbf{U}_{i_k}^{n+1} = \begin{cases} \mathbf{U}_{i_k}^n - \frac{\Delta t}{\Delta x}\left(\mathbf{F}^-_{i_k+\frac{1}{2}} - \mathbf{F}_k^\star\right) & \text{if} \quad \eta_k = -1 \\ \mathbf{U}_{i_k}^n - \frac{\Delta t}{\Delta x}\left(\mathbf{F}_k^\star - \mathbf{F}^+_{i_k-\frac{1}{2}}\right) & \text{if} \quad \eta_k = 1 \end{cases}, \tag{33}$$

which includes contributions from both its cell walls, of which one is computed regularly through the linearized flux, while the other is the exact flux through the wall associated with the junction. Figure 1 shows a schematic representation of a three-channel junction, with two parents ($\eta_1 = \eta_3 = 1$) and one child ($\eta_2 = -1$). Terminal cells, in a darker tone, must be computed using Equation (33), where the special flux $F_k^\star$ is associated with the junction walls in red. Rather than using an approximation, since a network of channels will contain many more inter-cell walls than junctions, the latter will be solved through an iterative method that calculates the exact values of the solution to the Riemann problem at each junction wall.

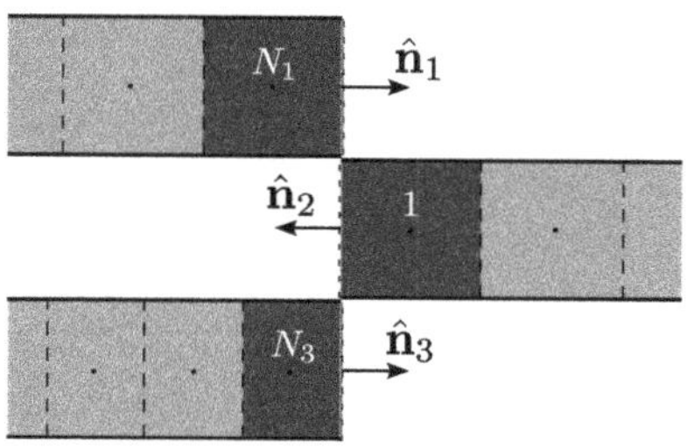


**Figure 1.** Schematic representation of a junction of three channels.

The Junction Riemann Problem is the set of $N_c$ initial value problems (IVPs),

$$\frac{\partial \mathbf{U}_k(x,t)}{\partial t} + \frac{\partial \mathbf{F}_k(x,t)}{\partial x} = \mathbf{S}_k(x, \mathbf{U}_k), \qquad \mathbf{U}_k(x,0) = \mathbf{U}_k^n, \tag{34}$$

where $k = 1, \dots, N_p$, along with $N_c$ coupling conditions:

$$\mathbf{\Phi}(\mathbf{U}_1, \dots, \mathbf{U}_{N_c}) = 0, \tag{35}$$

related to conservation of mass and, in this case, of energy. This yields a non-linear system of $2N_p$ equations that can be used to solve the $2N_p$ unknowns $\{u_1^\star, \dots, u_{N_c}^\star, A_1^\star \dots, A_{N_c}^\star\}$, which can be used to compute the inter-cell fluxes. In the subcritical regime, this system takes the following form:

$$\begin{cases} u_1^\star - u_1^n + \eta_1 \mathcal{B}_1 & = 0 \\ \vdots & \\ u_N^\star - u_N^n + \eta_N \mathcal{B}_N & = 0 \\ \sum_k (\eta_k u_k^\star A_k^\star) & = 0 \\ H_1^\star - H_2^\star & = 0 \\ \vdots & \\ H_1^\star - H_N^\star & = 0 \end{cases}, \tag{36}$$

where $\mathcal{B}_k$ is an auxiliary function that solves the rarefaction or shock wave produced as a solution to the Riemann problem and can be defined as follows:

$$\mathcal{B}_k(x) = \begin{cases} 2\sqrt{\frac{g}{B_k}(x - A_k^n)} & \text{if} \quad x \le A_k^n \\ (x - A_k^n)\sqrt{\frac{g}{2B_k}\left(\frac{x + A_k^n}{A_k^n x}\right)} & \text{if} \quad x > A_k^n \end{cases}. \tag{37}$$

Additionally, the $N_c$-th equation ensures the conservation of mass in the junction and equations $N_c + 1$ to $2N_c$ represent equality of the hydraulic head energy as defined in Equation (28) in all channels, taking channel 1 as the default reference. This does not mean that the energy of channel 1 is fixed, as $(u_1^\star, A_1^\star)$ can vary. The choice of energy conservation might seem controversial in a junction, where wall and angle effects seem important, and it certainly is in cases such as artificial ditches for irrigation or laboratory-scale channels. However, when the concern is on the large-scale wave propagation or transport phenomena in the whole network, or if walls of the junction have a reduced effect, as in river confluences, energy conservation is acceptable. Nevertheless, there is ample literature presenting other models where loss of energy and momentum due the geometry of the junction is taken into account.

System (36) can be solved by any iterative method suitable for a non-linear system of equations. An example is presented in [9,29] using Picard iterations, where a tentative solution $\left(u_k^\star, A_k^\star\right)^m$ is taken to the next iteration $m+1$ by means of inverting the Jacobian of the error vector.

This methodology is well known has been used in the past to solve networks of channels with one important limitation: the method fails for supercritical flows and even for relatively fast subcritcal flows (Fr $> 0.35$). This is due to the fact that the solution to each Riemann problem is performed under the assumption that only one wave is propagating into the channel domain from the junction, which is not true in the subcritical regime. Indeed, an out-flowing channel in the supercritical regime (SI $> 0$) does not fall under this assumption, given that the two waves forming the solution of the RP are at the exterior side of the wall.

In order to extend this method to all regimes, a series of modifications must be introduced, the details of which can be found in [29]. These take the form of a set of coefficients $\{\mathcal{D}_k, \mathcal{R}_k, \mathcal{T}_k, \mathcal{L}_k, \mathcal{Q}_k\}$ that take values 1 or 0 at each step $m$ of the iterative process. The coefficient $\mathcal{D}_k$ is defined as

$$\mathcal{D}_k{}^m = \begin{cases} 0 & \text{if} \quad \mathrm{SI}^n \geq 1 \text{ and, } \eta_k\left(Q_k^\star - Q_k^n\right) > 0 \text{ or } \left(A_k^\star - A_k^n\right) < 0 \\ 0 & \text{if} \quad \mathrm{SI}^n \geq 1 \text{ and } -\eta_k \mathcal{S}_k^{\star,m} < 0 \\ 1 & \text{otherwise} \end{cases}, \tag{38}$$

where $\mathcal{S}_k^{\star,m}$ is the speed of a tentative shock-type solution at iterative step $m$ at channel $k$, and is computed as follows:

$$\mathcal{S}_k^{\star,m} = \left(\frac{Q_k^\star - Q_k^n}{A_k^\star - A_k^n}\right)^m. \tag{39}$$

Then, $\mathcal{D}_k = 0$ occurs only in cases where the cell is initially in the supercritical regime and is tentatively being accelerated further, or if the tentative solution is a shock wave moving towards the junction.

Coefficients $\mathcal{T}_k$ and $\mathcal{R}_k$ are used to ensure that a rarefaction-type solution cannot be accelerated further towards a junction,

$$\mathcal{R}_k{}^m = \begin{cases} 1 & \text{if} \quad \mathrm{SI}^{\star,m} < 1 \text{ and } \mathrm{SI}^{\star,m+1} > 1 \text{ and } \left(A_k^{\star,m} - A_k^n\right) < 0 \\ 0 & \text{otherwise} \end{cases}, \tag{40}$$

and

$$\mathcal{T}_k{}^m = 1 - \mathcal{R}_k{}^m. \tag{41}$$

In this way, $\mathcal{R}_k$ takes a value of 1 right at the iteration when a tentative rarefaction-type solution $(A_k^{\star,m} - A_k^n)$ would enter the supercritical regime. Otherwise, $\mathcal{R}_k = 0$ and then $\mathcal{T}_k = 1$, the normal behavior.

The remaining coefficients, $\mathcal{L}_k$ and $\mathcal{Q}_k$, play their part in ensuring that channels of the junction whose behavior is affected by the previous coefficients cannot take part in energy conservation. If, by chance, the default reference channel for energy continuity is locked by $\mathcal{D}_k = 0$ or $\mathcal{T}_k = 0$, it cannot take part in energy conservation and thus another reference must be found. Hence, $\mathcal{L}_k$ is defined as

$$\mathcal{L}_k = \begin{cases} 1 & \text{if} \quad k = k_{\mathcal{L}} \geq 1 \\ 0 & \quad\text{otherwise} \end{cases}, \qquad k_{\mathcal{L}} = \min\{k \ : \ \mathcal{T}_k\mathcal{D}_k = 1\}, \tag{42}$$

so it takes a value of 1 only for the first channel that is not locked out of energy conservation, and therefore the reference energy can be defined as follows:

$$H^{\star}_{ref} = \sum_k \mathcal{L}_k H^{\star}_k. \tag{43}$$

Once the reference is chosen, every other ineligible channel must be excluded too. This is carried out through the $\mathcal{Q}_k$ coefficient, defined simply as follows:

$$\mathcal{Q}_k = \mathcal{T}_k\mathcal{D}_k(1 - \mathcal{L}_k). \tag{44}$$

By defining these coefficients, system (36) can be rewritten as follows:

$$\begin{cases} [u^{\bullet}_1 - u^n_1 + g_1\mathcal{B}(A^{\star}_1)]\mathcal{D}_1 & +[u^{\star}_1 - u^n_1](1 - \mathcal{D}_1) & = 0 \\ & \vdots & \\ [u^{\bullet}_N - u^n_N + g_N\mathcal{B}(A^{\star}_N)]\mathcal{D}_N & +[u^{\star}_N - u^n_N](1 - \mathcal{D}_N) & = 0 \\ \sum_k \eta_k\Big(\mathcal{D}_k(A^{\star}_k u^{\bullet}_k) & +Q^n_k(1 - \mathcal{D}_k)\Big) & = 0 \\ \left(H^{\star}_{ref} - H^{\star}_2\right)\mathcal{Q}_2 & +\delta A_1(1 - \mathcal{D}_1)\mathcal{L}_2 + \delta A_2(1 - \mathcal{D}_2) & = 0 \\ & \vdots & \\ \left(H^{\star}_{ref} - H^{\star}_N\right)\mathcal{Q}_N & +\delta A_1(1 - \mathcal{D}_1)\mathcal{L}_N + \delta A_N(1 - \mathcal{D}_N) & = 0 \end{cases}, \tag{45}$$

where $u^{\bullet}_k$ allows the velocity to be limited when a critical rarefaction occurs in any channel, and is defined as

$$u^{\bullet}_k = \eta_k\mathcal{R}_k c^{\star}_k + \mathcal{T}_k u^{\star}_k, \tag{46}$$

because SI $= 1$, the critical point, implies that $u = c$. It can be seen that all the equations of the system have two parts. The first deals with all cases where the terminal state is, in principle, not locked by $\mathcal{D}_k = 0$. The second part forces the evolved state to remain unchanged in the cases where the terminal cell cannot be affected due to it fulfilling the conditions of $\mathcal{D}_k = 1$. System (45) is again non-linear, but can be solved iteratively to find the exact solution of the JRP $\{u^{\star}_1, \ldots, u^{\star}_{N_c}, A^{\star}_1 \ldots, A^{\star}_{N_c}\}$ under all circumstances to compute $F^{\star}_k$ simply as

$$\mathbf{F}^{\star}_k = \begin{pmatrix} u^{\star}_k A^{\star}_k \\ u^{\star 2}_k A^{\star}_k + \frac{g A^{\star 2}_k}{2B^2_k} \end{pmatrix}, \tag{47}$$

which in turn completes the actualization of the terminal cell through Equation (33).

### 3.3. Two-Dimensional Numerical Method

The numerical method used to solve the 2-dimensional model given by the hyperbolic system (4) is also a first-order finite volume scheme, as laid out in Equation (9). The basic concepts are similar to the 1D scheme presented in this paper, so it will only be summarized here. Further information can be found in [31]. The Jacobian matrix, defined through the fluxes, is as follows:

$$\mathbf{J} = \frac{\partial \mathbf{F}}{\partial \mathbf{U}} n_x + \frac{\partial \mathbf{G}}{\partial \mathbf{U}} n_y = \begin{pmatrix} 0 & n_x & n_y \\ (c^2 - u^2) n_x - uv n_y & v n_y + 2u n_x & u n_y \\ (c^2 - v^2) n_y - uv n_x & v n_x & u n_x + 2v n_y \end{pmatrix}. \quad (48)$$

Its eigenvalues and eigenvectors can again be found in order to diagonalize the matrix

$$\lambda_1 = \mathbf{u} \cdot \mathbf{n} - c, \qquad \lambda_2 = \mathbf{u} \cdot \mathbf{n}, \qquad \lambda_3 = \mathbf{u} \cdot \mathbf{n} + c, \quad (49)$$

and

$$\mathbf{e}^1 = \begin{pmatrix} 1 \\ u - c n_x \\ v - c n_y \end{pmatrix}, \qquad \mathbf{e}^2 = \begin{pmatrix} 0 \\ -c n_y \\ c n_x \end{pmatrix}, \qquad \mathbf{e}^3 = \begin{pmatrix} 0 \\ u + c n_x \\ v + c n_y \end{pmatrix}. \quad (50)$$

Thus, the expression for the actualization of the 2D cell is analogous to that of the 1D scheme given in Equation (23): each cell $i$ with area $\Omega_i$ is updated according to the contribution of each wave $m$ at each of its $N_w$ walls (three for a triangular cell, four for a rectangular cell, etc.) of length $l_k$:

$$\mathbf{U}_i^{n+1} = \mathbf{U}_i^n - \frac{\Delta t}{\Omega_i} \left( \sum_{k=1}^{N_w} \sum_m (\tilde{\gamma}^m \mathbf{e}^m)_k l_k \right) + \Delta t \begin{pmatrix} 0 \\ \bar{S}_{fx} \\ \bar{S}_{fy} \end{pmatrix}_i, \quad (51)$$

where again $\tilde{\gamma}_k^m$ contains the contributions of the fluxes and source terms:

$$\tilde{\gamma}_k^m = (\tilde{\lambda}\tilde{\alpha} - \tilde{\beta})_k^m. \quad (52)$$

To ensure stability, the global time-step $\Delta t$ is limited, requiring the smallest out of all the meshes to be chosen, defined as:

$$\Delta t = \text{CFL} \min_{k,m} \left\{ \frac{\min\left\{ \frac{\Omega_i}{l_p} \right\}_{p=1,\dots,N_w}}{|\tilde{\lambda}_m^k|} \right\} \quad (53)$$

where, again, CFL $< 1$ must hold.

## 4. Computational Experiments

In this section, the 1D method is tested against a well-established 2D method that is often used to tackle large-scale simulations of free-surface flows through a series of numerical experiments involving transitory flow in junctions. It is worth noting that the reference 2D results have been obtained ad hoc by the authors using energy-balanced methods that can overcome all types of situations in shallow flows, not only under arbitrary geometries, but also independent of the rheological shear stress model selected. Reference [29] presents a series of academic test cases where a specific JRP configurations were solved using the 1D method and compared with the known exact solution. In contrast, the validation here is focused on practical cases.

### 4.1. Three-Channel Junctions

The first batch of experiments deals with a single junction of three channels, in which all of them are parallel. These types of configurations were proposed in [1] to test their 1D method based on the PSP. In principle, this particular geometry is the closest to the idea of the junction as a singular point. Figure 2 presents the configurations used in the tests, throughout which the width of parent channel 1 is kept constant at $B = 1$ m and the widths of children channels 2 and 3 are either $0.5B$ or $B$ depending on the case. Initially, the whole domain is set to the following conditions:

$$h(x,0) = 0.16 \text{ m}, \qquad hu(x,0) = 0 \text{ m}^2/\text{s}. \tag{54}$$

A discharge over time boundary condition is imposed on the parent channel, given by a Gaussian function

$$q(t) = 0.4e^{-0.5(t-3)^2}, \tag{55}$$

that evolves into a shock as it travels along the parent channel. A free boundary condition is imposed at the end of the children channels, but they are made sufficiently long as to avoid any effect of the outlets. Measurements are taken at positions $P1$, $P2$, and $P3$, all of them 2 m from the junction. In 1D, the measurement is taken directly at the cell. In 2D, however, the average value in the transversal direction is taken to smooth out local fluctuations.

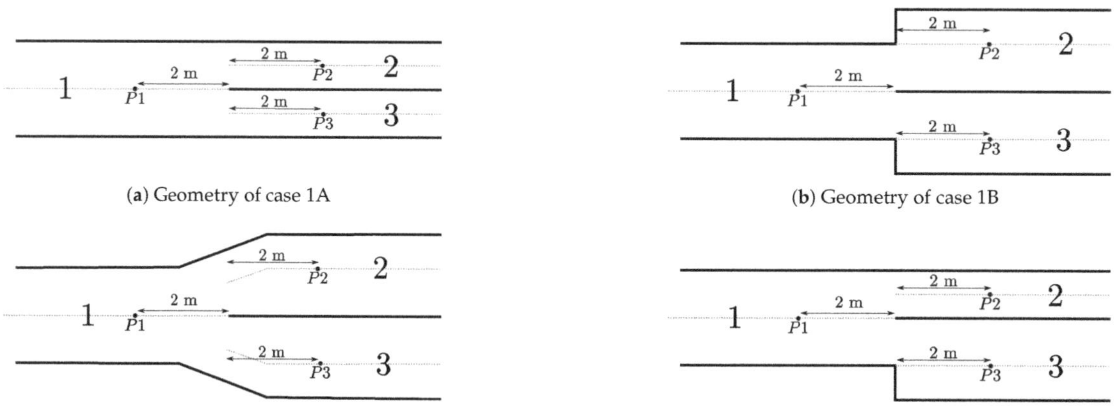


(**a**) Geometry of case 1A

(**b**) Geometry of case 1B

(**c**) Geometry of case 1C

(**d**) Geometry of case 1D

**Figure 2.** Geometries of 3-channel junctions tested.

Case 1A is the simplest case, where a parent channel seamlessly divides into two children channels of half the width. It is expected that the traveling wave will propagate without any change in shape. The time evolution of depth and the Froude number are pictured in Figure 3, where it can be seen that the coincidence is very good. The Froude numbers reach a peak value of 0.6 approximately, still in the subcritical regime but still relatively high for current methods used to solve junctions.

A snapshot of the channels at $t = 8$ s is shown in Figure 4, when the wave has already passed through the junction. Here, the discrepancies between the methods can be perceived more starkly, with the 1D solution producing a slower-moving wave.

This is due to the difference in propagation in a 1D discretized channel or in a 2D triangular unstructured mesh in a first-order method. To prove that the differences are not introduced by the junction, Figure 5 shows the propagation of the wave in the parent channel before it has reached the confluence. Here, the deformation can already be observed. The shock wave is not as sharp as in the 2D reference solution due to the unstructured mesh. As it propagates down the children channels, the difference in wave speed becomes more pronounced.

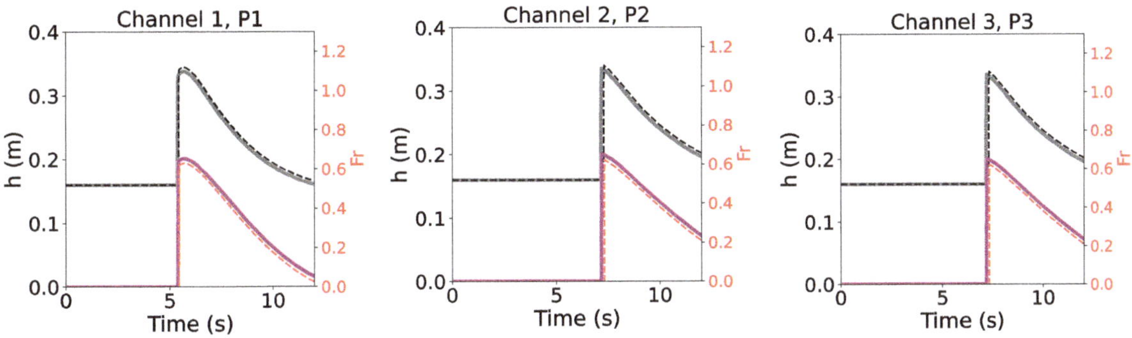


(**a**) Probe 1 (**b**) Probe 2 (**c**) Probe 3

**Figure 3.** Case 1A. Time evolution at probes *P*1, *P*2, and *P*3. The solid line (–) marks the reference 2D solution, while the dashed line (– –) marks the 1D solution. Depth (black) is pictured with reference to the left y-axis while the Froude (red) number is pictured with reference to the right y-axis.

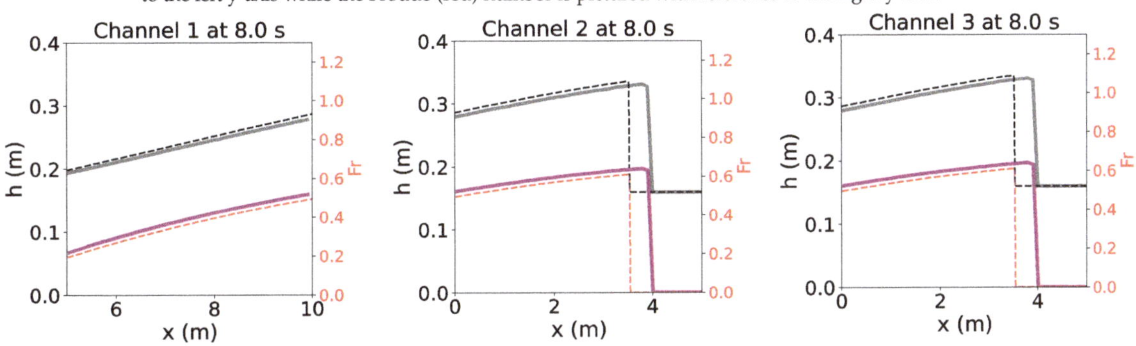


(**a**) Channel 1 (**b**) Channel 2 (**c**) Channel 3

**Figure 4.** Case 1A. State of the channels at $t = 8$ s. The solid line (–) marks the reference 2D solution, while the dashed line (– –) marks the 1D solution. Depth (black) is pictured with reference to the left y-axis while the Froude (red) number is pictured with reference to the right y-axis.

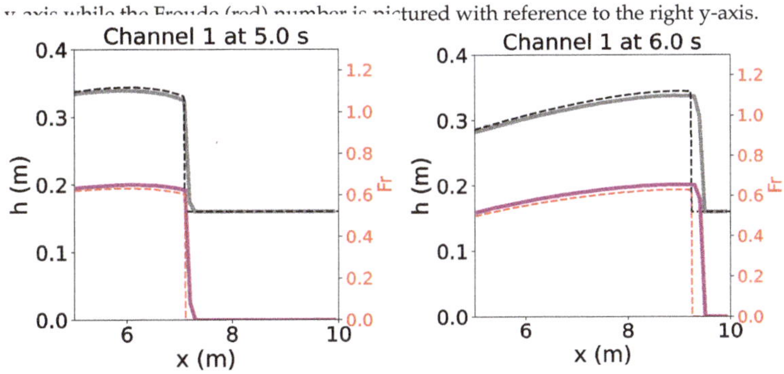


(**a**) Channel 1 at $t = 5$ s. (**b**) Channel 1 at $t = 6$ s.

**Figure 5.** Case 1A. (**a**): State of channel 1 at $t = 5$ s. (**b**): State of channel 1 at $t = 6$ s. The solid line (–) marks the reference 2D solution, while the dashed line (– –) marks the 1D solution. Depth (black) is pictured with reference to the left y-axis while the Froude (red) number is pictured with reference to the right y-axis.

Case 1B includes a sudden expansion of the parent channel into two children channels of the same width right at the junction. By the effect of the width source term, the parent channel will be accelerated, and the wave will be deformed. Again, the time evolution of depth and Froude number at the selected probes is shown in Figure 6. It can be observed that channel 1 is accelerated by the effect of the sudden expansion and a rarefaction propagates upstream, and this behavior is reproduced correctly in 1D. Downstream, the propagation of the shock wave is also well reproduced, albeit with slight delays in its arrival and in its maximum water surface elevation. Figure 7 provides further insight, as it pictures a snapshot of the channels at $t = 8$ s. Upstream, the parent channel is accelerated towards the junction, reaching the critical point exactly there. By the restrictions applied to the method, no further acceleration is allowed, as once the supercritical regime is reached, no perturbations can travel upstream. The reference solution in 2D is also accelerated towards the junction, but it does not quite reach the critical point there. Instead, its Froude number grows continuously until reaching a maximum in the children channels.

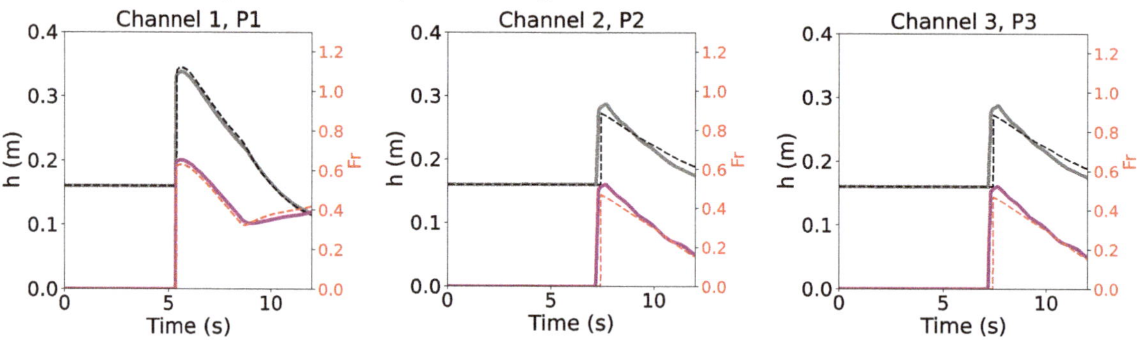


(**a**) Probe 1 (**b**) Probe 2 (**c**) Probe 3

**Figure 6.** Case 1B. Time evolution at probes *P*1, *P*2, and *P*3. The solid line (−) marks the reference 2D solution, while the dashed line (− −) marks the 1D solution. Depth (black) is pictured with reference to the left y-axis while the Froude (red) number is pictured with reference to the right y-axis.

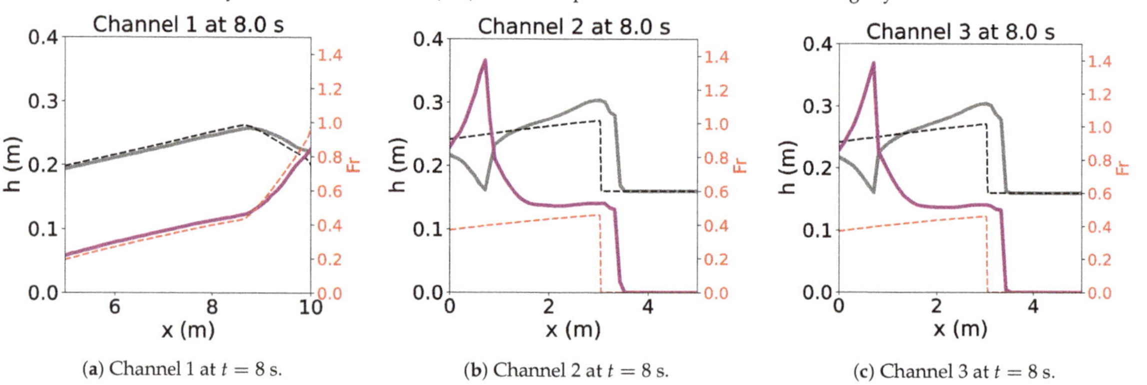


(**a**) Channel 1 at $t = 8$ s. (**b**) Channel 2 at $t = 8$ s. (**c**) Channel 3 at $t = 8$ s.

**Figure 7.** Case 1B. State of the channels at $t = 8$ s. The solid line (−) marks the reference 2D solution, while the dashed line (− −) marks the 1D solution. Depth (black) is pictured with reference to the left y-axis while the Froude (red) number is pictured with reference to the right y-axis.

Case 1C also considers a widening of the channel, but this time the expansion is linear, with the junction placed right at its middle. As a consequence, the acceleration to the supercritical regime occurs at the parent channel, as can be observed in Figures 8 and 9. The dynamics of the waves are correctly reproduced in 1D, both the rarefaction traveling upstream and the shockwave traveling downstream, albeit with the small delay that has been constantly observed. A noteworthy detail is the small bump observed in channels 2 and 3, not present in the 2D reference solution. This is a direct result of the supercritical-to-subcritical transition occurring upstream. To understand this issue, it is worth taking a look at the snapshot of the channels at $t = 11$ s, as pictured in Figure 10 with special attention at the parent channel in the vicinity of the junction.

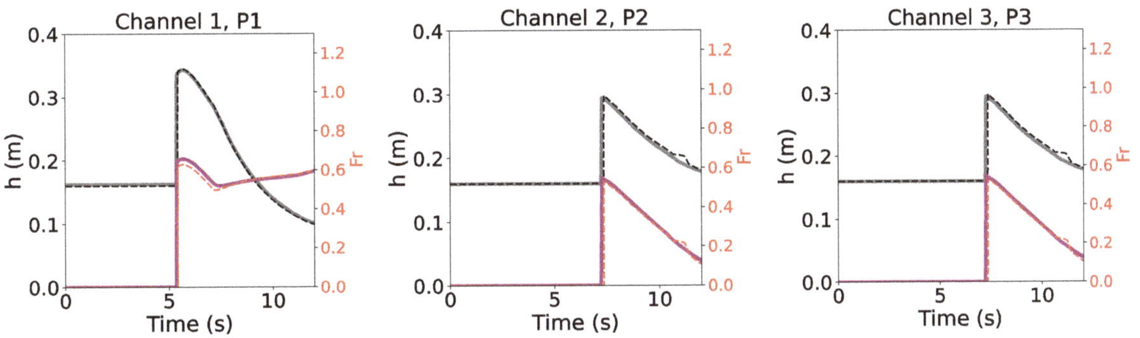


(a) Probe 1 (b) Probe 2 (c) Probe 3

**Figure 8.** Case 1C. Time evolution at probes $P1$, $P2$, and $P3$. The solid line (–) marks the reference 2D solution, while the dashed line (− −) marks the 1D solution. Depth (black) is pictured with reference to the left y-axis while the Froude (red) number is pictured with reference to the right y-axis.

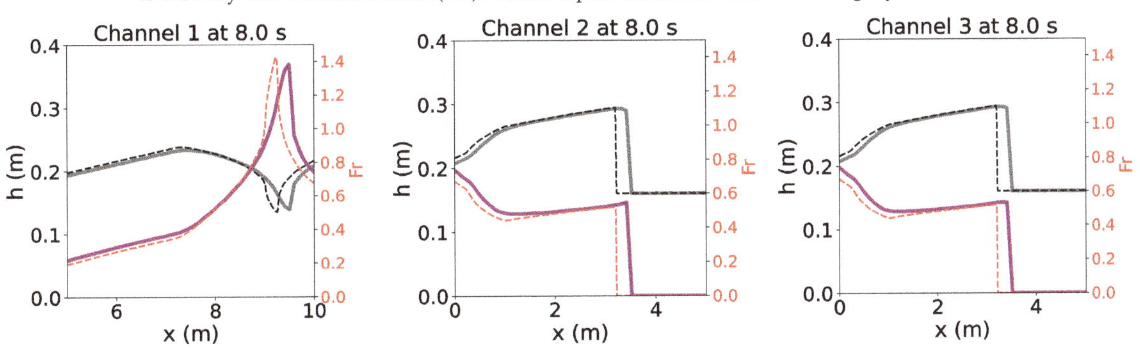


(a) Channel 1 at $t = 8$ s. (b) Channel 2 at $t = 8$ s. (c) Channel 3 at $t = 8$ s.

**Figure 9.** Case 1C. State of the channels at $t = 8$ s. The solid line (–) marks the reference 2D solution, while the dashed line (− −) marks the 1D solution. Depth (black) is pictured with reference to the left y-axis while the Froude (red) number is pictured with reference to the right y-axis.

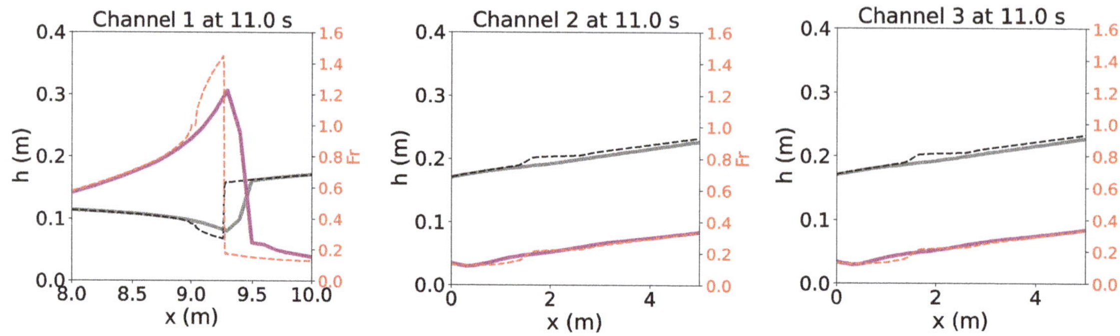


(a) Channel 1 at $t = 11$ s.
(b) Channel 2 at $t = 11$ s.
(c) Channel 3 at $t = 11$ s.

**Figure 10.** Case 1C. State of the channels at $t = 11$ s. The solid line (−) marks the reference 2D solution, while the dashed line (−−) marks the 1D solution. Depth (black) is pictured with reference to the left y-axis while the Froude (red) number is pictured with reference to the right y-axis.

While it is obvious that the 1D method is unable to exactly reproduce the shape of the 2D reference solution, it is remarkable that it reproduces the hydraulic jump. This is due to the JRP forcing a subcritical solution at the end of the supercritical 1D parent channel. The only possible solution to this configuration is the hydraulic jump pictured in the figure. The bump, or spike, appearing downstream is a known consequence of representing this sharp discontinuities in a finite-volume grid [32]. Regardless of the small bump, the fact that the 1D method can reproduce this feature is a remarkable result that has not been achieved by other methods, as it is a product of combining supercritical and subcritical regimes.

Case 1D is included to test the ability of the method to reproduce an asymmetrical junction. Again, the features of the reference 2D solution are reproduced by the 1D method, as observed in Figures 11 and 12, including the acceleration of the parent channel and its subsequent deceleration. Likewise, the two children channels behave appropriately, with differences more apparent in Figure 12.

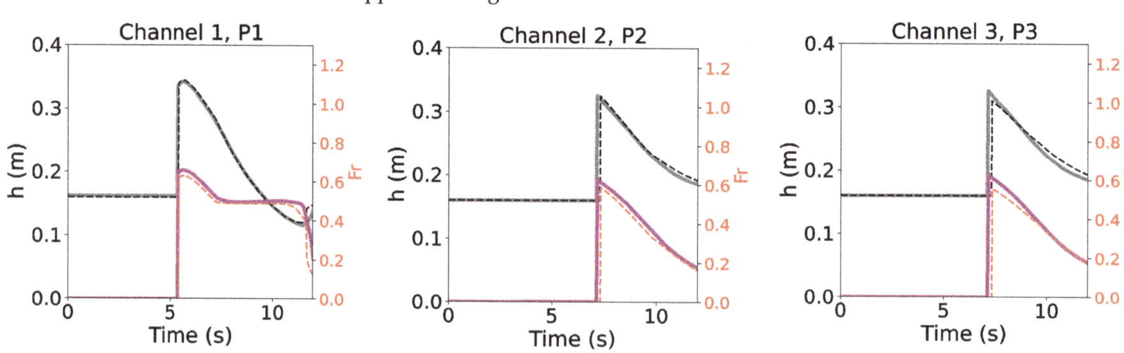


(a) Probe 1.
(b) Probe 2.
(c) Probe 3.

**Figure 11.** Case 1D. Time evolution at probes $P1$, $P2$, and $P3$. The solid line (−) marks the reference 2D solution, while the dashed line (−−) marks the 1D solution. Depth (black) is pictured with reference to the left y-axis while the Froude (red) number is pictured with reference to the right y-axis.

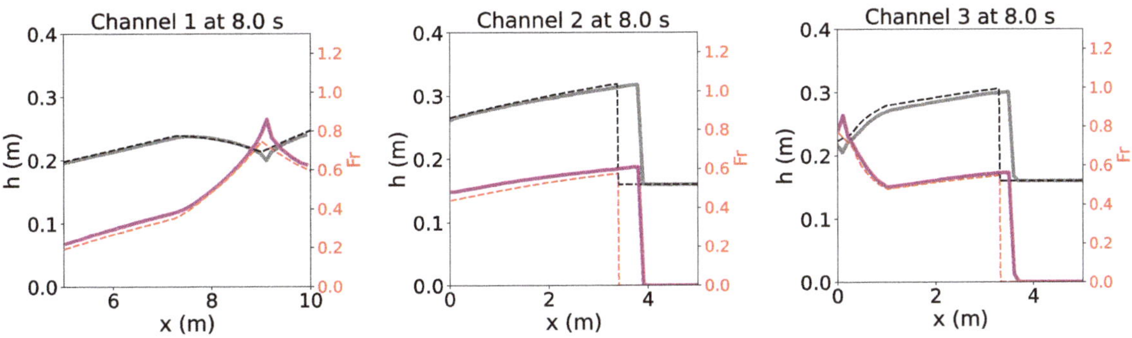


(**a**) Channel 1 at $t = 8$ s. (**b**) Channel 2 at $t = 8$ s. (**c**) Channel 3 at $t = 8$ s.

**Figure 12.** Case 1D. State of the channels at $t = 8$ s. The solid line (−) marks the reference 2D solution, while the dashed line (− −) marks the 1D solution. Depth (black) is pictured with reference to the left y-axis while the Froude (red) number is pictured with reference to the right y-axis.

The results with this basic junction configuration are promising. First, the method is robust and able.

### 4.2. Large Network

In order to test the capabilities of the model in a complex case, a large network of 24 channels and 16 junctions has been devised, inspired by a similar structure in [1]. The layout of the network in 2D is shown in Figure 13, with a detail of the fine meshing. The whole structure is 500 m long and all channels have a constant width of $B = 5$ m. There is a single inlet at channel 1, with a discharge over time boundary condition, and a single outlet at channel 24, with a free outflow boundary condition. In any case, channel 24 has been made long enough to avoid any effect of the outlet propagating back into the domain.

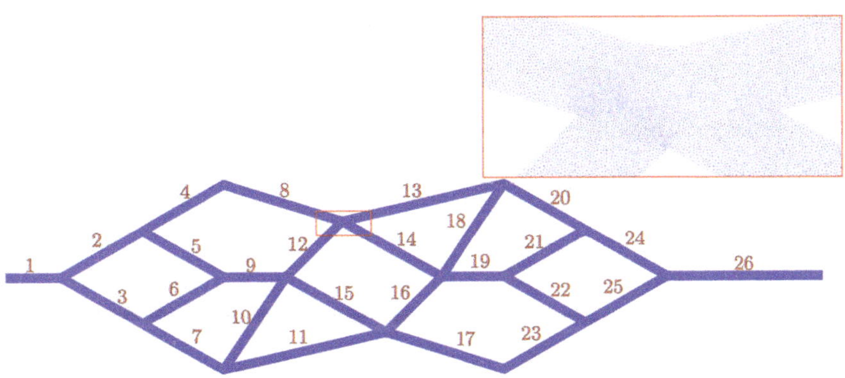


**Figure 13.** Two-dimensional mesh of the large network, with an example of the fine unstructured triangular cells used.

The boundary condition at channel 1 is given by a Gaussian function

$$q(t) = 4e^{-0.5(t-3)^2}, \tag{56}$$

which, combined with the initial conditions

$$h(x,0) = 0.2 \text{ m}, \qquad hu(x,0) = 0 \text{ m}^2/\text{s}, \tag{57}$$

ensures that the wave arrives at the first junction in the supercritical regime. This test serves to highlight the capabilities of the method, so the 1D solution is not expected to perfectly coincide with the 2D reference.

Indeed, Figure 14 show a collage of time evolution series at the center points of several channels. The 1D solution performs well for the first few channels, but as the wave propagates through the channels the discrepancies grow larger. However, the relative structures of the solutions are kept, meaning that the coupling conditions of the 1D method at the junctions are working correctly. It is then a matter of the propagation along the channels that is causing the delays observed. It is specially remarkable that the solutions at channels 12, 14, 15, and 16 keep a similar pattern, especially in the Froude number, even if displaced and dumped. These channels are located in the center part of the network; thus, they are subject to many oncoming waves and reflections. This is coherent with the observations made in the previous batch of test cases.

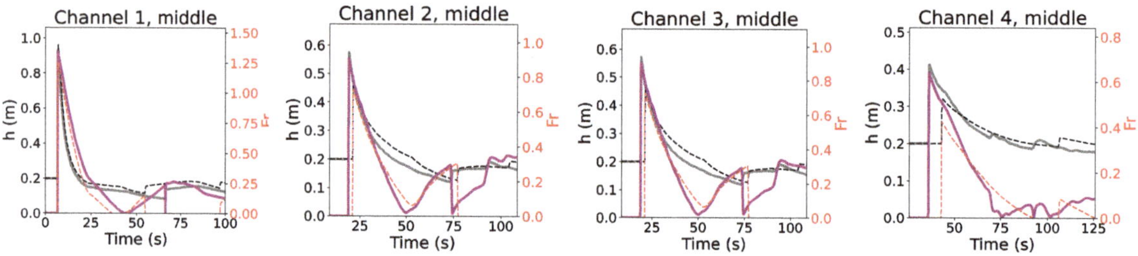


(**a**) Middle point of channel 1. (**b**) Middle point of channel 2. (**c**) Middle point of channel 3. (**d**) Middle point of channel 4.

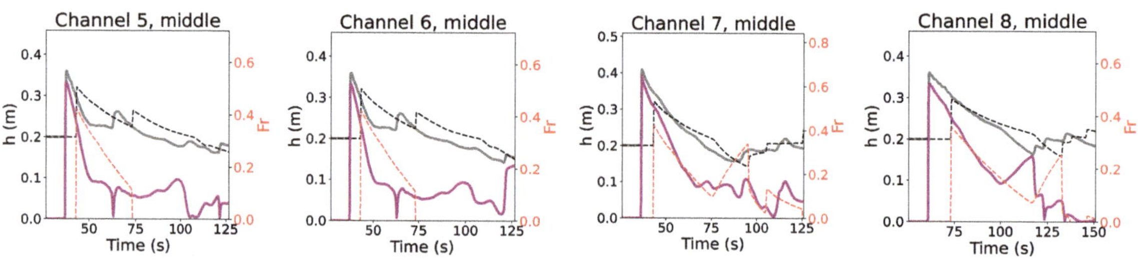


(**e**) Middle point of channel 5. (**f**) Middle point of channel 6. (**g**) Middle point of channel 7. (**h**) Middle point of channel 8.

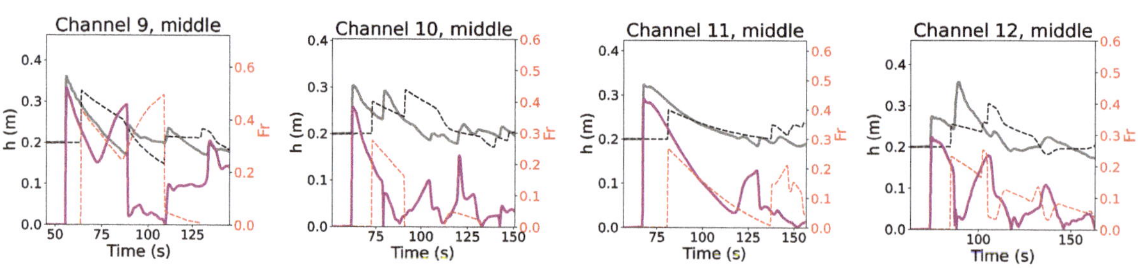


(**i**) Middle point of channel 9. (**j**) Middle point of channel 10. (**k**) Middle point of channel 11. (**l**) Middle point of channel 12.

**Figure 14.** *Cont.*

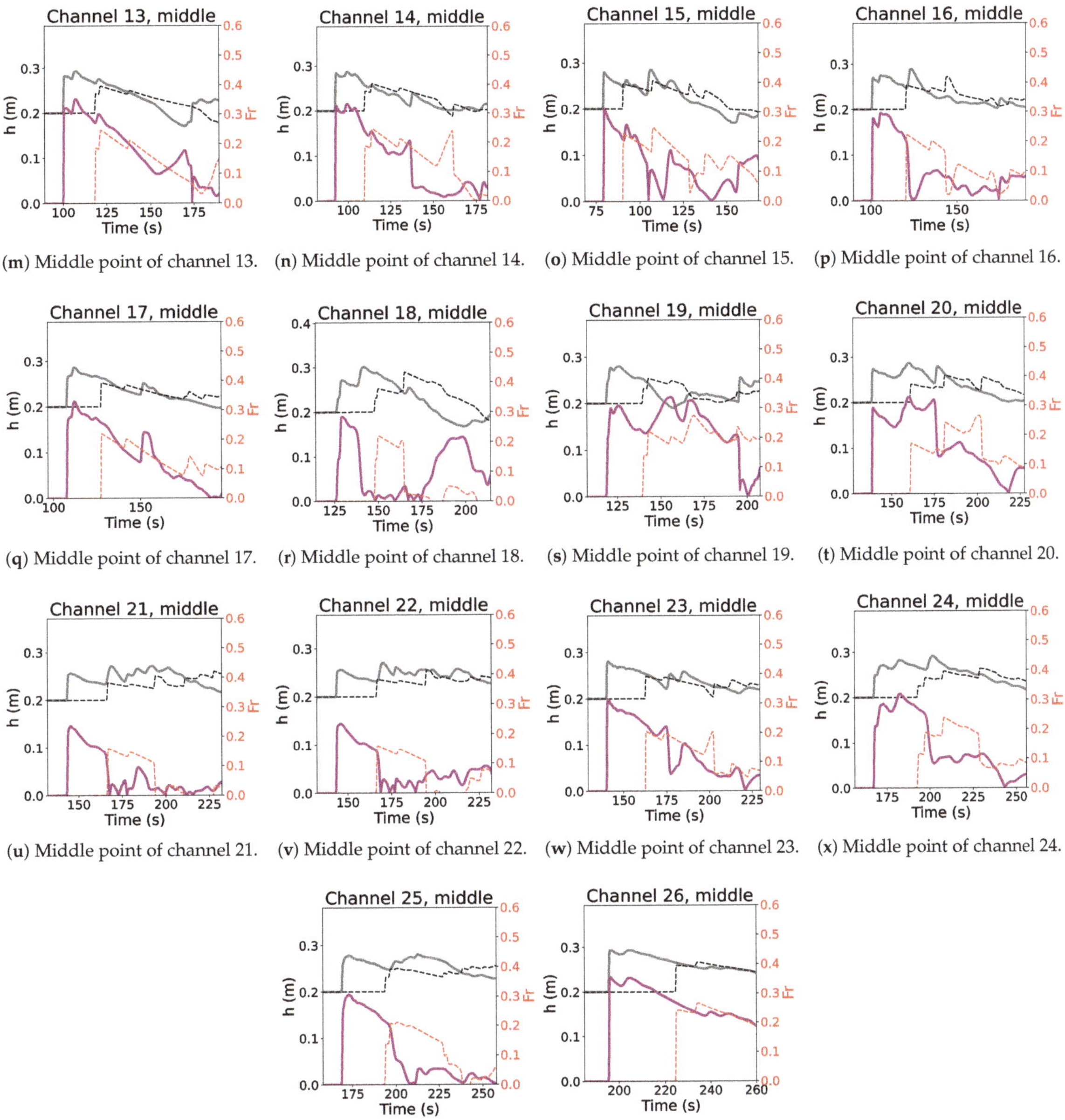


(**m**) Middle point of channel 13. (**n**) Middle point of channel 14. (**o**) Middle point of channel 15. (**p**) Middle point of channel 16.

(**q**) Middle point of channel 17. (**r**) Middle point of channel 18. (**s**) Middle point of channel 19. (**t**) Middle point of channel 20.

(**u**) Middle point of channel 21. (**v**) Middle point of channel 22. (**w**) Middle point of channel 23. (**x**) Middle point of channel 24.

(**y**) Middle point of channel 25. (**z**) Middle point of channel 26.

**Figure 14.** Time evolution of several probes located at the center of the channels of the large network. The solid line (−) marks the reference 2D solution, while the dashed line (− −) marks the 1D solution. Depth (black) is pictured with reference to the left y-axis while the Froude number (red) is pictured with reference to the right y-axis.

Likewise, the probe at Channel 26, the last channel of the network, shows that after going through 16 junctions, the wave This is a testament to the method's robustness.

## 5. Conclusions

This work shows that a fully one-dimensional formulation of channel networks that contemplates transitory states and supercritical transitions is possible through the Junction Riemann Problem and conservation of mass and energy.

By including some limiting coefficients in the coupling condition at the junction, the different combinations of regimes can be managed to achieve a coherent solution. This translates into a robust method that can handle complex networks with multiple junctions of several channels, independently of their geometry. Of course, given that energy conservation is assumed, the method will not take into account energy loss due to friction and reflections at the junction walls. This limitation can be addressed in future works. For now, the robustness of the method in the complex configurations tested is a promising result.

Given the computational efficiency of 1D methods as opposed to 2D, it can be concluded that this formulation of the JRP makes pure 1D methods a viable option for solving networks of channels independently of the regime of the flow.

**Author Contributions:** Conceptualization, J.M. (Javier Murillo) and P.G.-N.; methodology, J.M. (Juan Mairal) and J.M. (Javier Murillo); software, J.M. (Juan Mairal); validation, J.M. (Juan Mairal); formal analysis, J.M. (Juan Mairal); writing—original draft preparation, J.M. (Juan Mairal); writing—review and editing, J.M. (Juan Mairal), J.M. (Javier Murillo), and P.G.-N.; visualization, J.M. (Juan Mairal); supervision, J.M. (Javier Murillo) and P.G.-N.; project administration, P.G.-N.; funding acquisition, P.G.-N. All authors have read and agreed to the published version of the manuscript.

**Funding:** This work was supported by project PID2022-137334NB-I00 funded by MCIN/AEI/10.13039/501100011033 and by "ERDF/EU" of the Spanish Ministry of Science and Innovation and the European Regional Development Fund. It is also partially funded by the Government of Aragón, through the research grant T32_23R Fluid Dynamics Technologies.The authors belong to and are supported by the Aragón Institute of Engineering Research (I3A) of the University of Zaragoza. The first author is funded by the Spanish Ministry of Universities by grant FPU21/02876.

**Data Availability Statement:** The raw data supporting the conclusions of this article will be made available by the authors on request.

**Conflicts of Interest:** The authors declare no conflicts of interest. The funders had no role in the design of the study; in the collection, analyses, or interpretation of data; in the writing of the manuscript; or in the decision to publish the results

## References

1. Bellamoli, F.; Müller, L.O.; Toro, E.F. A numerical method for junctions in networks of shallow-water channels. *Appl. Math. Comput.* **2018**, *337*, 190–213. [CrossRef]
2. Herty, M.; Izem, N.; Seaid, M. Fast and accurate simulations of shallow water equations in large networks. *Comput. Math. Appl.* **2019**, *78*, 2107–2126. [CrossRef]
3. Bales, P.; Kolb, O.; Lang, J. Hierarchical modelling and model adaptivity for gas flow on networks. In Proceedings of the Computational Science–ICCS 2009: 9th International Conference, Baton Rouge, LA, USA, 25–27 May 2009; Proceedings, Part I 9; Springer: Berlin/Heidelberg, Germany, 2009; pp. 337–346.
4. Bermudez, A.; López, X.; Vázquez-Cendón, M.E. Numerical solution of non-isothermal non-adiabatic flow of real gases in pipelines. *J. Comput. Phys.* **2016**, *323*, 126–148. [CrossRef]
5. Bermúdez, A.; López, X.; Vázquez-Cendón, M.E. Treating network junctions in finite volume solution of transient gas flow models. *J. Comput. Phys.* **2017**, *344*, 187–209. [CrossRef]
6. Quarteroni, A.; Tuveri, M.; Veneziani, A. Computational vascular fluid dynamics: Problems, models and methods. *Comput. Vis. Sci.* **2000**, *2*, 163–197. [CrossRef]
7. Müller, L.O.; Blanco, P.J. A high order approximation of hyperbolic conservation laws in networks: Application to one-dimensional blood flow. *J. Comput. Phys.* **2015**, *300*, 423–437. [CrossRef]

8. Charlton, P.H.; Mariscal Harana, J.; Vennin, S.; Li, Y.; Chowienczyk, P.; Alastruey, J. Modeling arterial pulse waves in healthy aging: A database for in silico evaluation of hemodynamics and pulse wave indexes. *Am. J. Physiol.-Heart Circ. Physiol.* **2019**, *317*, H1062–H1085. [CrossRef] [PubMed]
9. Murillo, J.; García-Navarro, P. A solution of the junction Riemann problem for 1d hyperbolic balance laws in networks including supersonic flow conditions on elastic collapsible tubes. *Symmetry* **2021**, *13*, 1658. [CrossRef]
10. Murillo, J.; García-Navarro, P. Numerical coupling of 0D and 1D models in networks of vessels including transonic flow conditions. Application to short-term transient and stationary hemodynamic simulation of postural changes. *Int. J. Numer. Methods Biomed. Eng.* **2023**, *39*, e3751. [CrossRef] [PubMed]
11. Vallés, P.; Fernández-Pato, J.; Morales-Hernández, M.; Echeverribar, I.; García-Navarro, P. A 2D shallow water flow model with 1D internal boundary condition for subgrid-scale topography. *Adv. Water Resour.* **2024**, *189*, 104716. [CrossRef]
12. Gordillo, G.; Morales-Hernández, M.; García-Navarro, P. Solute transport control at channel junctions using adjoint sensitivity. *Mathematics* **2021**, *10*, 93. [CrossRef]
13. Cunge, J.A.; Holly, F.M.; Verwey, A. *Practical Aspects of Computational River Hydraulics*; Pitman Advanced Publishing Program: Boston, MA, USA, 1980.
14. Gurram, S.K.; Karki, K.S.; Hager, W.H. Subcritical junction flow. *J. Hydraul. Eng.* **1997**, *123*, 447–455. [CrossRef]
15. Hsu, C.C.; Wu, F.S.; Lee, W.J. Flow at 90 equal-width open-channel junction. *J. Hydraul. Eng.* **1998**, *124*, 186–191. [CrossRef]
16. Li, Q.; Liang, Q.; Xia, X. A novel 1D–2D coupled model for hydrodynamic simulation of flows in drainage networks. *Adv. Water Resour.* **2020**, *137*, 103519. [CrossRef]
17. Kesserwani, G.; Ghostine, R.; Vazquez, J.; Mosé, R.; Abdallah, M.; Ghenaim, A. Simulation of subcritical flow at open-channel junction. *Adv. Water Resour.* **2008**, *31*, 287–297. [CrossRef]
18. Rice, C.E. *Open Channel Junctions with Supercritical Flow*; ARS-United States Department of Agriculture, Agricultural Research Service (USA): Washington, DC, USA, 1985.
19. Kesserwani, G.; Ghostine, R.; Vazquez, J.; Ghenaim, A.; Mosé, R. One-dimensional simulation of supercritical flow at a confluence by means of a nonlinear junction model applied with the RKDG2 method. *Int. J. Numer. Methods Fluids* **2008**, *57*, 1695–1708. [CrossRef]
20. Ghostine, R.; Vazquez, J.; Terfous, A.; Mose, R.; Ghenaim, A. Comparative study of 1D and 2D flow simulations at open-channel junctions. *J. Hydraul. Res.* **2012**, *50*, 164–170. [CrossRef]
21. Sherwin, S.J.; Formaggia, L.; Peiro, J.; Franke, V. Computational modelling of 1D blood flow with variable mechanical properties and its application to the simulation of wave propagation in the human arterial system. *Int. J. Numer. Methods Fluids* **2003**, *43*, 673–700. [CrossRef]
22. Goudiaby, M.S.; Kreiss, G. A Riemann problem at a junction of open canals. *J. Hyperbolic Differ. Equ.* **2013**, *10*, 431–460. [CrossRef]
23. Elshobaki, M.; Valiani, A.; Caleffi, V. Junction Riemann problem for one-dimensional shallow water equations with bottom discontinuities and channels width variations. *J. Hyperbolic Differ. Equ.* **2018**, *15*, 191–217. [CrossRef]
24. Alcrudo, F.; Benkhaldoun, F. Exact solutions to the Riemann problem of the shallow water equations with a bottom step. *Comput. Fluids* **2001**, *30*, 643–671. [CrossRef]
25. Toro, E.F. *Riemann Solvers and Numerical Methods for Fluid Dynamics: A Practical Introduction*; Springer Science & Business Media: Berlin/Heidelberg, Germany, 2013.
26. Colombo, R.M.; Herty, M.; Sachers, V. On $2 \times 2$ conservation laws at a junction. *SIAM J. Math. Anal.* **2008**, *40*, 605–622. [CrossRef]
27. Murillo, J.; García-Navarro, P. Energy balance numerical schemes for shallow water equations with discontinuous topography. *J. Comput. Phys.* **2013**, *236*, 119–142. [CrossRef]
28. Murillo, J.; Navas-Montilla, A. A comprehensive explanation and exercise of the source terms in hyperbolic systems using Roe type solutions. Application to the 1D–2D shallow water equations. *Adv. Water Resour.* **2016**, *98*, 70–96. [CrossRef]
29. Mairal, J.; Murillo, J.; García-Navarro, P. The Junction Riemann Problem in 1d Shallow Water Channels Including Supercritical Flow Conditions. 2024. Available online: https://papers.ssrn.com/sol3/papers.cfm?abstract_id=5003532 (accessed on 17 February 2025 ).
30. Murillo, J.; García-Navarro, P. Improved Riemann solvers for complex transport in two-dimensional unsteady shallow flow. *J. Comput. Phys.* **2011**, *230*, 7202–7239. [CrossRef]
31. Echeverribar, I.; Morales-Hernández, M.; Brufau, P.; García-Navarro, P. 2D numerical simulation of unsteady flows for large scale floods prediction in real time. *Adv. Water Resour.* **2019**, *134*, 103444. [CrossRef]
32. Navas-Montilla, A.; Murillo, J. Overcoming numerical shockwave anomalies using energy balanced numerical schemes. Application to the Shallow Water Equations with discontinuous topography. *J. Comput. Phys.* **2017**, *340*, 575–616. [CrossRef]

fluids

MDPI

Article

# High $y^+$ Shear-Stress Turbulence Implementation for High Flux Isotope Reactor Narrow Channel Flows

**Emilian Popov [1], Nicholas Mecham [2,*] and Taylor Grubbs [2]**

[1] Oak Ridge National Laboratory (ORNL), Oak Ridge, TN 37830, USA; popove@ornl.gov
[2] Department of Nuclear Engineering, North Carolina State University (NCSU), Raleigh, NC 27695, USA; tegrubbs@ncsu.edu
* Correspondence: njmecham@ncsu.edu

**Abstract:** The research objective of this work was to improve the engineering predictions of the turbulence characteristics of flows in curved narrow channels. Such channel flows are commonly encountered in nuclear research and test reactors, with one of them being the high-flux isotope reactor (HFIR). Research reactors bear high heat fluxes, and the proper computing of turbulence is paramount for safe and reliable reactor operation. The study builds on the results of a previous direct numerical simulation of turbulence to inform a well-known Reynolds-averaged Navier–Stokes shear-stress turbulence model and improves its accuracy in simulating parallel channel flows. A new formulation of the loss term in the dissipation conservation equation is suggested. Combined with high wall distance computational grids, the new implementation provides a fast-running flow solution, suitable for engineering purposes. Model generalization for parallel channel flows, in a broader range of frictional Reynolds numbers, is suggested by introducing a new form of the model constants.

**Keywords:** CFD; RANS; channel flow; SST

## 1. Introduction

The HFIR, located at the Oak Ridge National Laboratory (ORNL), is a research and test reactor that provides very high neutron flux for a wide variety of applications. These applications range from neutron activation analysis, neutron scattering experiments, material irradiation, and isotope production. Engineering design and safety simulations are routinely used to calculate the kinematic and heat transfer properties of the coolant flow and to evaluate the reactor's safety margins. However, such calculations must be validated to ensure accuracy. Due to the restrictive geometry of the HFIR core (Figure 1) and the high coolant mass flux necessary to maintain operating temperatures, it is impossible to instrument the flow within the core itself. And it is prohibitively expensive to construct a test loop with similar conditions. To address these issues, a flow database of high-fidelity direct numerical simulation (DNS) was developed [1]. The DNS was computationally intense, resulting in a mesh size of 10 billion elements and necessitating the use of 50 thousand central processing units to solve the discretized Navier–Stokes equations. With the database in hand, the goal at this point of the research is to develop a Reynolds-averaged Navier–Stokes (RANS) turbulence model capable of delivering the accuracy of the DNS results at a fraction of the computational expense. The study explores a novel high $y^+$ wall function-based modeling approach and improves its prediction by modifying the loss term in the dissipation equation. Similar high-fidelity numerical simulations [2], although using different methods, were widely published before.

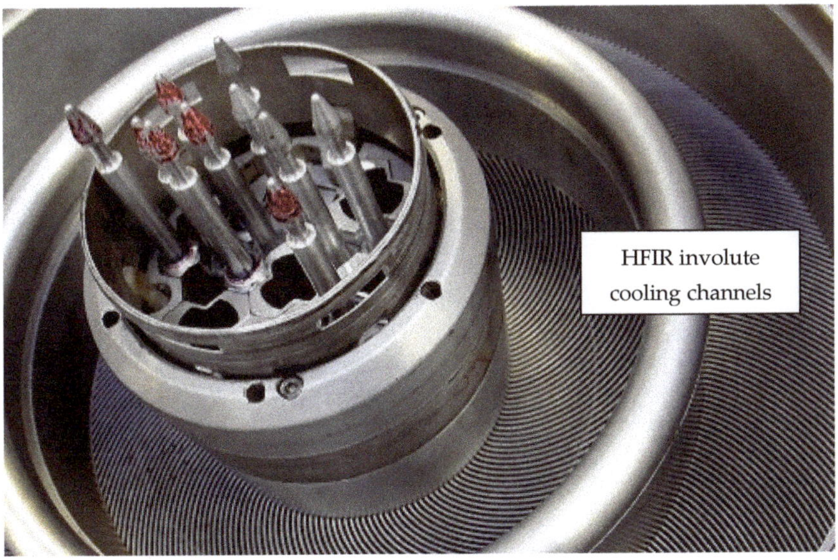


**Figure 1.** The HFIR core. The fuel plates are the curved elements arranged in two rings around the central flux trap. The coolant channels are the slots that separate the fuel plates. One of these channels is modeled in this study.

## 2. Problem Formulation

The code selected for this application was the popular open-source software OpenFOAM [3]. It has grown a large user community over the years because it is readily available, relative to commercial codes, and because of the broad spectrum of existing applications. OpenFOAM can simulate a wide range of problem scales up to a full resolution of turbulence (DNS). For this study, a RANS model was selected that can capture the effects of turbulence on mean flow parameters without necessitating a computationally intense simulation. The $k$-$\omega$ shear-stress transport (SST) model was chosen because of its reputation as a popular, generally accurate model for a wide variety of applications, ranging from aerospace to nuclear thermal hydraulics [4]. The problem was approached in three stages. First, a "blind" application of the RANS model to the HFIR geometry was conducted and compared with the database results to evaluate the model performance, as implemented originally in OpenFOAM. Second, the results were assessed, the discrepancies were determined, the model was reevaluated, and modifications were applied. Third, a series of analyses were performed to generalize the introduced changes for parallel channel flows.

## 3. Computational Domain and Grid

A domain representing a single HFIR involute cooling channel of around 120 hydraulic diameters in length was computed, which is about half of the physical size (Figure 1). The grid was composed of hexahedral elements with curved edges to capture the involute shape of the channel. For grid generation, a simple domain decomposition was applied with the *blockMesh* utility provided as part of the OpenFoam package. More information about the mesh generator can be found in the OpenFOAM User Guide [3]. This grid generation method provides full control over the element distribution in any direction. The geometry was specified using a combination of straight lines and high-order splines to match the curved shape of the HFIR channel. In total, 19 interpolation points were used to define each spline, and all arc lengths were verified to $\pm 0.01$ mm of the nominal dimensions. A comparison of the DNS [1] and RANS grids is provided in Table 1. The computing load reduction is evident from this comparison.

**Table 1.** DNS cyclic grid details compared with the RANS grid expressed in relative units.

| Problem | Reference | $Re_\tau$ | $Lx/\delta$ | $Lz/\delta$ | $L^+x$ | $L^+z$ | $\Delta x^+$ | $\Delta z^+$ | $\Delta y_w{}^+$ | $\Delta y_c{}^+$ | $Ny$ |
|---|---|---|---|---|---|---|---|---|---|---|---|
| RANS | Present | 960 | 480 | 37.6π | 437,745 | 113,556 | 2155 | 2155 | 20 | 129 | 21 |
| Database | [1] | 960 | 2.6π | 37.6π | 7682 | 113,556 | 12 | 12 | 0.5 | 15 | 184 |

### *3.1. Grid Node Positioning and Boundary Conditions*

The grid is significantly longer in the streamwise direction, equivalent to 480δ, because it is intended for an inlet–outlet type of problem and is more representative of the real HFIR channel. The first-to-the-wall layer thickness for both the gapwise and side walls was placed outside of the viscous sublayer. This feature defines the problem as a high $y^+$ RANS problem. Placing the first element outside of the viscous layer saves computational resources and significantly simplifies the engineering modeling of HFIR flow. Otherwise, the layer must be resolved, which imposes an additional computing burden on the simulations, making them often unaffordable in engineering types of simulations.

The plots shown in Figure 2, based on the DNS database [1], illustrate the principle. The plots show that the linear (red) dependency between the nondimensional velocity and distance meets the log profile at around $y^+ = 10$. The first node of the RANS grid was placed at $y^+ = 20$ from both walls. The mesh had 532,875 hexagonal cells distributed in streamwise, gapwise, and spanwise directions, as follows: 203 × 21 × 125. The mesh was graded to provide increased resolution near the walls, as is depicted in Figure 3. The boundary conditions applied to this problem are listed in Table 2. The inlet velocity was taken from the HFIR database [1]. The maximum aspect ratio of the grid is 177.

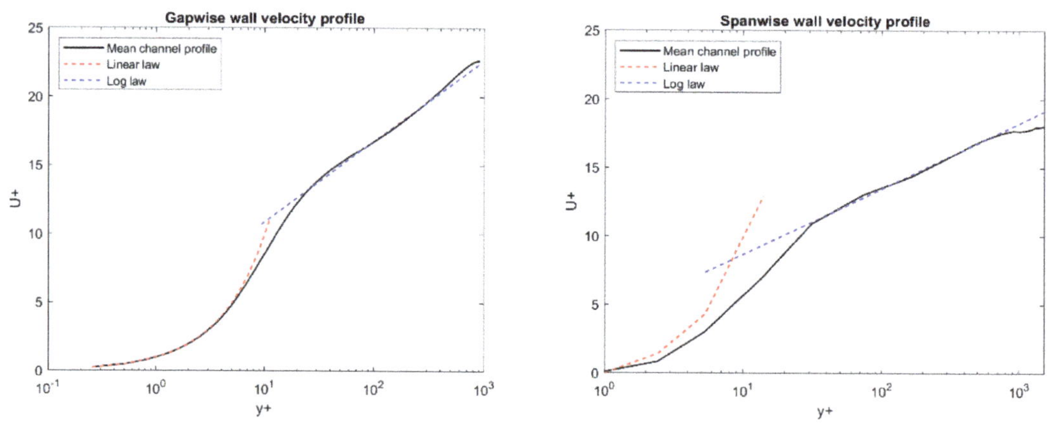


**Figure 2.** DNS velocity profiles illustrating extent of viscous sublayer on gapwise wall (**left**) and spanwise wall (**right**) from [1].

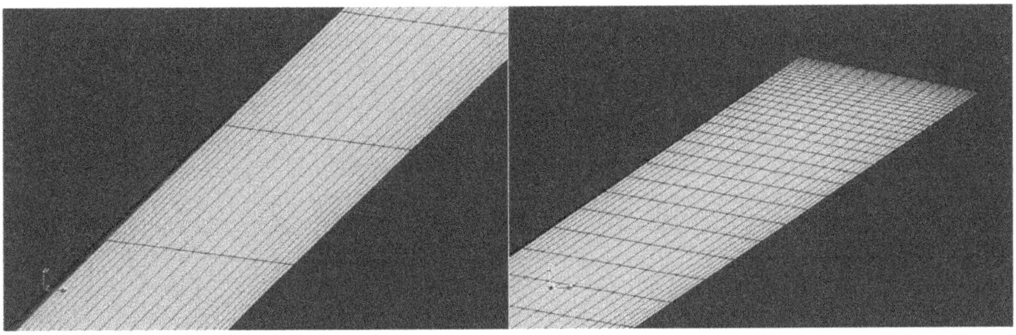

**Figure 3.** RANS grid., channel center (**left**), and channel corner (**right**).

**Table 2.** Boundary conditions for the RANS problem.

| **Face** | $k\left[\frac{m^2}{s^2}\right]$ | $\nu_t\left[\frac{m^2}{s}\right]$ | $\omega\left[\frac{1}{s}\right]$ | $p\left[\frac{m^2}{s^2}\right]$ | $U\left[\frac{m}{s}\right]$ |
|---|---|---|---|---|---|
| Inlet | Turbulent Intensity Kinetic Energy | Zero Gradient | Turbulent MixingLength Frequency | Zero Gradient | Fixed Value |
| Outlet | Zero Gradient | Zero Gradient | Zero Gradient | Fixed Value | Pressure Outlet |
| Walls | kqRWallFunction | BlendedWallFunction | omegaWallFunction | Zero Gradient | No Slip |

### 3.2. *Grid Convergence*

A grid convergence study was conducted to ensure that the grid resolution is properly selected and the estimated uncertainty of the spatial discretization is acceptable. The study was carried out using the grid convergence index (*GCI*) theory outlined in Roache [5]. As the *GCI* approaches 0%, the numerical result theoretically approaches the exact solution. *GCI* is computed by Equation (1):

$$GCI = \frac{F_s|f_{i+1} - f_i|}{f_i(r^p - 1)}, \tag{1}$$

where $F_s$ is a factor and is equal to 1.25 when comparing more than three meshes, $r$ is the uniform refinement ratio chosen to be 1.2, $p$ is the order of accuracy of the solver equal to 2, and $f$ is a figure of merit to be compared between meshes $i$ and $i+1$. Given the focus of the present study on turbulence modeling, the bulk velocity profile shape factor $R_s$ from the relation in Equation (2) was selected as the figure of merit.

$$U_{CL}^+ - U^+ = \frac{1}{2}R_s\left(1 - \frac{y}{\delta}\right)^2, \tag{2}$$

where $U_{CL}$ is the centerline velocity and the superscript + denotes wall units. The meshes selected for the convergence study are summarized in Table 3, with the computed $R_s$ and *GCI* plotted in Figure 4. It is apparent that GCI converges to a very low value with increased refinement, indicating that, as the cell size approaches zero, the exact solution is obtained. Owing to the good convergence exhibited starting with the third mesh, the 532,875-cell mesh was chosen for performing the analysis. Note that the database results indicated that $R_s$ should be 14.2, as opposed to the value of approximately 8.18 to which the RANS model converges, which is one of the initially observed discrepancies.

**Table 3.** Summary of meshes used in the grid convergence study.

| Cells Across the Gap | Mesh Size | $R_s$ | *GCI* |
|---|---|---|---|
| 15 | 181,395 | 8.33317 | - |
| 18 | 323,190 | 8.26321 | 2.39% |
| 21 | 532,875 | 8.21931 | 1.51% |
| 26 | 951,600 | 8.19906 | 0.70% |
| 31 | 1,629,360 | 8.18836 | 0.37% |

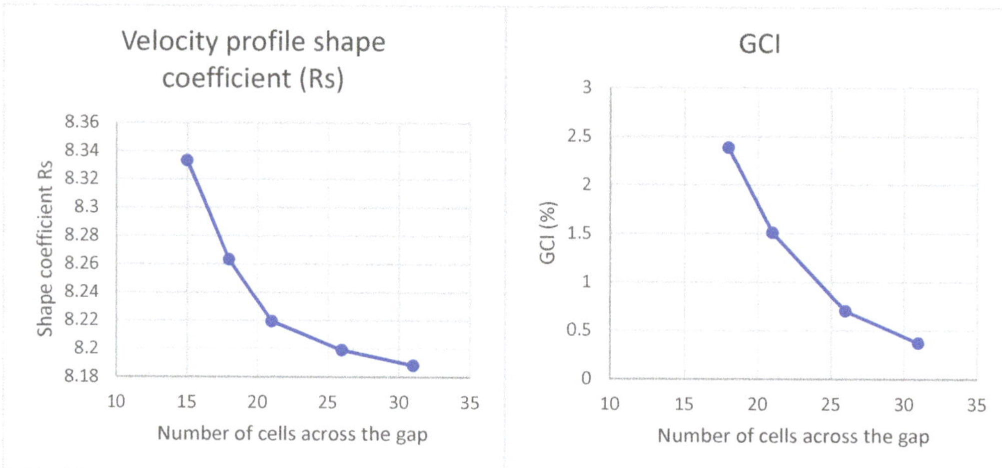


**Figure 4.** Results of grid convergence study. Velocity shape factor $R_s$ (**left**) and computed GCI (**right**).

## 4. Menter's SST RANS Model

This RANS model was selected because it is one of the most popular among the engineering computational fluid dynamics (CFD) applications. The motivation for developing the $k$-$\omega$ SST model was to combine the advantages of the standard $k$-$\varepsilon$ and $k$-$\omega$ models [4]. Specifically, the advantage of the $k$-$\varepsilon$ is its insensitivity to free-stream dissipation values, whereas the $k$-$\omega$ model is much more accurate in the boundary layer by comparison with the test data. The idea behind the SST model is to switch from one model to the other outside the boundary layer. The basic approach is to multiply the $k$-$\varepsilon$ equations by a blending function $F_1$, multiply the $k$-$\omega$ equations by $(1 - F_1)$, and add the two together. Additional considerations regarding the formulation of this function are covered in the following sections.

### *4.1. Turbulent Kinetic Energy (TKE) Equations*

The base $k$-$\omega$ SST model implementation in OpenFOAM [3] defines the $k$ equation as shown in Equation (3):

$$\frac{\mathrm{D}k}{\mathrm{D}t} = \widetilde{P}_k - \beta^* k\omega + \frac{\partial}{\partial x_i}\left[(\nu + \sigma_k \nu_t)\frac{\partial k}{\partial x_i}\right] \tag{3}$$

where $k$ is the turbulent kinetic energy (TKE), $\omega$ is the specific dissipation rate, $\widetilde{P}_k$ is the modified production term, ν is the kinematic viscosity, $\sigma_k$ is a constant, and $\nu_t$ is the turbulent viscosity. The implementation follows from Menter's work [6].

Note that the $F_1$ blending function does not make an explicit appearance in this equation since the standard $k$-$\varepsilon$ and $k$-$\omega$ models implement similar $\kappa$ equations that differ only by a constant ($\sigma_k$). The left side of Equation (3) represents the material derivative of turbulent kinetic energy. The right-side terms are, from left to right, turbulence production, turbulence dissipation, and turbulence viscous diffusion.

The dissipation term requires more attention. Such an implementation presumes that only a fraction of the specific dissipation rate is used to compute the dissipation of TKE. This fraction is well-known as the coefficient beta star ($\beta^* = 0.09$). By assuming this, the specific dissipation rate $\omega$ formulation inherently becomes $\omega = \varepsilon/(\beta^* k)$. The rationale for this conclusion is based on the work of Wilcox [7]. Such a postulate is reflected later in this study, when the results are compared with the database. The specific dissipation rate $\omega$ is reduced by a factor of $\beta^*$.

The production term is implemented with a limit of ten times the dissipation term to prevent turbulence build-up in stagnation regions, as shown in Equation (4). Note that, most of the time, the limiting value is not reached.

$$\widetilde{P}_k = \min(\nu_t G; c1k\omega) \tag{4}$$

The dissipation term is simply a product of the turbulent kinetic energy and the specific dissipation rate with a coefficient equal to $\beta^*$ [8].

The SST model, as implemented in OpenFOAM, employs 12 coefficients applied to the basic conservation equation and to the blending functions. They are given in Table 4 with their default values.

**Table 4.** SST model coefficients and their default values.

| Coefficient | Value | Usage |
|---|---|---|
| alphaK1 | 0.85 | TKE diffusion Equation (5) |
| alphaK2 | 1.0 | TKE diffusion Equation (5) |
| alphaOmega1 | 0.5 | Omega diffusion Equation (18) |
| alphaOmega2 | 0.856 | Omega diffusion Equation (18) |
| beta1 | 0.075 | Beta coefficient Equation (17) |
| beta2 | 0.0828 | Beta coefficient Equation (17) |
| betaStar | 0.09 | Equation (3) and blending functions |
| gamma1 | 5/9 | Alpha coefficient Equation (16) |
| gamma2 | 0.44 | Alpha coefficient Equation (16) |
| a1 | 0.31 | Turbulent viscosity Equation (6) |
| b1 | 1.0 | Not used |
| c1 | 10.0 | TKE production Equation (4) |

The coefficient $\sigma_k$ is computed with the $F_1$ blending function to fully switch from the standard *k*-ε model to the standard *k*-ω model at the boundary layer because the two models use different coefficients. This implementation is shown in Equation (5). According to this formulation, the coefficient $\sigma_k$ varies in the range of 0.85–1.0.

$$\sigma_k = alphaK1 \cdot F_1 + alphaK2(1 - F_1) \tag{5}$$

All coefficients in the *k*-ω SST model are blended in this manner [4]. The eddy viscosity is computed via a second blending function $F_2$, as shown in Equation (6), to satisfy the framework of an eddy viscosity model without overpredicting shear stress in large adverse pressure gradients [4]:

$$\nu_t = \frac{a1 \cdot k}{\max(a1\omega; SF_2)}, \tag{6}$$

where $S$ is the square root of $S2$, where $S2 = 2S_{ij}S_{ij}$ is defined by Equation (7), and $S_{ij}$ is the symmetry part of the velocity gradient tensor or the strain rate tensor.

$$S2 = 2\left\|\frac{\partial U_i}{\partial x_j} + \frac{\partial U_j}{\partial x_i}\right\|^2 = 2\left(\frac{\partial U_i}{\partial x_j} + \frac{\partial U_j}{\partial x_i}\right)\cdot\left(\frac{\partial U_i}{\partial x_j} + \frac{\partial U_j}{\partial x_i}\right) \tag{7}$$

By definition, the turbulent viscosity, used as a substitute for the Reynolds stress tensor in the RANS modeling paradigm, is formulated by Equation (8). The database contains all the necessary information for the computation of the true turbulence viscosity, including the Reynolds stress components and the mean velocity derivatives.

$$-\overline{u_i'u_j'} = \nu_t\left(\frac{\partial U_i}{\partial x_j} + \frac{\partial U_j}{\partial x_i}\right), \tag{8}$$

where $u'$ denotes fluctuating velocity components, the overbar denotes averaging, and $U$ capital is a mean velocity.

### 4.2. Blending Functions

The SST model blends both the TKE and the dissipation rate conservation equations by two blending functions, $F_1$ and $F_2$. They require elevated attention because the formulations depend strongly on $k$ and $\omega$ and their distribution across the channel.

#### 4.2.1. $F_1$ Blending Function

The blending function $F_1$ is defined in Equations (9) and (10) [3]:

$$F_1 = \tanh\left(\arg_1^4\right) \tag{9}$$

$$\arg_1 = \min\left\{\min\left[\max\left(\frac{\sqrt{k}}{\beta^*\omega y}; \frac{500\nu}{y^2\omega}\right); \frac{3.424k}{CD_{k\omega}y^2}\right]; c1\right\} \tag{10}$$

The meaning behind the terms in Equation (10), as defined in the original paper [4], is worth discussing. The term $\frac{\sqrt{k}}{\beta^*\omega y}$ represents the turbulence length scale divided by the distance to the wall—$y$. According to Menter, this term should be around 2.5 in the log layer and approach 0 toward the outer boundary layer edge. The term $\frac{500\nu}{y^2\omega}$ is supposed to maintain the value of $F_1$ at one within the viscous sublayer. This is because it exhibits a roughly constant value near the wall and decays to zero in the log region. The term $\frac{3.424k}{CD_{k\omega}y^2}$ is proposed as an additional safeguard against a freestream-dependent solution. The term $CD_{k\omega}$ is defined in Equation (11).

$$\mathrm{CD}_{k\omega} = \max\left(\gamma\frac{1}{\omega}\frac{\partial k}{\partial x_i}\frac{\partial \omega}{\partial x_i}; 10^{-10}\right) \tag{11}$$

The components and the overall value of $\arg_1$ are plotted in Figure 5 for the fully developed turbulent flow in the HFIR channel (right plot) and by using data from the database (left plot). According to the SST model, the blending function $F_1$ takes a value of 1 throughout the whole channel gap, since $\arg_1$ is above the threshold of approximately 1.3 (Figure 6, left). This implies that the boundary layer thickness for HFIR flow would be larger than the channel half-width, and therefore, only the $k$-$\omega$ model would be triggered. For most of the channel width, Equation (10) takes the value of the first term, except very close to the wall and in a narrow region around $y/\delta$~$0.6$. The database result shows a different response. It predicts that some mixing between $k$-$\varepsilon$ and $k$-$\omega$ should occur in the range $y/\delta$ = 0.8–1.0. This difference is mostly driven by the denominator value of term 1 in Equation (10) (Figure 6, right plot), which strongly depends on the specific dissipation.

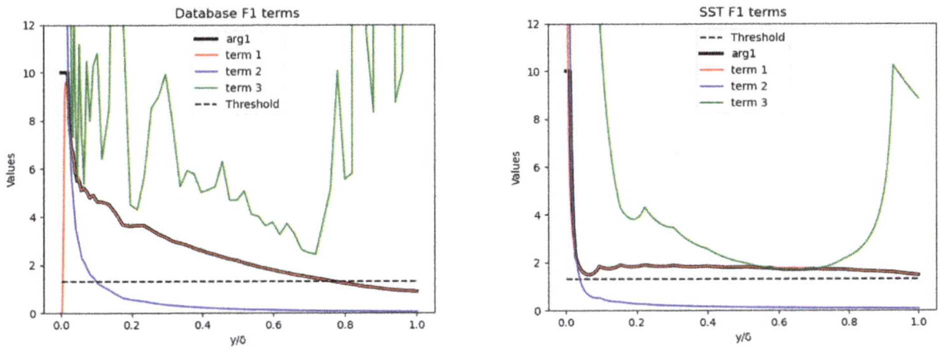


**Figure 5.** $F_1$ component distribution in the gap. The left plot shows $F_1$ based on the database results, and the right plot shows $F_1$ as computed in the SST model. The threshold value of $arg_1$, at which $F_1$ becomes noticeably less than 1, is shown as a dashed line.

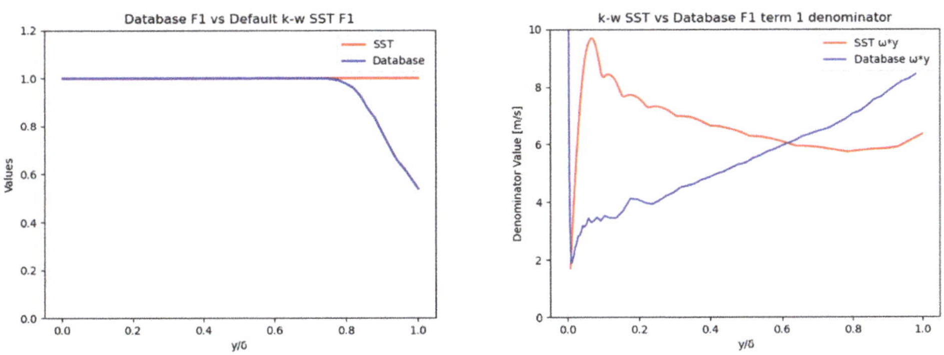


**Figure 6.** $F_1$ discrepancy between SST and database (**left**) and the corresponding difference in the denominator of term 1 (**right**).

The review of the $F_1$ blending function revealed that mainly the $k$-$\omega$ model should be used for simulating the HFIR flow. The database predicted that some partial (up to 50%) mixing with the $k$-$\varepsilon$ model should happen in the mid-channel. This mixing would only slightly affect the TKE conservation equation, where the factor $\sigma_\kappa$ may vary only in the range of 0.85–0.92. And this factor applies only to the diffusion term of TKE.

4.2.2. $F_2$ Blending Functions

The blending function $F_2$ is defined in Equations (12) and (13) [3]:

$$F_2 = \tanh\left(\arg_2^2\right) \tag{12}$$

$$\arg_2 = \min\left\{\max\left(\frac{2\sqrt{k}}{\beta^*\omega y};\frac{500\nu}{y^2\omega}\right);100\right\} \tag{13}$$

Equation (13) contains similar terms to Equation (10), except for the additional safeguard against freestream dependence, where term 1 is in fact doubled compared with the Equation (10) term 1. It comes as no surprise then that $F_2$ takes a value of one throughout the channel in both the database and SST implementation. This function has limited use only in the expression for the turbulent viscosity given by Equation (6). Both functions are shown in Figure 7.

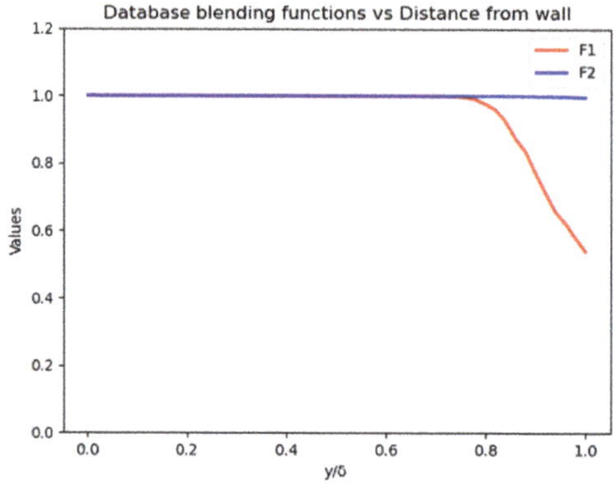


**Figure 7.** $F_1$ and $F_2$ distributions from wall to channel center.

### 4.3. Specific Dissipation Rate Equation

The base *k-ω* SST model implementation in OpenFOAM [3] defines the $\omega$ equation as shown in Equation (14):

$$\frac{D\omega}{Dt} = \alpha S^2 - \beta\omega^2 + \frac{\partial}{\partial x_i}\left[(\nu + \sigma_\omega \nu_t)\frac{\partial \omega}{\partial x_i}\right] + \gamma \frac{1}{\omega}\frac{\partial k}{\partial x_i}\frac{\partial \omega}{\partial x_i} \tag{14}$$

where $S$ is defined in Equation (7), $\nu_t$ is defined in Equation (6), and $F_1$ is defined in Equation (9). The blending coefficients $\alpha$, $\beta$, $\sigma_\omega$, and $\gamma$ are defined in Equations (16)–(19).

Equation (14) is to some extent simplified to reduce the complexity of the low-order terms. Based on evaluations by the authors, after a review of the Menter papers [4,6,8], a more comprehensive form of Equation (14) should look like Equation (15):

$$\frac{D\omega}{Dt} = \alpha S^2 - \beta\omega^2 + \frac{\partial}{\partial x_i}\left[(\nu + \sigma_\omega \nu_t)\frac{\partial \omega}{\partial x_i}\right] + \gamma \frac{1}{\omega}\frac{\partial k}{\partial x_i}\frac{\partial \omega}{\partial x_i} + \gamma \frac{\partial^2 k}{\partial {x_i}^2} + \gamma \frac{1}{k}\left(\frac{\partial k}{\partial x_i}\right)^2 \tag{15}$$

Two extra $\kappa$-derivative terms appear on the right side. However, Menter in [4] explicitly says that these terms are small compared with the $\omega$ terms and do not play a major role in the overall balance. Further, in this work, Equation (14) is considered comprehensive enough to allow for a transition between the *k-ω* and *k-ε* models by varying only the blending function $F_1$, which is embedded in the free coefficients preceding each of the terms, as shown in Equations (16)–(19).

$$\alpha = gamma1 \cdot F_1 + gamma2(1 - F_1) \tag{16}$$

$$\beta = beta1 \cdot F_1 + beta2(1 - F_1) \tag{17}$$

$$\sigma_\omega = alphaOmega1 \cdot F_1 + alphaOmega2(1 - F_1) \tag{18}$$

$$\gamma = 2 \cdot alphaOmega2(1 - F_1) \tag{19}$$

It was shown earlier that, for HFIR flow, $F_1$ may vary in the range of 0.5–1.0 and only in the mid-channel. It is interesting to evaluate how these variations affect the SST model predictions of the main turbulence properties. This evaluation is carried out below.

## 5. Baseline SST Model Performance

Initial simulations were executed with the default implementation of the SST model in OpenFOAM. These simulations were intended to provide insights regarding the accuracy of the SST model when it is "blindly" applied to the HFIR problem. Such blind simulations, when the RANS models are used as developed, are commonly encountered in engineering practices.

Because the viscous sublayer was not resolved in the RANS model, the accuracy in the calculation of the first node velocity becomes increasingly important. The method for calculation of the first node velocity in the SST model was examined, and the results are presented briefly in Section 5.1.

### *5.1. First Node Velocity*

When the first-to-the-wall node resides outside of the viscous layer, the SST model estimates the node velocity using the logarithmic law approximation. The velocity is computed as shown by Equation (20) [3].

$$U^+ = \frac{1}{\kappa}\ln\left(Ey^+\right), \tag{20}$$

where $k$ is the von Karman constant equal to 0.41 and $E$ is a roughness coefficient equal to 9.8 for smooth walls. $U^+$ and $y^+$ are the nondimensional velocity and wall distance computed by Equation (21).

$$U^+ = \frac{U_1}{u_\tau} \text{ and } y^+ = \frac{y u_\tau}{\nu} \tag{21}$$

To use Equation (21) for computing the dimensional first node velocity $U_1$, the frictional velocity $u_\tau$ must be known. In OpenFOAM, $u_\tau$ is iteratively calculated. The initial guess is based on the velocity gradient:

$$u_\tau = \sqrt{(\nu_t + \nu)\left|\nabla \vec{U}\right|} \tag{22}$$

Subsequent iteration over the following equations occurs until they converge to a specified error and iteration tolerance, as shown by Equation (23):

$$u_\tau = \sqrt[4]{\left(\frac{\left|\vec{U_1}\right|}{y^+}\right)^4 + \left(\frac{\kappa\left|\vec{U_1}\right|}{\ln\left(Ey^+\right)}\right)^4} \tag{23}$$

Once the frictional velocity is computed, the first node velocity can be calculated from the logarithmic law. The database provides results for the full velocity profile, including the viscous layer. These results were used to benchmark the first node velocity in the SST model. The velocity computed in the database was fit with a logarithmic expression, which is given by Equation (24).

$$U^+ = \frac{1}{0.39}\ln\left(y^+\right) + 4.28 \tag{24}$$

Table 5 summarizes the results for the frictional and first node velocities from the SST model, the database, and Equation (24), which approximates the database result. The velocity was also computed for both (convex and concave) involute channel walls for completeness.

**Table 5.** Comparison of friction velocity and first node velocity.

| Case | $u_\tau$ [m/s] | $U_1$ Convex [m/s] | $U_1$ Concave [m/s] |
|---|---|---|---|
| Blind SST | 0.82 | 10.36 | 10.18 |
| Database | 0.85 | 10.07 | 10.04 |
| Equation (24) | 0.85 | 10.17 | 10.17 |
| Equation (24) in SST | 0.85 | 10.21 | 10.07 |

As seen in Table 5, the values for $u_\tau$ are acceptably close, keeping in mind that this is an overall quantity. Regarding the $U_1$ calculation, a test was implemented to evaluate the sensitivity of the entire model, including the blending functions, to the first node velocity. Equation (24) was implemented in the SST model, replacing the original Equation (22), and the result is provided in the last row of Table 5. The computed values for $U_1$ are close. Replacing the log expression in the SST model gives little advantage, and it is not worth the programming effort. It can be concluded that the first node velocity calculation, as implemented, is accurate enough for properly simulating the HFIR velocity at that location.

### 5.2. Blind Model Application

The mean velocity, turbulent viscosity, and main turbulence properties in the channel were computed with the SST model, without any model modification, and are plotted in Figure 8 alongside the database results. It is apparent that the default SST model overpredicts the TKE in the center of the channel and strongly underpredicts it next to the wall, which flattens the velocity profile. The TKE distributions were integrated over the profiles shown in Figure 8b, and the values are very close (database TKE = 0.00153 versus SST TKE = 0.00154). These numbers demonstrate that the total produced turbulence is the same, but the distribution in the channel is different. The RANS grid node spacing, in the range of the peak TKE, has at least 3–5 elements, which eliminates the possibility of smearing the TKE and confirms that it is, in fact, underpredicted by the SST model.

A large discrepancy is observed in the dissipation rate. The plot (Figure 8c) is on a logarithmic scale and in actual units, and the result shows a significant overprediction by the SST model. This discrepancy, combined with the differences in the computation of TKE, leads to a totally different turbulent (eddy) viscosity distribution (Figure 8d). To eliminate any ambiguity, the turbulent viscosity was additionally calculated by simply dividing $k$ by $\omega$ (or $k^2/\varepsilon$), and the result is plotted in Figure 8d with the black line. As expected, and as shown by the plot, the two denominator formulations are almost identical and produce the same $\nu_t$ distribution.

The database value for $\nu_t$ was calculated from the database values for $k$ and $\varepsilon$ plotted in Figure 8b,c. However, the true formulation of the turbulent viscosity, originating in the replacement of the Reynolds stresses in the TKE conservation equation, is given by Equation (8), which is another method of calculating the turbulent viscosity. This was plotted and added to the figure in a light-blue color. The mean velocity derivative tensor and the Reynolds stresses were taken from the database. The result shows values that are lower than the database result based on $\kappa$ and $\varepsilon$, but the profile still has two expressed humps. This result can be further explained by the slightly lower $\varepsilon$ calculated by the database (Figure 8c).

To further verify that the discrepancy between the database and the SST model is, in fact, true, data from Moser [9] were added to Figure 8. Because these data were produced for a distinct set of conditions, they were renormalized for the conditions in the database/SST case. All compared properties deviate, but the $\nu_t$ profile resembles the profile computed by the database, thus confirming its validity.

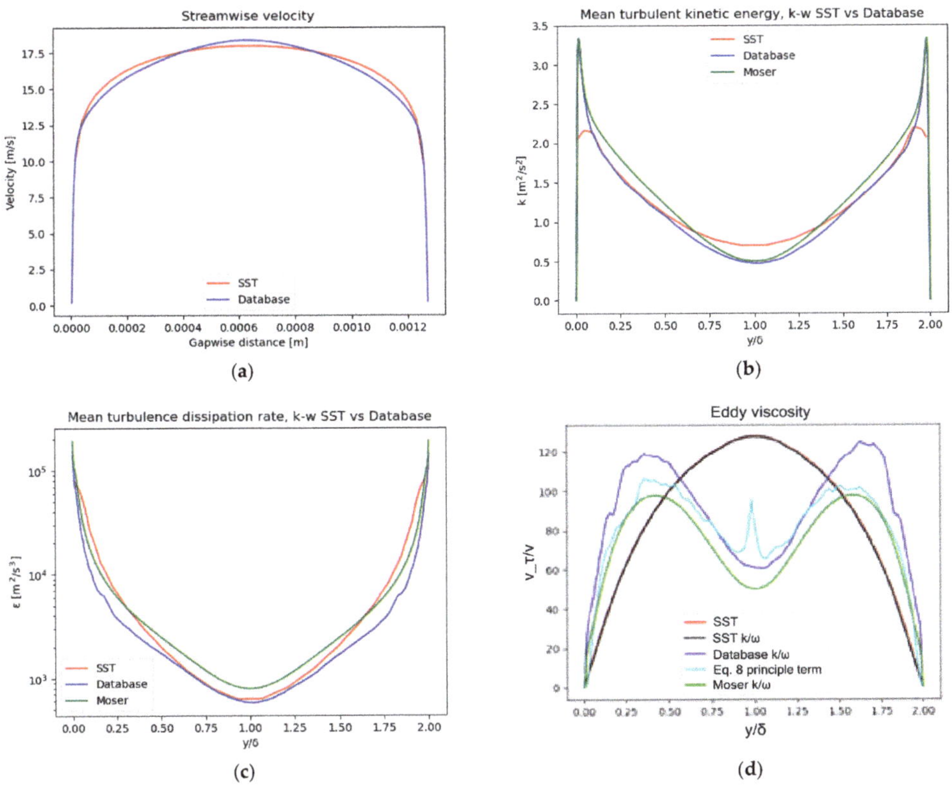


**Figure 8.** Comparison of (**a**) mean velocity, (**b**) TKE, (**c**) dissipation, and (**d**) eddy viscosity profiles computed in the database, in the default SST implementation (red SST) and by simply dividing the $k^2$ from (**b**) with $\varepsilon$ from (**c**) (black). Data of Moser [9] are plotted in green.

It is apparent that the result is not satisfactory, and a modification is needed. Turbulent viscosity is the sole property returned from the turbulence model to the Navier–Stokes solver, and it has a major impact on the model's accuracy. Below, the SST model implemented in OpenFOAM is analyzed to determine the most effective method for correction by using the DNS database previously generated.

### *5.3. SST Sensitivity to Model Factors*

To improve the result, model factor sensitivity was attempted by varying only a single factor at a time. The factor values used in the sensitivity are summarized in Table 6, with their ranges based on the variation of the $F_1$ function. These ranges were expanded by ±50% to amplify the effect.

**Table 6.** Summary of coefficient values used in parametric studies.

| Coefficient | Default Value and Blending Range | Values Used for Sensitivity |
|---|---|---|
| $\beta^*$ | 0.09 | 0.081, 0.09, 0.099 |
| $\sigma_k$ | 0.85–0.92 | 0.443, 0.885, 1.328 |
| $\alpha$ | 0.56–0.50 | 0.265, 0.53, 0.795 |
| $\beta$ | 0.075–0.079 | 0.0385, 0.077, 0.1155 |
| $\sigma_w$ | 0.5–0.68 | 0.295, 0.59, 0.885 |
| $\gamma$ | 0.856–1.712 | 0.856, 1.712, 2.568 |

Multiple runs were performed to evaluate the effects of the model's factors. The results are not reported in this manuscript for the sake of brevity. This simple range-sweep type of run demonstrated that the SST model prediction cannot be qualitatively improved by varying only a single free model coefficient. In fact, the default values are often the best fit for the properties of interest. To some extent, the results can be improved by adjusting the diffusion coefficients of both TKE and $\omega$, but such adjustments must be very substantial.

Major discrepancies are seen in the wall proximity. It is evident that the first node value in the SST for $\kappa$ is much lower than in the database. The latter is true for all coefficients, including their variations. The specific dissipation rate response shows the opposite—the SST value of $\omega$ is higher at the first node. This poses the question of how these first node values for $\kappa$ and $\omega$ are computed in the SST model. The bounding values of the turbulence properties are discussed in Section 6.

## 6. Data-Informed SST Model

Before going into the details of the conservation equations, a closer look was taken at the bounding values for $\kappa$ and $\omega$. The bounding values are the values in the first cell (face) next to the wall. In the current setup, this cell was chosen to be in the log range. This quest stemmed from the coefficients' sweep sensitivity discussed in the previous section.

### 6.1. Bounding Values for Major Turbulence Properties

The SST model solves for two turbulence properties, TKE and omega. When the first cell resides in the logarithmic layer, the bounding (or the first cell value) must be computed and imposed at this cell. The OpenFoam documentation does not provide information regarding these formulations. The following discussion is based on the SST model as coded, plus open literature searches.

#### 6.1.1. Omega Bounding Value

The bounding value for $\omega$ is set according to Equation (25). The first term is derived from a solution of an equation that applies to the viscous layer, whereas the second term is valid away from the wall. These expressions correspond to Equations (32) and (30) in the Wilcox work [7]. Equation (25) shows an option of "binomial blending", which is the default in OpenFOAM implementation. There are other blending options; one of them is "stepwise", in which the terms are applied separately as a function of $y^+$. The blending concept is still to be investigated. But, because it has little justification, further in this analysis, the stepwise option was chosen. With this assumption, only the second term in Equation (25) applies, and its derivation is discussed below.

$$\omega_1 = \left( \left( \frac{6\nu}{\beta y_1^2} \right)^2 + \left( \frac{\sqrt{k_1}}{\beta^{*0.25} \cdot kappa \cdot y_1} \right)^2 \right)^{1/2}, \tag{25}$$

where the coefficients' default values are in Table 4, and kappa is the von Karman constant *kappa* = 0.41

Following Wilcox [7], this term results from the solution of the system of equations formulated according to Equation (26).

$$\begin{aligned} 0 &= \frac{\partial}{\partial y}\left[\upsilon_T \frac{\partial u}{\partial y}\right] \\ 0 &= \upsilon_T \left[\frac{\partial u}{\partial y}\right]^2 - \beta^* \omega k + \sigma^* \frac{\partial}{\partial y}\left[\upsilon_T \frac{\partial k}{\partial y}\right] \\ 0 &= \gamma \left[\frac{\partial u}{\partial y}\right]^2 - \beta \omega^2 + \sigma \frac{\partial}{\partial y}\left[\upsilon_T \frac{\partial \omega}{\partial y}\right] \end{aligned} \tag{26}$$

This system is formulated for a layer of the flow "far enough from the wall to render the molecular viscosity negligible relative to eddy viscosity, yet close enough for convective effects to be negligible relative to the rate at which the turbulence is being created and destroyed", as stated in the Wilcox paper. When the velocity is substituted with the log law expression, $u = \frac{u_\tau}{kappa}\ln\left(\frac{yu_\tau}{v}\right)$, and the corresponding derivatives are taken, the resulting solution yields expressions for $k$ and $\omega$:

$$k = \frac{u_\tau^2}{(\beta^*)^{\frac{1}{2}}} \text{ and } \omega = \frac{u_\tau}{(\beta^*)^{\frac{1}{2}} kappa \cdot y} \tag{27}$$

Replacing the frictional velocity in those expressions produces the second term in Equation (25). It must be noted that the expression for $k$ serves only to compute the frictional velocity in the SST model according to Equation (29). The database-computed bounding value of $\omega$ is about two-times lower, suggesting some of the coefficients in Equation (27) are not fully applicable to the HFIR channel flow.

#### 6.1.2. TKE Bounding Value

The SST model provides two options to set the $\kappa$ bounding value ($k_1$) for the high and low $y^+$ given by Equation (28). The high $y^+$ option is further used in these analyses.

$$k_1 = \begin{cases} \left(\frac{-0.416\ln(y_1^+)}{0.41} + 8.366\right)\left((\beta^*)^{0.25}\sqrt{k_1}\right)^2 & , y^+ > 11.4 \\ 664.82\left(\frac{1}{(y_1^+ + 11)^2} + \frac{2y_1^+}{1331} - \frac{1}{121}\right) & , y^+ \leq 11.4 \end{cases} \tag{28}$$

Equation (28) has two components in the brackets. The first one is a function of the cell distance to the wall, which is used to decrease a constant of 8.366. The second is the frictional velocity squared, as computed in the SST model and shown in Equation (29).

$$u_\tau = (\beta^*)^{0.25}\sqrt{k_1} \tag{29}$$

Equation (28) is derived through a fit to the so-called "$v^2$-$f$" model. A detailed explanation can be found in the work by Kalitzen [10]. In that work, the authors solve for the dimensionless $k^+ = \frac{k}{u_\tau^2}$ and do not explicitly define $u_\tau$. The implementation must use the result obtained by Wilcox in Equation (27). Because the frictional velocity is defined by the TKE, the expression becomes implicit. This, together with the overall solution for TKE, which is by default iterative, allows little flexibility for using this expression to modify the TKE bounding value. Instead, the focus was placed on the TKE production term bounding value $G_1$, which is discussed in Section 6.1.3.

#### 6.1.3. TKE Production Bounding Value

The production terms are the first terms on the equation's right side. The TKE production is given by Equation (4). It has a hard limiter of ten times the dissipation rate. The production is defined after averaging the Navier–Stokes equations as the product of the tensor ($G$) and the turbulent viscosity. The $G$ formulation is given by Equation (30).

$$G = \frac{\partial u_i}{\partial x_j}\left(\frac{\partial u_i}{\partial x_j} + \frac{\partial u_j}{\partial x_i}\right) \tag{30}$$

The bounding value of $G_1$ is computed by Equation (31). In the viscous layer, it is zero.

$$G_1 = \begin{cases} 0 & , y_1^+ < 11.4 \\ (\nu_{t,1} + \nu)|\mathbf{n}\cdot(\nabla \mathbf{u}_1)|(\beta^*)^{0.25}\frac{\sqrt{k_1}}{kappa \cdot y_1} & , y_1^+ \geq 11.4 \end{cases} \tag{31}$$

The expression for the log layer can be evaluated if the frictional velocity is known. The first two terms translate to the shear stress at the wall, which can be approximated with $\tau_w/\rho = u_\tau{}^2$. And the last term in the numerator is again the frictional velocity given by Equation (28). Substituting in Equation (31) enables rewriting the $G_1$ bounding expression as

$$G_1 \approx \frac{u_\tau{}^3}{kappa \cdot y_1} = \frac{\left(\upsilon_\tau \frac{\partial u}{\partial y}\right)^{\frac{3}{2}}}{kappa \cdot y_1} \tag{32}$$

Accounting for the first node positioning, the estimate for $G_1$ is around twice as high as the resulting value from Equation (31).

In the preceding section, the bounding value of $\omega$ was determined to be too large in the default SST model. If this value is reduced by a constant multiplicative factor of about 0.4, the SST model generates a better prediction of the dissipation near the wall. However, this reduction of near-wall dissipation generates a larger TKE at the wall. It was found that this result could be corrected by simply reducing the value of $G_1$ by the same amount, about 0.4. This result suggests that, to properly compute the $k$ and $\omega$ bounding values, another factor can be applied to the computation of frictional velocity.

The major turbulence properties were computed and compared with the database after the bounding values for $\omega$ and $G_1$ were corrected according to the above derivations. They are depicted in Figure 9. The model is still clearly deficient at predicting the correct profiles for nearly all quantities. The work in this section and the sensitivity study in Section 5.3 demonstrate that better accuracy cannot be obtained by just modifying the model constants and bounding values. This finding suggests that the form of the model equations is inconsistent or at least is not fully capturing the physics involved.

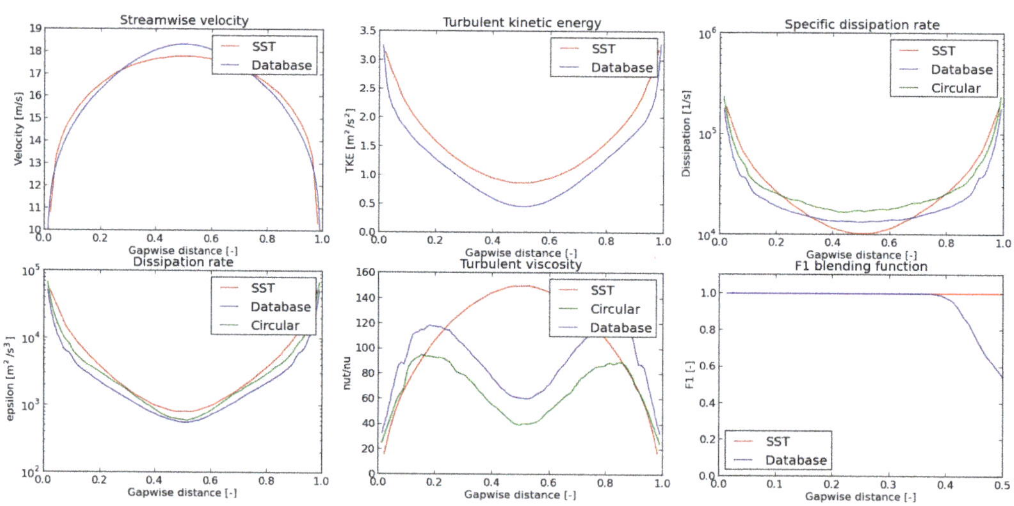


**Figure 9.** Major turbulence properties after the bounding values for k and ω have been corrected to match the database.

## 6.2. Specific Dissipation Rate (ω) Production and Loss

The production of $\omega$ is given by Equation (7) and represents the trace of the mean strain rate squared. The production of dissipation is expected to occur in the wall's proximity, where the maximum shear is taking place. This is well-confirmed by plotting S2 in Figure 10. Practically no dissipation is produced in the bulk of the flow. Such a formulation of the production term was deemed credible and likely is not the cause of improper $\omega$ prediction by the model.

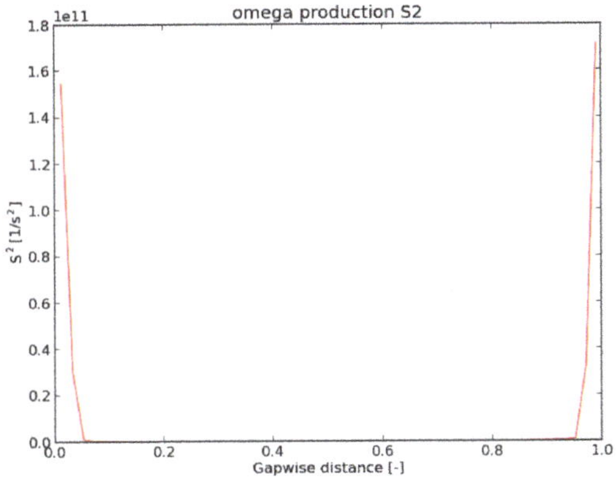


**Figure 10.** Omega production distribution in channel.

The loss term is formulated in Equation (14) as omega squared. To understand this loss term formulation, one must go back to the exact dissipation equation and the reasoning used in the original $k$-$\varepsilon$ model. The well-known form of the $k$-$\varepsilon$ dissipation equation originates in the thesis by Hanjalic [11]. By subtracting the RANS momentum equation from the instantaneous Navier–Stokes equation, differentiating with respect to $x_l$, multiplying by $2\nu\frac{\partial u_i'}{\partial x_l}$, and then, time averaging, an exact equation for the dissipation

$\epsilon = \nu\overline{\frac{\partial u_i'}{\partial x_l}\frac{\partial u_i'}{\partial x_l}}$ can be derived (Equation (33)):

$$\frac{\partial \epsilon}{\partial t} + U_k\frac{\partial \epsilon}{\partial x_k} = -2\nu\frac{\partial U_i}{\partial x_k}\left(\overline{\frac{\partial u_i'}{\partial x_l}\frac{\partial u_k'}{\partial x_l}} + \overline{\frac{\partial u_l'}{\partial x_i}\frac{\partial u_l'}{\partial x_k}}\right) - 2\nu\frac{\partial^2 U_i}{\partial x_k \partial x_l}\overline{u_k'\frac{\partial u_i'}{\partial x_l}} - 2\nu\overline{\frac{\partial u_i'}{\partial x_k}\frac{\partial u_i'}{\partial x_l}\frac{\partial u_k'}{\partial x_l}} - \frac{\partial}{\partial x_k}\left(\overline{u_k'\epsilon'} - \nu\frac{\partial \epsilon}{\partial x_k}\right) - 2\nu\overline{\left(\frac{\partial^2 u_i'}{\partial x_k \partial x_l}\right)^2} - \frac{1}{\rho}\frac{\partial}{\partial x_i}\overline{\left(\frac{\partial p'}{\partial x_l}\frac{\partial u_i'}{\partial x_l}\right)}, \tag{33}$$

where $u_i'$ and $p'$ refer to the fluctuating velocity and pressure, and $U_i$ refers to the mean velocity.

Each of these terms on the right side represent forms of either production or destruction of dissipation and are explained in Hanjalic's thesis. The terms on the right side do not have a closed form, and so, the dissipation equation cannot be solved directly. Multiple assumptions are made to simplify Equation (33) into the form found in the well-known $k$-$\varepsilon$ model (Equation (34)).

$$\frac{D\epsilon}{Dt} = \frac{\partial}{\partial x_i}\left[(\nu + \nu_T)\frac{\partial \epsilon}{\partial x_i}\right] + c_1\frac{\epsilon}{k}\nu_T\left(\frac{\partial U_i}{\partial x_j} + \frac{\partial U_j}{\partial x_i}\right)\frac{\partial U_i}{\partial x_j} - c_2\frac{\epsilon^2}{k} \tag{34}$$

Of interest is the term corresponding to the destruction of dissipation due to viscosity, as shown in Equation (35). Hanjalic [11] simply states that this term "may plausibly be expressed in approximate form as"

$$2\nu^2\overline{\left(\frac{\partial^2 u_i'}{\partial x_j \partial x_k}\right)^2} = c_2\frac{\epsilon^2}{k} \tag{35}$$

The original $k$-$\omega$ model is not significantly different from the $k$-$\varepsilon$ model. The difference is that Wilcox simply solves for the specific dissipation instead of the true dissipation. By defining the specific dissipation as $\omega = \frac{\epsilon}{\beta^* k}$ and noting that $\frac{D\omega}{Dt} = \frac{1}{\beta^* k}\frac{D\epsilon}{Dt} - \frac{\omega}{k}\frac{Dk}{Dt}$, the

$k$-$\omega$ model can be derived directly from the standard $k$-$\varepsilon$ model [12]. However, after this variable transformation is performed, there are additional terms that are proportional to the gradient of $k$ (as in Equation (15)). If these terms are ignored, one arrives at the original $k$-$\omega$ model by Wilcox. When Wilcox changes variables from $\epsilon \to \omega$, the following transformation occurs to the loss term:

$$c_2 \frac{\epsilon^2}{k} \to \beta\omega^2 \tag{36}$$

Changing the variables from $\epsilon \to \omega$ changes each term's dimensions to $1/\mathrm{s}^2$. The form of the $\omega$ equation really is heavily dependent upon the form of the $\varepsilon$ equation and the assumptions that go into modeling each term. It is likely that the loss term was posited based purely on a dimensional analysis. This is explicitly stated in the book by Rebollo and Lewandowski [13]. The authors state that this loss term could be the source of a totally new closure equation but that "a large number of numerical experiments conducted since the work by Launder and Spalding suggest that two closure equations are more than enough, and an additional equation do not bring greater accuracy. This is why we may close [the loss term] using dimensional analysis".

After variables are changed from $\varepsilon$ to $\omega$. The $\omega^2$ form of the loss, present in the $k$-$\omega$ and SST models, appears. By reviewing the original literature, one can conclude that the form of the $\omega$ loss is derived based solely on a dimensional analysis. This conclusion cannot really be trusted on physical grounds, since it involves no derivation from fundamental theories. Its only justification is that it balances the units of other terms in the equation ($1/\mathrm{s}^2$ in the case of the $\omega$ equation, and $\mathrm{m}^2/\mathrm{s}^4$ in the $\varepsilon$ equation). The $k$-$\omega$ and $k$-$\varepsilon$ turbulence models are widely used because of their empirically observed value. If the interest is only in the empirical results, then new expressions for the loss can be suggested, which are not necessarily physical as long as they generate accurate results.

The real implementation of the loss term is implicit and means that $\omega$ is decreased by a factor equal to itself, i.e., to $\omega$, which produces the square. This is the major factor leading to the parabolic shape of $\omega$ seen in Figure 9. It is evident that the loss is excessive in the mid-channel and insufficient next to the wall when applied to HFIR. Suggesting another dependency for the loss as a linear function of $\omega$ and $S$ produces the desired results. The general form of that function is given by Equation (37).

$$\beta\left(a\omega^b + c\sqrt{S2}\right) \tag{37}$$

The best result for application to the HFIR was achieved with $a = 1 \times 10^{-5}$, $b = c = 2$. In order for this new form to have correct units, the constant $a$ must have units of $\mathrm{s}^{b-1}$. This formulation improves the result primarily because of S2 term inclusion. A plausible physical explanation could be because the dissipation is generated and destroyed in the same vicinity of the wall proximity.

After the implementation of these modifications to the SST model, runs were performed for the estimated limiting HFIR Reynolds (Re) numbers of a minimum of 74,000 and a maximum of 95,000. The database flow conditions were chosen for the minimum Reynolds flow number. Both runs are plotted in Figure 11, with circles for the low Re and crosses for the high Re number. The Re number mostly affects the TKE and the viscosity distributions. In general, the lower Re number run better matches with the database, which is an indication of relevance.

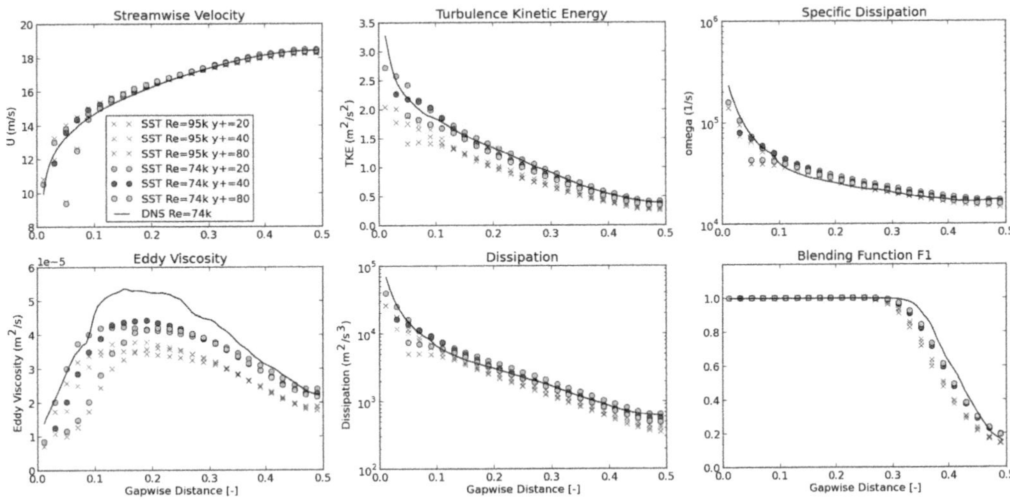


**Figure 11.** Modified SST model results for a range of HFIR-specific Reynolds numbers and meshes with different $y^+$. The SST model implementation includes the modified sink term, with $\omega_1$ and $G_1$ decreased by the same factor.

The effect of the mesh was also tested by running the new SST implementation with different meshes, each with a different $y^+$ for the first node off the wall. Table 7 describes the mesh parameters. The higher $y^+$ meshes use the same total number of cells and the same number of cells across the gap (21), with uniform spacing (no grading). This arrangement results in slightly higher bulk resolution. These results are also shown in Figure 11. The meshes with $y^+ = 80$ deviate from the database in the wall proximity but show fair agreement toward the center of the channel. The best results occur for the meshes with a $y^+ \approx$ 20–40. A lower $y^+$ might also improve performance because the $k$-$\omega$ model is valid within the viscous sublayer, but determining whether that is so was not an objective because it would incur extra computational cost.

**Table 7.** Mesh specifications for new SST model tests.

| Mesh Name | $y_1^+$ | $y_1(m)$ | $y_1/\delta$ | $\Delta y^+$ in Bulk |
|---|---|---|---|---|
| $y^+ = 20$ | 20 | $1.39 \times 10^{-5}$ | $2.19 \times 10^{-2}$ | 46.3–132 |
| $y^+ = 40$ | 39.4 | $2.74 \times 10^{-5}$ | $4.31 \times 10^{-2}$ | 87.9 |
| $y^+ = 80$ | 80.4 | $5.59 \times 10^{-5}$ | $8.80 \times 10^{-2}$ | 79.3 |

Based on this assessment, it can be concluded that the new SST implementation properly simulates the turbulence of HFIR flow in the entire range of Re numbers with moderate-to-low computational resources and can be used for engineering modeling purposes.

### *6.3. SST Model Generalization for Channel Flows*

In the sections above, changes to the SST model were proposed after analyzing the derivations for certain wall functions and the loss term in the specific dissipation equation. Specifically, it was observed that the default bounding value for TKE production ($G_1$) was low and the bounding value for specific dissipation ($\omega_1$) was high when the HFIR channel flow was simulated. These values were corrected by reducing each bounding value by the same constant. In the specific dissipation equation, the loss term was modified according to Equation (37). All these changes together produced a model that better captured HFIR profiles of the streamwise velocity and other turbulence quantities.

These modifications were made based on the database simulation at a single HFIR Reynolds number. Therefore, the introduced changes were specific to the dataset used. To further extend the model to general channel flows, more conditions were analyzed at different Reynolds numbers.

The next section reviews an additional study that was carried out to help generalize the model. The model was evaluated on a simple parallel channel geometry and compared against the DNS results reported by Lee and Moser [9] (referred to as "LM" hereinafter). From these tests, additional relationships were derived that helped generalize the model to a wider range of conditions.

A series of tests was performed using a simple parallel channel geometry (Figure 12). The geometry is identical to that of the LM, with a streamwise length of about $8\pi$, a spanwise length of $3\pi$, and a wall-to-wall distance of $2\delta$.

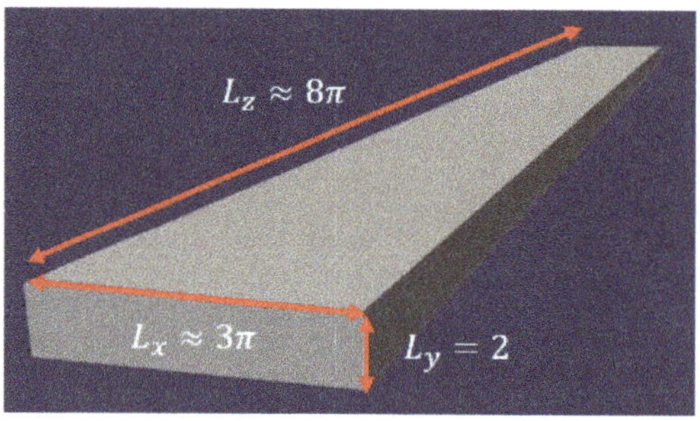


**Figure 12.** Parallel-plate geometry for channel flow testing. Units are in half-gap width—δ.

Periodic boundary conditions were applied in the spanwise ($x$) and streamwise ($z$) directions. The OpenFOAM mean velocity force option was used to keep the average streamwise velocity constant. This was accomplished by dynamically adjusting the pressure gradient, which was the same method used by the LM. Also following the LM, only the kinematic viscosity was varied to reproduce the corresponding frictional Reynolds ($\mathrm{Re}_\tau$) number. In the following, the "baseline SST" refers to the original Menter's formulation, and the "new SST" refers to the modified implementation. For each Reynolds number, the mesh was adjusted to keep the $y^+$ in the range of 20–30.

The new model, with the modified loss term given by Equation (37) and the new bounding values for $\omega$ and $G$, was first tested on the $Re_\tau = 1000$ case, which is the closest to the HFIR $Re_\tau$ of 960. This result is shown in the top four plots of Figure 13. Clearly, one or both constants are somehow not dependent on just $Re_\tau$.

Attention was given to the constants used in Equation (37). Constant $a$ should have units of time in order to balance with the rest of the dissipation equation. An option for providing dimension to constant $a$ is to divide by the shear rate at the wall: $\frac{du}{dy}\Big|_{y=0} = \frac{u_\tau^2}{\nu}$. For the HFIR case, it was found that an $a$ of about $1 \times 10^{-5}$ gave the right distribution for omega. The constant $12\frac{\nu}{u_\tau^2} = 9.37 \times 10^{-6}$ s $\approx 1 \times 10^{-5}$ s gives the right magnitude and units for this term. For the LM $Re_\tau = 1000$ case, this constant is equal to about 0.24 s, because of the viscosity difference. This value is vastly different from that of the HFIR case, even though both cases have nearly the same frictional Reynolds number. Application of this new constant $a$ formulation produces the desired result, as shown in the bottom four plots of Figure 13.

This final implementation was then tested for each of the $Re_\tau$ cases reported by the LM, and the results are shown in Figure 14. The constants used are shown in Table 8. The values of viscosity and friction velocity are the exact values reported by LM. For each run, the amount of the $\sqrt{S2}$ contribution (constant $c$ in Equation (37)) was modified slightly to provide a better fit to the DNS results, but this modification has not been heavily fine-tuned. This contribution increases slightly with $Re_\tau$, but constant $a$ changes more dramatically between the cases. In general, the new model obtains a closer fit to the DNS results. Both the original and the new SST models perform worst for the lower $Re_\tau$ cases. It is interesting to note that the new model improves with higher Reynolds number flows.

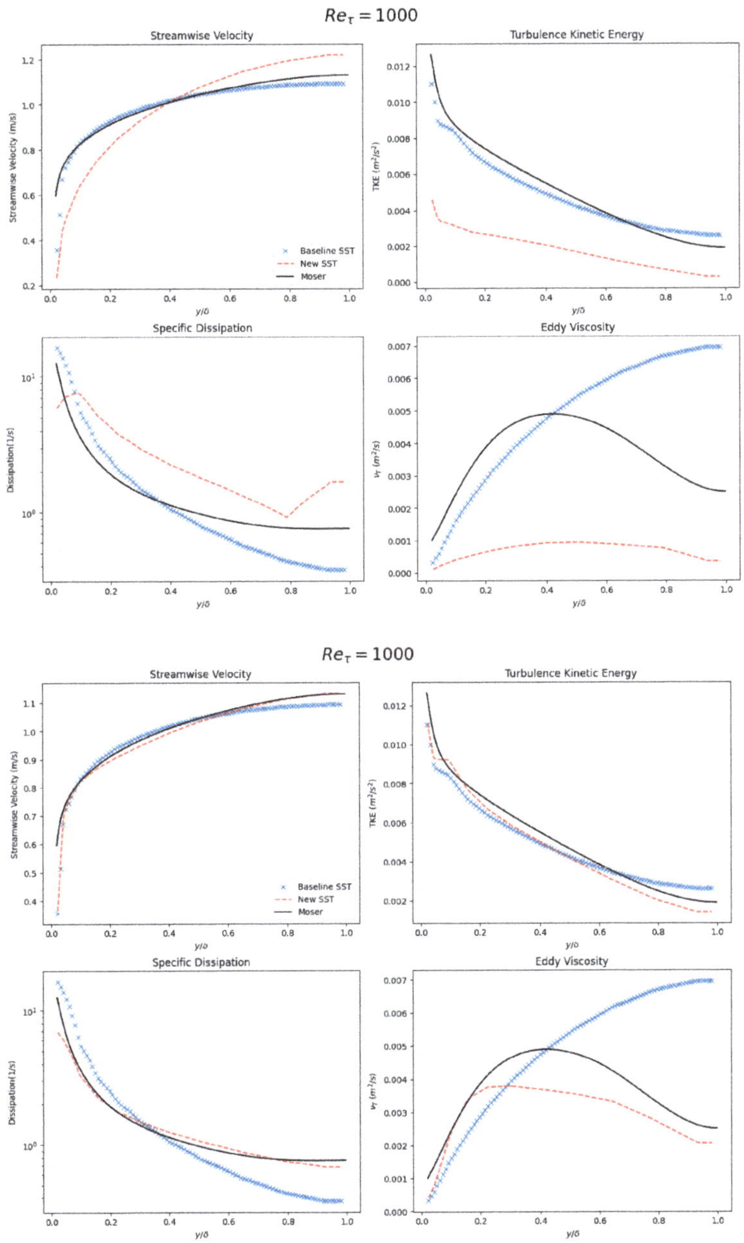


**Figure 13.** Application of new model to the LM case of $Re_\tau = 1000$. Top four plots are before and bottom four are after the constant modification.

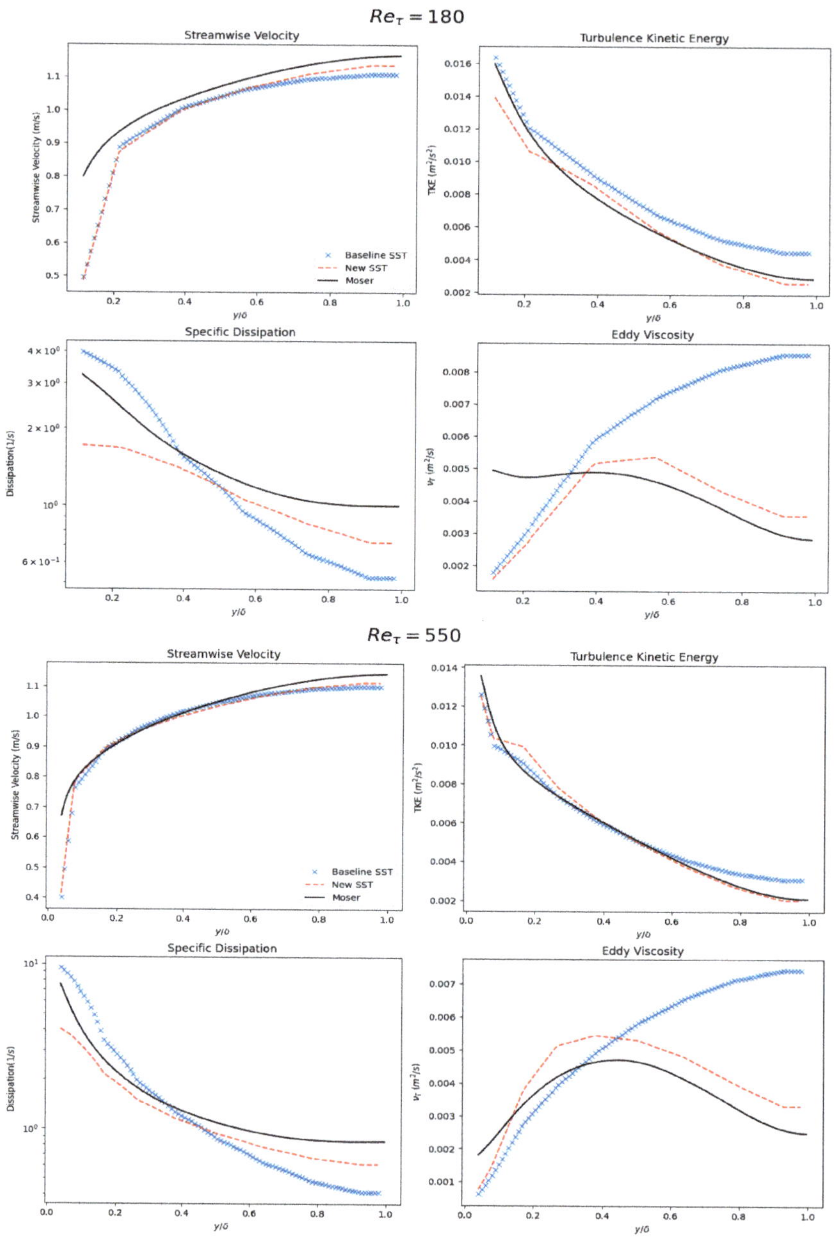


**Figure 14.** *Cont.*

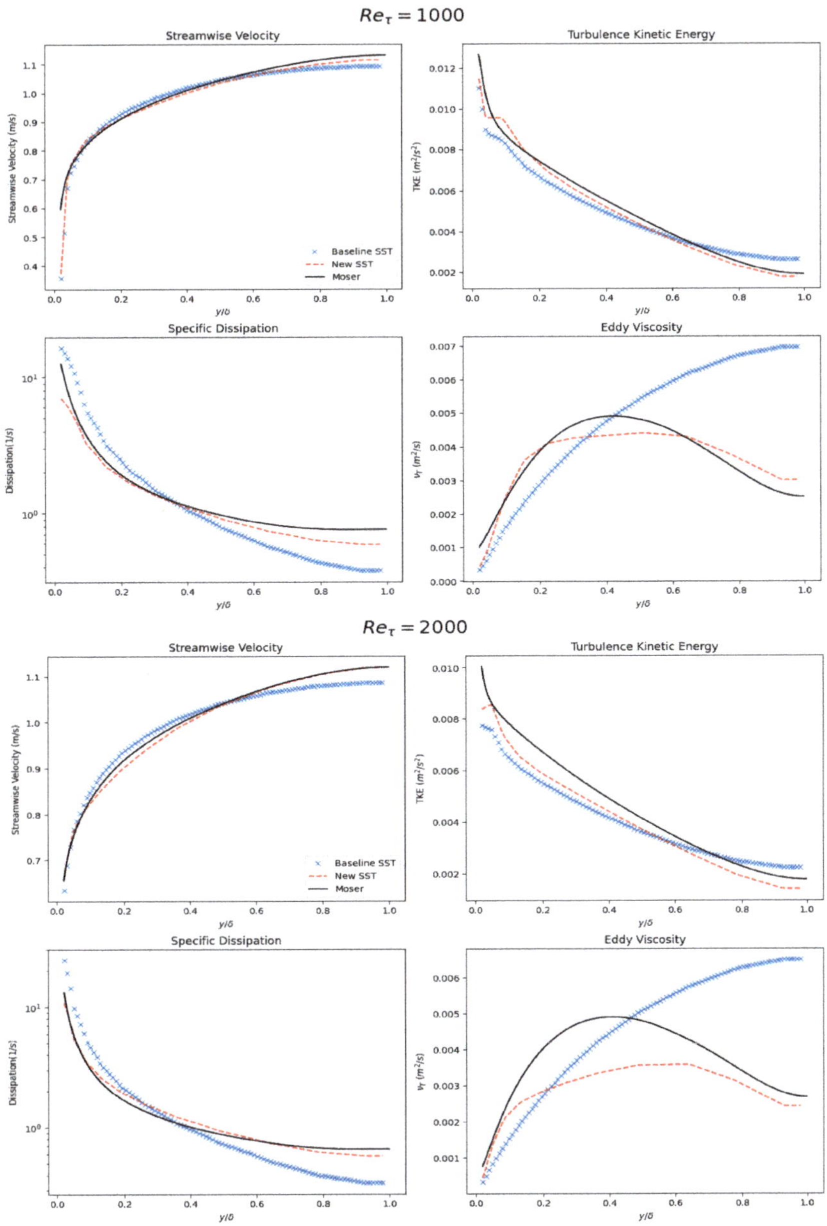


**Figure 14.** *Cont.*

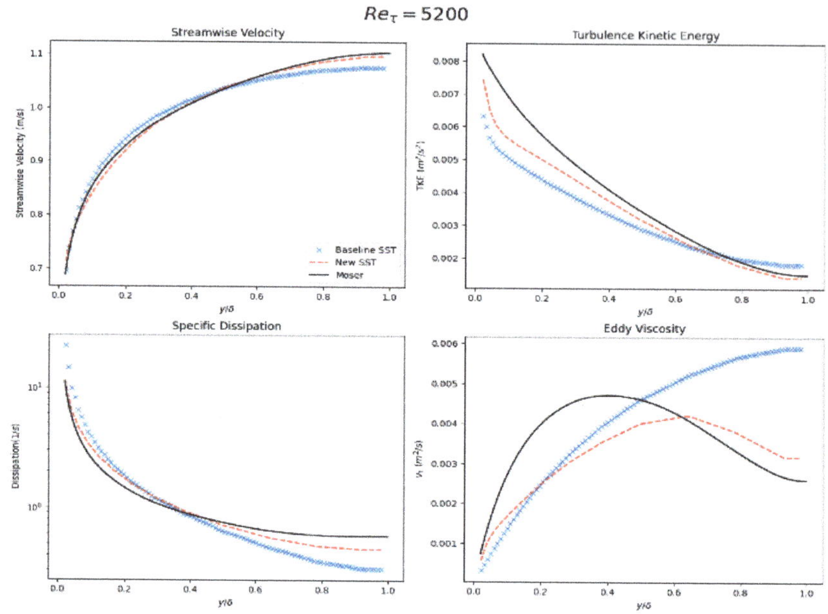


**Figure 14.** Results of parallel channel tests. The new SST model is shown by the red dashed line. The original OpenFOAM SST model is shown by the blue crosses, and Moser data are shown by the black solid line.

**Table 8.** Constants used in new SST model on each Lee and Moser case and HFIR.

| $Re_\tau$ | $Re$ | $\nu$ ($m^2$/s) | $u_\tau$ (m/s) | $a = 12\frac{\nu}{u_\tau^2}$ (s) | $c$ |
|---|---|---|---|---|---|
| 180 | 2857 | $3.5 \times 10^{-4}$ | $6.37309 \times 10^{-2}$ | 1.034 | 0 |
| 550 | 10,000 | $1.0 \times 10^{-4}$ | $5.43496 \times 10^{-2}$ | 0.406 | 2 |
| 1000 | 20,000 | $5.0 \times 10^{-5}$ | $5.00256 \times 10^{-2}$ | 0.24 | 2.5 |
| 2000 | 43,478 | $2.3 \times 10^{-5}$ | $4.58794 \times 10^{-2}$ | 0.131 | 2.75 |
| 5200 | 125,000 | $8.0 \times 10^{-6}$ | $4.14872 \times 10^{-2}$ | 0.056 | 3.25 |
| 960 (HFIR) | 74,000 | $5.64 \times 10^{-7}$ | 0.85 | $1.0 \times 10^{-5}$ | 2 |

One specific of the model is the need for the frictional velocity to be known beforehand. For many flow configurations, it can be easily estimated using the simple wall shear-stress correlation

$$\tau_w = \frac{1}{8} f_D \rho \, U^2,$$

where $f_D$ is the Darcy–Weisbach friction factor calculated via the Colebrook–White equation and $U$ is the average streamwise velocity. This calculation gives a friction velocity estimate of simply $u_\tau = \sqrt{\frac{1}{8} f_D U^2}$, which provides a result close to the true friction velocity reported by LM. Using this estimate of $u_\tau$ instead of the true values results in a negligible change in the new SST model's performance, and it is not reported here. In the future, the new implementation of the SST model derived here should be assessed against more data for channel flows (including experimental) to confirm its validity. It can be tried equally, as well on other channel geometries, to evaluate performance.

## 7. Conclusions

This work brings information obtained by high-resolution studies of HFIR flow turbulence to practical use for simplified, but still sufficiently accurate, engineering simulations. A well-known RANS model of turbulence was studied and modified to allow for the implementation of fast, computationally efficient algorithms in any CFD computer code. To achieve high performance, relatively coarse meshing was used, but relevance was demonstrated on meshes with varying $y^+$ between 20 and 80. The runs were performed on a few (4–8) compute cores with meshes of around 0.5 M elements, and each run takes less than 5 min to execute. It was found that introducing changes in the calculation of the properties' bounding values, and in the dissipation conservation loss term, fundamentally improved the model prediction for channel flows. The accomplishments in this work can be further developed and fine-tuned for specific applications, beyond HFIR. The focus was on the major factors leading to qualitative discrepancies. Those were identified, and methods for correction were offered. The results can be used to compute industrial channel flow turbulence in general, reactor thermal hydraulics, and other relevant applications. Additional validation work might be required to complement that performed in this study.

This study lays a foundation for research on channel flow heat transfer. Proper prediction of turbulence is the key to accurate simulation of the complex thermal phenomena occurring in the narrow HFIR channels, including heat transfer coefficients and channel pressure losses, which are critical for safe reactor operation. The research results have improved the understanding of the flow characteristics and built a foundation to improve reactor performance in the process of transitioning to low-enriched fuels.

**Author Contributions:** Conceptualization, E.P.; methodology, all; software, N.M. and T.G.; validation, all; writing—original draft preparation, E.P.; writing—review and editing, E.P.; visualization, T.G. and N.M. All authors have read and agreed to the published version of the manuscript.

**Funding:** This research was funded by Office of Material Management and Minimization (M3) of the US Department of Energy's National Nuclear Security Administration, grant number DE-AC05-00OR22725.

**Data Availability Statement:** The data presented in this study are available on request from the ORNL-RRD (Research Reactor Division).

**Acknowledgments:** The authors would like to acknowledge the support for this work provided by the Office of Material Management and Minimization (M3) of the US Department of Energy's National Nuclear Security Administration.

**Conflicts of Interest:** The authors declare no conflict of interest.

## Abbreviations

| | |
|---|---|
| CFD | computational fluid dynamics |
| DNS | direct numerical simulation(s) |
| GCI | grid convergence index |
| HFIR | high-flux isotope reactor |
| ORNL | Oak Ridge National Laboratory |
| RANS | Reynolds-averaged Navier–Stokes |
| SST | Shear-stress transport (model) |
| TKE | turbulent kinetic energy |
| $\delta$ | half gap size used as characteristic length |
| $x$ | streamwise direction |
| $y$ | wall direction |
| $z$ | spanwise direction |
| $y^+$ | nondimensional wall distance ($y^+ = y\, u_\tau / \nu$) |
| $Lx/\delta$, $Lz/\delta$ | domain size in gap units |

| | |
|---|---|
| $L^+x, L^+z$ | domain size in wall units |
| $\Delta x^+, \Delta z^+$ | grid size in wall units |
| $\Delta y_w{}^+, \Delta y_c{}^+$ | normal-to-wall grid size, wall units, at the wall w and in the bulk c |
| $Ny$ | resolution in y direction |
| $u, v, w$ | instantaneous velocity components in the lab (orthogonal) coordinate system |
| $U, V, W$ | mean values of velocity components |
| $u', v', w'$ | fluctuating components of velocity |
| $\partial u_i/x_i$ | velocity derivatives tensor |
| $\partial u'_i/x_i$ | fluctuating velocity derivatives tensor |
| $u_\tau$ | frictional velocity |
| $Re$ | flow Reynolds number |
| $Re_\tau$ | frictional Reynolds number |
| $k$ | turbulent kinetic energy |
| $\nu_t$ | eddy viscosity |
| $\omega$ | specific dissipation rate |
| $p$ | pressure |
| $\nu$ | kinematic viscosity |

## References

1. Popov, E.; Bolotnov, I.; Mecham, N. *Numerical Database of HFIR Flow Turbulence, ORNL TM2022/2814*; Oak Ridge National Laboratory: Oak Ridge, TN, USA, 2022.
2. Raisee, M.; Vahedi, N.; Rostamzadeh, A. Prediction of Gas Flow Through Short and Long 2-D Micro and Nano-Channels Using a Generalized Slip Model. In Proceedings of the ASME 2008 6th International Conference on Nanochannels, Microchannels, and Minichannels, Darmstadt, Germany, 23–25 June 2008.
3. OpenCFD Ltd. OpenFOAM: User Guide. 2019. Available online: https://www.openfoam.com/ (accessed on 8 February 2023).
4. Menter, F. Two-Equation Eddy-Viscosity Turbulence Models for Engineering Applications. *Am. Inst. Aeronaut. Astronaut. J.* **1994**, *32*, 1598–1605.
5. Roache, J. Verification of Codes and Calculations. *Am. Inst. Aeronaut. Astronaut.* **1998**, *36*, 696–702. [CrossRef]
6. Menter, F.; Kuntz, M.; Langtry, R. Ten Years of Industrial Experience with the SST Turbulence Model. *Turbul. Heat Mass Transf.* **2003**, *4*, 625–632.
7. Wilcox, D. Reassessment of the Scale-Determining Equation for Advanced Turbulence Models. *AIAA J.* **1988**, *26*, 1299–1310. [CrossRef]
8. Menter, F.; Esch, T. Elements of industrial heat transfer predictions. In Proceedings of the 16th Brazilian Congress of Mechanical Engineering, Uberlândia, Brazil, 26–30 November 2001; pp. 117–127.
9. Lee, M.; Moser, R. Direct numerical simulation of turbulent channel flow up to $Re_\tau \approx 5200$. *J. Fluid Mech.* **2015**, *774*, 395–415.
10. Kalitzin, G.; Medic, G.; Laccarino, G.; Durbin, P. Near-wall behavior of RANS turbulence models and implications for wall functions. *J. Comput. Phys.* **2005**, *204*, 265–291.
11. Hanjalic, K. Two-dimensional Asymmetric Turbulent Flow is Ducts. Ph.D. Thesis, University of London, London, UK, 1970.
12. Bredberg, J.; Peng, S.; Davidson, L. An improved *k*-$\omega$ turbulence model applied to recirculating flows. *Int. J. Heat Fluid Flow* **2002**, *23*, 731–743. [CrossRef]
13. Rebollo, T.; Lewandowski, R. *Mathematical and Numerical Foundations of Turbulence Models and Applications*; Birkhauser: New York, NY, USA, 2014.

fluids

MDPI

*Article*

# CFD-Based Transient Analysis for the Detection and Characterisation of Extended Partial Blockages in Pipes

**Nuno M. C. Martins [1,2,*], Dídia I. C. Covas [1], Bruno Brunone [2], Silvia Meniconi [2] and Caterina Capponi [2]**

1 CERIS, Instituto Superior Técnico, Universidade de Lisboa, Av. Rovisco Pais, 1049-001 Lisboa, Portugal
2 Department of Civil and Environmental Engineering, The University of Perugia, Via G. Duranti 93, 06123 Perugia, Italy
* Correspondence: nunomiguelmartins@tecnico.ulisboa.pt

**Abstract**

Partial blockages in pressurised pipe systems present significant challenges for precise detection, characterisation, and ongoing monitoring. Transient test-based techniques, which utilise sharp but small pressure waves, have shown considerable potential due to their safety and diagnostic capabilities. This paper investigates the transient response of an extended partial blockage—an evolution of a discrete partial blockage that protrudes longitudinally—an increasingly complex condition which has a greater impact on the behavior of pipe systems. Through Computational Fluid Dynamics simulations, the interaction of pressure waves with extended partial blockages of different severity and lengths is examined to assess the resulting pressure response. The results confirm that the pressure signature, generated by extended partial blockages, differs markedly from those of discrete partial blockages. In particular, the magnitudes of the first and second pressure peaks enable accurate characterisation of the severity and extent of the extended partial blockage. These results demonstrate that transient test-based techniques can play a significant role in managing water pipe systems, facilitating more targeted maintenance interventions. Broader implementation of these techniques could enable water utilities to reduce energy consumption, maintain water quality with lower chlorine dosing, and prevent the progression of partial blockages to total pipeline blockage.

**Keywords:** extended partial blockages; hydraulic transients; pressure wave propagation; fault detection

## 1. Introduction

Pressurised pipe systems are fundamental components of modern infrastructures, enabling the reliable transport of vital fluids—such as potable water, natural gas, industrial chemicals, and crude and refined petroleum products—across vast geographical areas. These systems underpin a range of sectors from municipal water supply and energy distribution to industrial activities [1].

In the context of global climate change, the increasing frequency and severity of extreme events—such as droughts and floods—make water demand more uncertain and jeopardise its availability. As a consequence, water system managers face intensifying operational challenges. These include increasing resource scarcity, high energy requirements, growing urban populations, ageing and costly infrastructures, more stringent regulatory frameworks, and increased environmental concerns [2]. Recent reviews highlight that these stressors are accelerating infrastructure deterioration and intensifying the hydraulic vulnerability of pipe systems [3,4]. From the mentioned perspective of water resource scarcity, as an alarming example, the situation of small lakes without natural emissaries—but with a relevant role for the local water supply and a sort of climate sentinel—is emblematic, as in many cases they are at risk of survival, e.g., [5,6].

To address the above challenges, system managers need to improve methods used for the monitoring, diagnosis, prognosis, supervision, and optimisation of water-related systems in alignment with sustainability objectives [7,8]. These issues are compounded by the operational vulnerabilities of pressurised pipe systems, which must maintain uninterrupted flow under demanding conditions, yet are susceptible to faults—most notably leaks, pipe wall deterioration, and partial blockages (PBs)—that compromise continuity, safety, and efficiency. Among these, PBs are recognised as particularly insidious, as they may persist undetected for long periods, progressively degrading system performance while offering little or no external indication of their presence [9,10].

PBs arise from many faults. Physical and chemical deposition processes lead to the accumulation of sediments, mineral scale, biofilms, or corrosion products on the internal pipe wall. In oil and gas pipelines, waxes, paraffin, hydrates, and asphaltenes can precipitate, particularly in subsea environment with low ambient temperature [11]. In these contexts, hydrate and wax blockages remain among the critical flow assurance problems, motivating advanced numerical approaches for prediction and mitigation [12,13]. Air pockets in undulating profiles and partially closed in-line valves can produce similar effects from a hydraulic point of view [14]. All these obstructions narrow the effective pipe cross-section and then increase head losses. Accordingly, they reduce the carrying capacity of gravity pipelines on the one hand and increase the energy demand in elevatory mains on the other hand [15,16]. In addition, PBs may interfere with the dynamics of biofilm formation in the pipe wall with possible microbial proliferation and water quality deterioration [17]. In metallic and concrete pressurised pipelines, due to the possibility of sediments adhering to the internal pipe wall, PBs are predominantly concentric (Figure 1) [9,18]. On the contrary, in polymeric pipes, in most cases, sediments accumulate at the bottom of the pipe.

The reasons outlined above contrast with PBs, which is a priority for water system managers.

(a) (b)

**Figure 1.** Partial blockage (PB) from sediment deposition, reducing the inner section by (**a**) 50%, and (**b**) 90%.

Without external evidence, the most common approaches used for PBs detection include localised intrusive inspections, insertion of probes, or extensive video and ultrasonic surveys [19–22]. While technically robust, such methods are costly, labour-intensive, service-disruptive, and often impractical in buried or inaccessible sections of pipe systems. To overcome these constraints, attention has turned towards non-invasive diagnostic methods capable of detecting and characterising PBs with minimal operational impact.

Among the most promising of these are the transient test–based techniques (TTBTs), which exploit controlled hydraulic transients—those characterised by sharp but small-amplitude pressure waves [23]—to detect and characterise anomalies in a pipeline. When a pressure wave interacts with an anomaly, a part of it is reflected back, whereas the other is transmitted beyond. The characteristics of the reflected pressure wave, i.e., its time of occurrence at the pressure measurement section, entity, and shape, allow it to be located and facilitate the identification of its type and severity without interior pipe access. TTBTs are non-destructive and well-suited to buried or otherwise inaccessible pipes; they are also cost-effective, as only the pressure signal must be acquired at some selected sections. Moreover, the whole procedure can be executed on demand by trained utility personnel, avoiding additional expense and the waiting list of external companies [24,25]. Pressure signal (i.e., the pressure time-history) acquired during transients can be analysed using different approaches e.g., [26–28], as discussed in [29–32].

With regard to PBs, a distinction must be made between *discrete*, DPBs, and *extended*, EPBs, partial blockages. The first type, resulting from the deposition of sediment on the internal pipe wall at an early stage in terms of longitudinal length, includes features (e.g., partially closed in-line valves) that behave like an orifice from the hydraulic point of view (Figure 2a). The second type is the result of the progressive longitudinal development of DPBs along the pipe axis (Figure 2b). Accordingly, the transient response of DPBs and EPBs is remarkably different [33].

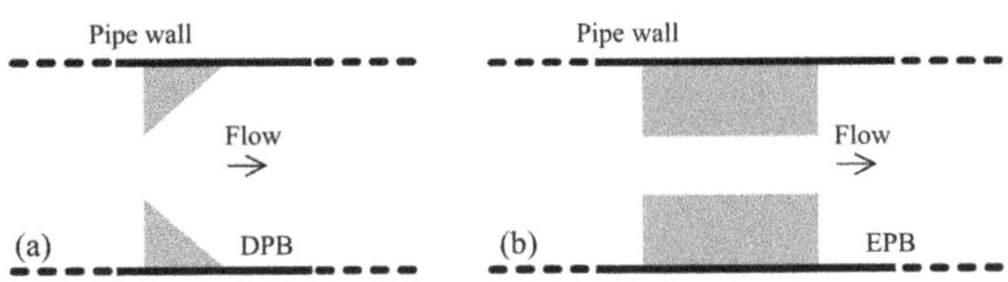


**Figure 2.** Partial blockage classification: (**a**) discrete partial blockage (DPB), and (**b**) extended partial blockage (EPB).

In a previous paper [34], the interaction of pressure waves with a sharp-edged orifice was analysed by a Computational Fluid Dynamics (CFD) model. Scenarios examined by numerical experiments, i.e., layouts with different ratios between the diameter of the orifice and the one of the pipe, confirmed the following key features: (i) The first reflected pressure wave encapsulates the information required for DPB detection and localisation; (ii) The amplitude of the reflected wave increases as the DPB severity increases, allowing its characterisation; and (iii) The characteristics of the acquired pressure signal depend significantly on the relative position of the DPB, and location of the measurement section and the one where the transient is generated. Numerical tests were executed in laminar conditions for two reasons. The first reason is that, because of the small value of the mean flow velocity that characterises such a regime, a small amount of overpressure is generated even for very fast maneuvers. As a matter of fact, within TTBTs fast maneuvers are requested as sharp pressure waves improve significantly their performance in terms of both fault localisation and evaluation. At the same time, small overpressures limit the risk of damage to pipe systems and issues caused by fatigue, as system managers must carry out integrity checks quite frequently, particularly in aging systems. The second reason why the laminar regime is of interest is that pressure waves generated "to explore" the pipe overlap the pre-transient pressure oscillations. As shown in [23], the smaller the value of the pre-transient Reynolds number, the more visible the travelling pressure waves.

This paper, which focuses on EPBs, is structured as follows. A brief description of the CFD model and the system layout and executed numerical tests is given in Section 2. The obtained results are described and discussed in Section 3. Conclusions are listed in Section 4.

## 2. Materials and Methods

### 2.1. Computational Fluid Dynamics (CFD) Model

Numerical simulations have been executed using OpenFOAM (OF) 2106, an open-source CFD software package [35,36]. This tool has been chosen for its robust capabilities in handling transient scenarios and its ability to accurately simulate the intricate interactions between pressure waves and different PB configurations [34,37–43]. OF is based on the Navier–Stokes equations, derived from the conservation principles of mass and momentum. In the absence of mass sources and sinks, the principle of mass conservation is expressed by the following continuity equation:

$$\frac{D\rho}{Dt} + \frac{\partial u_k}{\partial x_k} = 0 \tag{1}$$

where $\rho$ = fluid density, $u_k$ = velocity component, $t$ = time, and $x_k$ = spatial co-ordinate (with $k = 1, 2, 3$).

The principle of momentum conservation, in terms of the Cauchy stress tensor, $\sigma$, is expressed by the Cauchy equation:

$$\rho \frac{Du_i}{Dt} = \frac{\partial \sigma_{ij}}{\partial x_j} + \rho f_i \tag{2}$$

where $f$ = forces per unit mass (e.g., gravity). For isotropic, homogeneous, and Newtonian fluids, the stress tensor is expressed in terms of the velocity field by

$$\sigma_{ij} = -p\,\delta_{ij} - \frac{2}{3}\mu\,\delta_{ij}\frac{\partial u_k}{\partial x_k} + \mu\left(\frac{\partial u_i}{\partial x_j} + \frac{\partial u_j}{\partial x_i}\right) \tag{3}$$

where $p$ = thermodynamic pressure (equal to static pressure for motionless fluids), $\delta_{ij}$ = Kronecker delta, and $\mu$ = fluid dynamic viscosity.

The governing equations are discretised using the finite volume method [44], implemented within the computational framework. The domain consists of a structured mesh of hexahedral cells [45], with local refinement applied to critical regions such as near-wall zones and the vicinity of the EPB.

For the mesh generation (Figure 3), the `snappyHexMesh` tool of the OF framework is employed. The process begins with a base cuboid mesh defined according to the three-dimensional (3D) discretisation of the mesh size parameters proposed in [46]: mesh size in the axial direction, described in terms of number of diameters, $\eta_a = 0.57$; mesh size in the circumferential direction described in terms of the perimeter, $\eta_c = 0.009$; and the ratio between the first layer thickness and theoretical viscous sublayer in the radial direction, $\eta_r = 0.03$. The initial mesh is subsequently refined and adapted to conform to the geometry of the fluid domain, corresponding to a pipe featuring an EPB. The procedure involves successive sculpting and refinement steps until the mesh accurately represents the physical configuration, resulting in meshes from 2 to 9 million points. Consequently, the fluid domain is divided into smaller, non-overlapping subdomains. As highlighted below, this mesh strategy ensures adequate resolution [47] of steep velocity and pressure gradients in the flow field.

Appropriate initial and boundary conditions are imposed. As pre-transient conditions, the fluid is presumed to be at rest ($u_{k,0} = 0$). At the pipe wall, zero relative velocity happens with respect to the fluid (in the simulation, water, with $\rho = 1000$ kg/m$^3$, has been considered) enforcing the no-slip boundary condition. Moreover, as discussed below, where the system layout is described in more detail, a given pressure increase characterises the pipe inlet, determining the corresponding velocity increase. On the contrary, no outlet is assumed at the opposite end section of the pipe where there is a dead end.

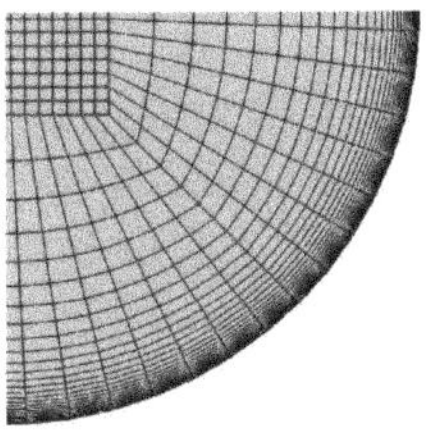

**Figure 3.** Structured hexahedral efficient mesh.

The numerical scheme employs the `cavitatingFoam` solver, a pressure-based solver with a PIMPLE (merged PISO-SIMPLE) algorithm to handle compressibility effects, which is crucial for modelling pressure wave propagation accurately. In particular, the solver uses the Homogeneous Equilibrium Model to determine the compressibility of the liquid according to the selected barotropic equation of state. Temporal discretisation is achieved through a transient scheme, providing stability and precision in time-dependent simulations. Given the laminar flow regime under consideration, turbulence modelling is not incorporated. The convergence of the numerical solution is evaluated by monitoring the residuals that arise during each iteration. In particular, convergence studies have been conducted to optimise mesh resolution and time-step sizes, striking a balance between computational efficiency and solution accuracy [46]. Convergence is achieved when the residuals fall below the threshold of $10^{-7}$.

The above numerical approach has been validated extensively with regard to both steady- and unsteady-state conditions. As the solution of a CFD model depends greatly on the accuracy with which the flow field is simulated, particular attention has been paid to the description of the DPB and EPB geometries with the adoption of an appropriate spatial mesh. To verify the adequacy of the geometry and mesh adopted, the model results were compared with the laboratory experimental data available in [48–54] under steady-state conditions, with satisfactory results [41,55]. Subsequently, the model's ability to appropriately capture the characteristics of transients in pressurized flows was verified [34,42,56].

### *2.2. System Layout and Parameters*

Numerical experiments were carried out in a 3D fluid domain (Figure 4). The setup included a constant pressure head reservoir maintained at a pressure head, $h_r$, equipped with a valve—hereafter referred to as the connection valve—that was initially closed. In fact, at the inlet section, such a valve was connected to a pipe—with an inner diameter, $D$, and length, $X$ (=3 m)—initially at a pressure head, $h_0$, with the subscript 0 indicating the pre-transient conditions. As discussed in detail in [34], such a setup is an idealised version of the Portable Pressure Wave Maker (PPWM) device, refined at the Water Engineering Laboratory (WEL) of the University of Perugia, Italy [23]. At the opposite section of the pipe, there is a dead end.

However idealised, such a layout is not too far from the configuration that real TTBT pipe systems must assume before transient tests to maximise the performance of the method. In fact, closing a valve and then creating a dead end isolates the pipe under study from the other parts of the system. This prevents the entry of pressure waves that are not related to the transient test, which would complicate the subsequent analysis of the pressure signal and then the detectability of the anomalies. The values chosen for $D$ (=0.02 m), $h_r$ (=42.39 m), and $h_0$ (=40.86 m) imply the establishment of a laminar flow during the transient generated by the opening of the connection valve. In fact, as mentioned previously, the smaller the value of the Reynolds number, $Re$ (=$UD/\nu$, with $U$ = mean flow velocity, and $\nu$ = fluid kinematic viscosity), the more effective the identification of the pressure waves reflected by anomalies with respect to the pressure oscillations that characterise the flow. The EPB, with a diameter $d$, starts at $x_{1,bl} = X/2$ and extends over a length $L$. The value of the pressure wave speed, $a$ (=300 m/s)—which is significantly smaller than that of elastic pipes—has been chosen to increase the pipe characteristic time, $\tau = 2X/a$, rather than increasing the pipe length, which would have increased the computation time. The pressure signal, $h$, is acquired at the measurement section located at $x^*_{1,M} = x_{1,M}/X = 0.20$.

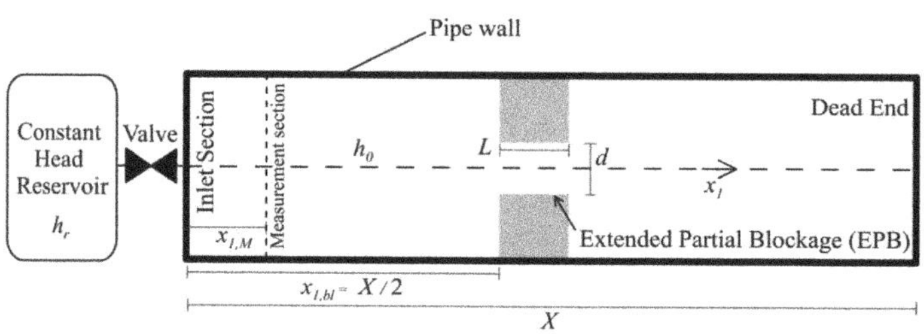


**Figure 4.** Layout schematic representation: a constant-head reservoir connected through a valve to a pressurized pipe, with an EPB located at mid-length.

In the assumed pre-transient conditions—i.e., with the fluid at rest—background disturbances are negligible, ensuring that subsequent pressure variations originate solely from the imposed transient.

The connection valve opening (Figure 5) is characterised by a dimensionless manoeuvre duration $T^*_V = t_V/\tau = 0.10$, with $t_v$ being the manoeuvre duration. The considered manoeuvre profile is based on experimental measurements for a ball valve [42].

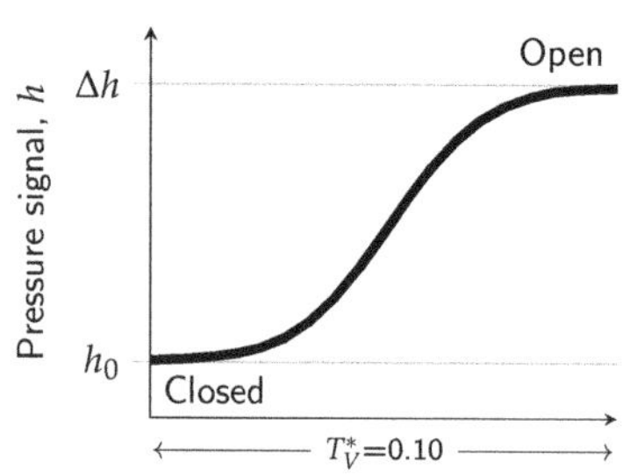


**Figure 5.** Pressure signal at the inlet section, illustrating the valve-opening manoeuvre and resulting pressure surge.

As $h_r > h_0$, a pressure wave $\Delta h = h_r - h_0$ is generated; such a pressure wave propagates along the pipe, crossing the EPB and reaching the dead end. The mean induced velocity, $U$ (=$(g\,\Delta h)/a$) determines the establishment of laminar conditions with $Re$ = 1000.

A comprehensive series of 3D CFD simulations was executed for a wide range of EPB geometries. In particular, for the orifice diameter ratio, $\beta = d/D$, the following values were considered: 5%, 6.25%, 7.5%, 8.75%, 10%, 11.25%, 12.5%, 13.75%, 15%,

17%, 19%, 21.4%, 25%, 35%, 42.5%, 50%, 69.7%, 75%, and 90%; the EPB length ratio, $\phi = L/d$, assumed the following values: 0.5, 1, 2, 5, 10, and 25. Accordingly, in total, 114 numerical simulations were executed—taking an average of 16 h each for a 16-processor unit—allowing a comprehensive exploration of how variations in EPB severity and length affect the transient response.

## 3. Results and Discussion

### 3.1. Role of EPB Severity in the Transient Response

To analyse the role of the diameter ratio, $\beta$, in the EPB transient response, as a first example, the pressure signal at $x^*_{1,M}$, the smallest length of the EPB ($\phi = 0.50$) has been considered for the above-mentioned range of the diameter ratio, $\beta$ (Figure 6); where $T^* = t/\tau$ is the dimensionless time. All cases are characterised by an initial sharp step increase in the pressure signal from $h_0$ to $\Delta h$. Successively, different patterns can be observed according to the value of $\beta$. Precisely, for the smallest value of $\beta$ (=0.05, blue line), the transient response is similar to the one observed for the DPB with the same $\beta$ [34] with the largest peak reaching near $2\Delta h$, which corresponds to the dead end scenario. The intermediate value of $\beta$ (=0.19, green line) shows a much more moderate response, with attenuated amplitude pressure changes. Finally, for the highest value of $\beta$ (=0.90, red line), the effect of the EPB is almost negligible. A notable feature characterises the second pressure rise that results from the overlapping of pressure waves due to the EPB and dead end for the largest value of $\beta$. The implications of this feature extend beyond theoretical understanding to practical system management. Specifically, the observed pressure amplification phenomenon suggests that EPBs, which are associated with close dead ends, may subject pipeline systems to pressure exceeding design specifications based on conventional wave reflection principles. This observation carries implications for system operation and safety protocols.

With regard to the EPB localization, it can be obtained by considering the time interval taken for a pressure wave to travel from the measurement section to the EPB and return. As an example, the time instants $T^*_1$ and $T^*_2$, indicated in Figure 6, allow EPB localisation. Specifically, the calculated distance, $x_c$, of the EPB from the measurement section, $x^*_M$, is given by $x_c = X\Delta T^* = X(T^*_2 - T^*_1)$.

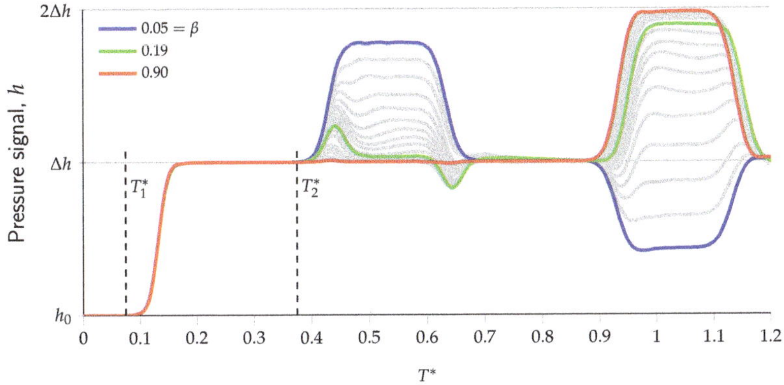


**Figure 6.** Pressure signal at $x^*_{1,M}$ for $\phi$ = 0.50 and different values of $\beta$ (pressure signals for $\beta$ = 0.05, 0.19, and 0.90 are highlighted in bold, whereas those for the intermediate values of $\beta$ are shown in gray).

For the same range of diameter ratio, $\beta$, as a second example, Figure 7 shows the pressure signals concerning the largest EPB length ($\phi$ = 25). Such pressure signals demonstrate distinctive characteristics, which are particularly evident in the case of $\beta = 0.05$, where the extended nature of the EPB and the associated damping effects are observable. These features provide valuable indicators for EPB characterisation through pressure signal analysis.

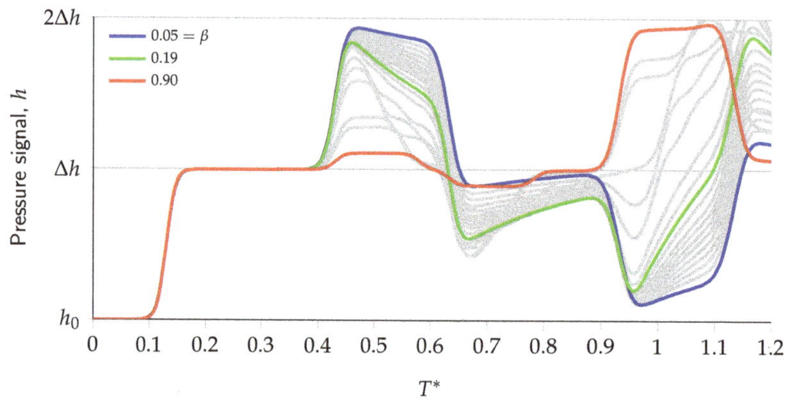


**Figure 7.** Pressure signal at $x^*_{1,M}$ for $\phi = 25$ and different values of $\beta =$ (pressure signals for $\beta$ = 0.05, 0.19 and 0.90 are highlighted in bold, whereas those for the intermediate values of $\beta$ are in gray).

### *3.2. Role of EPB Length in the Transient Response*

The pressure signal at the measurement section, $x^*_{1,M}$, for the diameter ratio, $\beta = 0.05$ is presented in Figure 8, illustrating a range of values of $\phi$ (= 0.50, 1, 2, 5, 10 and 25).

The pressure signal reveals distinct characteristics for different EPB lengths, with all cases enabling precise location identification. The most significant contrast appears between the smallest and largest value of $\phi$ (0.5 and 25, represented by solid and dashed black lines, respectively). The smallest EPB length generates a pressure response similar to DPB with the same value of $\beta$, both in amplitude and duration. Conversely, the largest EPB length produces substantially higher amplitude peaks and exhibits increased decay.

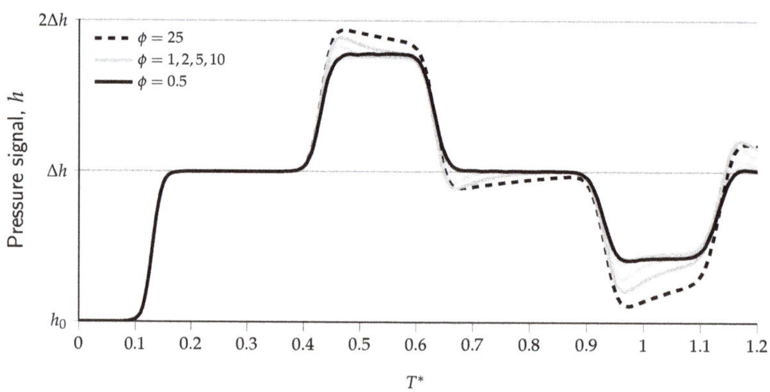


**Figure 8.** Pressure signal at $x^*_{1,M}$ for $\beta = 0.05$ and different values of $\phi =$ (pressure signals for $\phi = 0.50$ and 25 highlighted in black, whereas those for the intermediate values of $\phi$ are in gray).

Intermediate cases, depicted by grey lines ($\phi = 1, 2, 5,$ and 10), demonstrate a progressive evolution in pressure signal characteristics. These traces reveal a systematic increase in peak amplitude and through depth as the EPB length increases, establishing a continuous relationship between blockage length and pressure signal behaviour. This response pattern provides fundamental insights into the relationship between blockage length and transient pressure characteristics, with longer EPBs producing progressively larger pressure changes. The systematic nature of these changes suggests a direct correlation between the length of the EPB and its corresponding transient response, offering valuable diagnostic potential for EPB characterisation in pipeline systems [57].

The pressure signal for $\beta = 0.10$ (Figure 9) shows overall patterns similar to those observed in the previous case, but with some notable distinctions. The transient response maintains the ability to identify EPB location while exhibiting characteristic differences across the range of the EPB length ($\phi = 0.5$ to 25). The smallest length ($\phi = 0.50$, a continuous black line) generates pressure peaks comparable to DPB, through with reduced amplitude compared to the case for $\beta = 0.05$. The largest length ($\phi = 25$, dashed black line) maintains its distinct higher amplitude response, but the magnitude of these fluctuations is notably different from the previous case. The intermediate cases (gray lines) display a systematic progression in pressure response characteristics, through the overall pressure variations are less pronounced than in the $\beta = 0.05$ scenario.

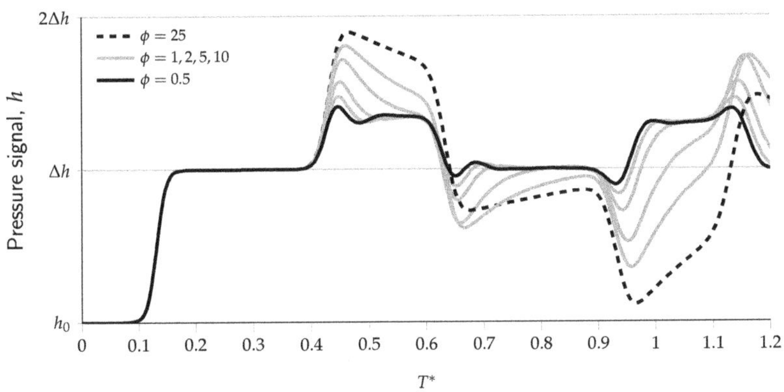


**Figure 9.** Pressure signal at $x^*_{1,M}$, for $\beta = 0.10$ and different values of $\phi =$ (pressure signals for $\phi = 0.50$ and 25 are highlighted in bold, whereas those for the intermediate values of $\phi$ are in gray).

In Figure 10, the pressure signal for $\beta = 0.15$ shows a notable evolution compared to the previous cases ($\beta = 0.05$ and 0.10). This larger diameter ratio reveals distinct behavioural patterns across the range of blockage length, whilst maintaining the fundamental ability to identify blockage locations through pressure signals. For this increased diameter ratio, the smallest length, $\phi = 0.5$, exhibits a markedly different response pattern, with reduced peak amplitude compared to the cases with smaller values of $\beta$. The pressure signal shows a more moderate initial peak, suggesting a fundamental shift in the pressure wave–EPB interaction dynamics. The largest length ($\phi = 25$) generates the highest amplitude response, through the magnitude of these fluctuations is notably attenuated compared to the previous cases.

Intermediate EPB lengths ($\phi = 1, 2, 5$, and 10) display a systematic progression in their pressure response, with each successive increase in length producing incrementally larger amplitude variations. However, these variations are less pronounced than in the cases with smaller diameter ratios, indicating that the larger flow area moderates the pressure wave interactions. This moderating effect is particularly evident in the secondary reflections.

These observations confirm that the influence of EPB length on the transient response becomes more subtle as the diameter ratio increases, suggesting that detection and characterisation methodologies may need to be adjusted accordingly for larger diameter ratios. The above analysis confirms the importance of considering both the EPB length and diameter ratio for understanding and interpreting transient pressure responses in pipeline systems.

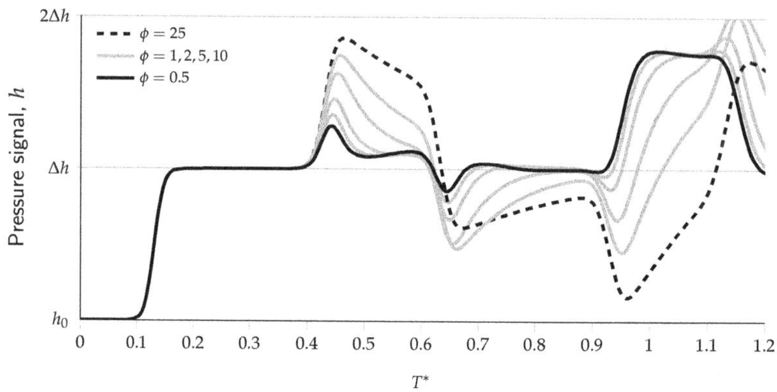


**Figure 10.** Pressure signal at $x^*_{1,M}$, for $\beta = 0.15$ and different values of $\phi =$ (pressure signals for $\phi = 0.50$ and 25 are highlighted in bold, whereas the other signals are shown in gray).

## 4. Conclusions

Extended partial blockages (EPBs) represent the longitudinal progression of discrete partial blockages (DPBs) and create substantial operational challenges for pressurised pipe systems through altered flow dynamics and potential system compromise. This paper investigates the transient response of EPBs using three-dimensional Computational Fluid Dynamics (CFD) simulations based on compressible Navier–Stokes equations implemented in OpenFOAM. Extending upon previous DPB investigations, this analysis examines transient pressure wave interactions with EPBs at mid-pipe locations featuring different orifice diameter ratios and longitudinal extensions.

The results reveal distinct pressure signature patterns for EPBs compared with DPBs. The first reflected pressure wave from an EPB shares qualitative similarities with the DPB response, with its amplitude being inversely proportional to the EPB diameter ratio. However, significant differences arise in the second pressure peak, which can exceed the maximum pressure increase expected under traditional wave reflection theory, sometimes exceeding the inserted pressure wave two times over ($2\Delta h$). This feature, due to the interaction between multiple reflected waves within the EPB region and dead end, carries important implications for structural safety and operation. Moreover, the EPB length significantly influences the transient pressure response. Longer EPBs generate higher amplitude peaks and more complex pressure changes, highlighting the need for refined detection algorithms.

The presented results contribute to improving monitoring strategies for water utilities by advancing non-invasive diagnostic capabilities. Transient test-based techniques leveraging pressure wave reflections can effectively differentiate between DPBs and EPBs whilst providing quantitative estimates of their severity through analysis of reflection magnitude and temporal characteristics. These enhanced diagnostic capabilities enable water system managers to implement optimised maintenance practices through targeted interventions, achieving significant energy savings by preserving hydraulic efficiency and mitigating the risks of progressive EPB development into full obstruction. Furthermore, the improved characterisation methodology supports proactive asset management strategies, reducing operational costs and enhancing network reliability through early detection and timely remediation of partial blockages before they compromise system performance or require emergency interventions.

**Author Contributions:** Conceptualization, N.M.C.M., D.I.C.C., B.B., S.M. and C.C.; Methodology, N.M.C.M., D.I.C.C., B.B., S.M. and C.C.; Software, N.M.C.M., D.I.C.C., B.B., S.M. and C.C.; Validation, N.M.C.M., D.I.C.C., B.B., S.M. and C.C.; Formal analysis, N.M.C.M., D.I.C.C., B.B., S.M. and C.C.; Investigation, N.M.C.M., D.I.C.C., B.B., S.M. and C.C.; Resources, N.M.C.M., D.I.C.C., B.B., S.M. and C.C.; Data curation, N.M.C.M., D.I.C.C., B.B., S.M. and C.C.; Writing—original draft, N.M.C.M., D.I.C.C., B.B., S.M. and C.C.; Writing—review & editing, N.M.C.M., D.I.C.C., B.B., S.M. and C.C.; Visualization, N.M.C.M., D.I.C.C., B.B., S.M. and C.C.; Supervision, N.M.C.M., D.I.C.C., B.B., S.M. and C.C.; Project administration, N.M.C.M., D.I.C.C., B.B., S.M. and C.C.; Funding acquisition, S.M. and C.C. All authors have read and agreed to the published version of the manuscript.

**Funding:** This research has been jointly supported by the University of Perugia "Fondi di ricerca di Ateneo-edizioni 2021 e 2022"; by the Ministry of University and Research (MUR) under the Project of Relevant Interest PRIN2022 "Hybrid Transient-Machine Learning Approach for Anomaly Detection and Classification in Water Transmission Mains (TANDEM)" (CUP: J53D23002110006). Finally, the Authors would like to thank the European Commission, MUR (Italy) (CUP: J93C23002030006, C57J24000010001), Fapesc (Brazil), and FCT (Portugal) for funding in the framework of the collaborative international consortium MORE4WATER financed under the 2022 Joint call of the European Partnership 101060874-Water4all.

**Institutional Review Board Statement:** Not applicable.

**Informed Consent Statement:** Not applicable.

**Data Availability Statement:** The original contributions presented in this study are included in the article. Further inquiries can be directed to the corresponding author.

**Conflicts of Interest:** The authors declare no conflicts of interest.

## References

1. Li, C.; Zhang, Y.; Jia, W.; Hu, X.; Song, S.; Yang, F. Blockage detection techniques for natural gas pipelines: A review. *Gas Sci. Eng.* **2024**, *122*, 205187 . [CrossRef]
2. Rosińska, W.; Jurasz, J.; Przestrzelska, K.; Wartalska, K.; Kaźmierczak, B. Climate change's ripple effect on water supply systems and the water-energy nexus—A review. *Water Resour. Ind.* **2024**, *32*, 100266 . [CrossRef]
3. Fan, Q.; Zhan, J.; Hu, W.; Huang, Z. Effects of orifice size on the efficiency of a heaving oscillating water column. *Eng. Appl. Comput. Fluid Mech.* **2022**, *16*, 666–676. [CrossRef]
4. Li, D.; Wang, C.; Ju, P.; Han, G.; Jin, D.; Wen, J. Effect of the number of orifice plate stages on cavitation characteristics. *J. Phys. Conf. Ser.* **2024**, *2752*, 012168. [CrossRef]
5. Woolway, R.I.; Kraemer, B.M.; Lenters, J.D.; Merchant, C.J.; O'Reilly, C.M.; Sharma, S. Global lake responses to climate change. *Nat. Rev. Earth Environ.* **2020**, *1*, 388–403. [CrossRef]
6. Bongioannini Cerlini, P.; Saraceni, M.; Silvestri, L.; Meniconi, S.; Brunone, B. Monitoring the water mass balance variability of small shallow lakes by an ERA5-Land reanalysis and water level measurement-based model. An application to the Trasimeno Lake, Italy. *Atmosphere* **2022**, *13*, 949. [CrossRef]
7. Yakhot, V.; Bailey, S.C.C.; Smits, A.J. Scaling of global properties of turbulence and skin friction in pipe and channel flows. *J. Fluid Mech.* **2010**, *652*, 65–73. [CrossRef]
8. Walski, T.; Chase, D.V.; Savic, D.; Grayman, W.; Beckwith, S.; Koelle, E. *Advanced Water Distribution Modeling and Management*; Haestad Press: Waterbury, CT, USA, 2003.
9. Duan, H.F.; Lee, P.J.; Che, T.C.; Ghidaoui, M.S.; Karney, B.W.; Kolyshkin, A.A. The influence of non-uniform blockages on transient wave behavior and blockage detection in pressurized water pipelines. *J. Hydro-Environ. Res.* **2017**, *17*, 1–7. [CrossRef]
10. Duan, H.F.; Lee, P.J.; Kashima, A.; Lu, J.; Ghidaoui, M.S.; Tung, Y.K. Extended blockage detection in pipes using the system frequency response: Analytical analysis and experimental verification. *J. Hydraul. Eng.* **2013**, *139*, 763–771. [CrossRef]
11. Olajire, A.A. Review of wax deposition in subsea oil pipeline systems and mitigation technologies in the petroleum industry. *Chem. Eng. J. Adv.* **2021**, *6*, 100104. [CrossRef]
12. Duan, X.; Shi, B.; Wang, J.; Song, S.; Liu, H.; Li, X.; Chen, Y.; Liao, Q.; Gong, J.; Chen, S.; et al. Simulation of the hydrate blockage process in a water-dominated system via the CFD-DEM method. *J. Nat. Gas Sci. Eng.* **2021**, *96*, 104241. [CrossRef]
13. Duan, X.; Shi, B.; Ruan, C.; Jia, J.; Song, S.; Liao, Q.; Chen, Y.; Chen, S.; Wang, J.; Yao, H.; et al. Quantitative assessment of hydrate blockage risk in pipelines based on reliability theory. *J. Nat. Gas Sci. Eng.* **2022**, *98*, 104345. [CrossRef]
14. Escarameia, M. *Air Problems in Pipelines—A Design Manual*; HR Wallingford Ltd.: Oxford, UK, 2005; p. 49.

15. Mays, L.W. *Water Distribution Systems Handbook*; Mays, L.W., Ed.; McGraw-Hill: New York, NY, USA, 2000.
16. Boulos, P.; Lansey, K.; Karney, B. *Comprehensive Water Distribution Systems Analysis Handbook for Engineers and Planners*, 2nd ed.; MWH Soft: Pasadena, CA, USA, 2006.
17. Soini, S.; Koskinen, K.; Vilenius, M.; Puhakka, J. Effects of fluid-flow velocity and water quality on planktonic and sessile microbial growth in water hydraulic system. *Water Res.* **2002**, *36*, 3812–3820. [CrossRef] [PubMed]
18. Likitha, J.; Savitha, M.; Karthik, M.; Kavya, P.; Usha, G. CFD analysis on different shapes of concentric orifice plate for turbulent flow. *Int. J. Eng. Res. Technol.* **2018**, *7*, 421–426.
19. Zain, R.; Terry, A.; Dahing, N.S.; Yusoff, N.; Yahya, R.; Dahing, L.N.S.; Rahman, M.; Mustapha, I.; Hassan, N.; Chik, E.; et al. Technical aspect on procedure of gamma-ray pipeline inspection. In Proceedings of the Nuclear Technical Convention 2015 (NTC 2015), Bangi, Malaysia, 3–5 November 2015; pp. 1–5.
20. Rogers, L. Pipeline blockage location by strain measurement using a ROV. In Proceedings of the Offshore Technology Conference, Houston, TX, USA, 1–4 May 1995. [CrossRef]
21. Qunli, W. Reconstruction of blockage in a duct from single spectrum. *Appl. Acoust.* **1994**, *41*, 229–236. [CrossRef]
22. Qunli, W.; Fricke, F. Estimation of blockage dimensions in a duct using measured eigenfrequency shifts. *J. Sound Vib.* **1989**, *133*, 289–301. [CrossRef]
23. Brunone, B.; Capponi, C.; Meniconi, S. Design criteria and performance analysis of a smart portable device for leak detection in water transmission mains. *Measurement* **2021**, *183*, 109844. [CrossRef]
24. Meniconi, S.; Brunone, B.; Tirello, L.; Rubin, A.; Cifrodelli, M.; Capponi, C. Transient tests for checking the Trieste subsea pipeline: Toward field tests. *J. Mar. Sci. Eng.* **2024**, *12*, 374. [CrossRef]
25. Meniconi, S.; Brunone, B.; Tirello, L.; Rubin, A.; Cifrodelli, M.; Capponi, C. Transient tests for checking the Trieste subsea pipeline: Diving into fault detection. *J. Mar. Sci. Eng.* **2024**, *12*, 391. [CrossRef]
26. Kim, S. Inverse transient analysis for a branched pipeline system with leakage and blockage using impedance method. *Procedia Eng.* **2014**, *89*, 1350–1357. [CrossRef]
27. Kim, D.S.; Shin, S.; Choi, G.B.; Jang, K.H.; Suh, J.C.; Lee, J.M. Diagnosis of partial blockage in water pipeline using support vector machine with fault-characteristic peaks in frequency domain. *Can. J. Civ. Eng.* **2017**, *44*, 707–714. [CrossRef]
28. Mohapatra, P.K.; Chaudhry, M.H. Frequency responses of single and multiple partial pipeline blockages. *J. Hydraul. Res.* **2011**, *49*, 263–266. [CrossRef]
29. Guo, X.; Zhao, L.; Cao, Z.; Deng, J. The development and application of transient-based leak detection methods in pressurized liquid pipelines: A review. *Measurement* **2025**, *257*, 118867. [CrossRef]
30. Ayati, A.H.; Haghighi, A.; Lee, P.J. Statistical review of major standpoints in hydraulic transient-based leak detection. *J. Hydraul. Struct.* **2019**, *5*, 1–26. [CrossRef]
31. Colombo, A.F.; Lee, P.; Karney, B.W. A selective literature review of transient-based leak detection methods. *J. Hydro-Environ. Res.* **2009**, *2*, 212–227. [CrossRef]
32. Xu, X.; Karney, B. An overview of transient fault detection techniques. In *Modeling and Monitoring of Pipelines and Networks: Advanced Tools for Automatic Monitoring and Supervision of Pipelines*; Verde, C., Torres, L., Eds.; Applied Condition Monitoring; Springer International Publishing: Berlin/Heidelberg, Germany, 2017; pp. 13–37. [CrossRef]
33. Brunone, B.; Ferrante, M.; Meniconi, S. Discussion of "Detection of partial blockage in single pipelines" by P. K. Mohapatra, M. H. Chaudhry, A. A. Kassem, and J. Moloo. *J. Hydraul. Eng.* **2008**, *134*, 872–874. [CrossRef]
34. Martins, N.M.C.; Covas, D.I.C.; Brunone, B.; Meniconi, S.; Capponi, C. Transient response of discrete partial blockages in laminar pressurized flow: Insights into pressure and velocity variations by Computational Fluid Dynamics. *Phys. Fluids* **2024**, *36*, 106112. [CrossRef]
35. Jasak, H. OpenFOAM: Open Source CFD in Research and Industry. Technical Report, International Workshop on Coupled Methods in Numerical Dynamics, 2009. Available online: https://openfoam.org (accessed on 1 January 2025).
36. Weller, H.G.; Tabor, G.; Jasak, H.; Fureby, C. A tensorial approach to computational continuum mechanics using object-oriented techniques. *Comput. Phys.* **1998**, *12*, 620–631. [CrossRef]
37. Yang, L.; Fu, H.; Liang, H.; Wang, Y.; Han, G.; Ling, K. Detection of pipeline blockage using lab experiment and computational fluid dynamic simulation. *J. Pet. Sci. Eng.* **2019**, *183*, 106421. [CrossRef]
38. Denton, J.D.; Dawes, W.N. Computational Fluid Dynamics for turbomachinery design. *Proc. Inst. Mech. Eng. Part J. Mech. Eng. Sci.* **1998**, *213*, 107–124. [CrossRef]
39. Fay, J.A. *Introduction to Fluid Mechanics*; MIT Press: London, UK, 1994.
40. White, F.M. *Viscous Fluid Flow*, 2nd ed.; McGraw-Hill: Columbus, OH, USA, 1991.
41. Martins, N.M.C.; Covas, D.I.C.; Meniconi, S.; Capponi, C.; Brunone, B. Hydrodynamics of laminar pipe flow through an extended partial blockage by CFD. *J. Hydroinform.* **2023**, *25*, 2268–2280. [CrossRef]
42. Martins, N.M.C.; Covas, D.I.C.; Capponi, C.; Meniconi, S.; Brunone, B. Unified approach for damping rate of transient laminar flow: Experiments, Computational Fluid Dynamics, one-dimensional, and global models. *J. Fluids Eng.* **2024**, *146*, 1–32. [CrossRef]

43. Chaudhry, M.H. *Applied Hydraulic Transients*, 3rd ed.; Springer: New York, NY, USA, 2014. [CrossRef]
44. Acharya, S.; Baliga, B.R.; Karki, K.; Murthy, J.Y.; Prakash, C.; Vanka, S.P. Pressure-Based Finite-Volume Methods in Computational Fluid Dynamics. *J. Heat Transf.* **2007**, *129*, 407–424. [CrossRef]
45. Ushakova, O.V. *Advances in Grid Generation*; Nova Science Publishers Inc.: Hauppauge, NY, USA, 2007.
46. Martins, N.M.C.; Carriço, N.J.G.; Ramos, H.M.; Covas, D.I.C. Velocity distribution in pressurized pipe flow using CFD: Accuracy and mesh analysis. *Comput. Fluids* **2014**, *105*, 218–230. [CrossRef]
47. Oh, H.W. *Applied Computational Fluid Dynamics*; IntechOpen: London, UK, 2012.
48. Zampaglione, D. Sul moto di una corrente laminare attraverso diaframmi. *L'Energia Elettr.* **1969**, *66*, 821–828. (In Italian)
49. Johansen, F.C. Flow through pipe orifices at low Reynolds numbers. *Proc. R. Soc. London Ser.* **1997**, *126*, 231–245. [CrossRef]
50. Hasegawa, T.; Suganuma, M.; Watanabe, H. Anomaly of excess pressure drops of the flow through very small orifices. *Phys. Fluids* **1997**, *9*, 1–3. [CrossRef]
51. Sahin, B.; Ceyhan, H. Numerical and experimental analysis of laminar flow through square-edged orifice with variable thickness. *Trans. Inst. Meas. Control* **1996**, *18*, 166–174. [CrossRef]
52. Jankowski, T.A.; Schmierer, E.N.; Prenger, F.C.; Ashworth, S.P. A series pressure drop representation for flow through orifice tubes. *J. Fluids Eng.* **2008**, *130*, 1–7. [CrossRef]
53. Kiljanski, T. Discharge coefficient for free jets from orifices at low Reynolds number. *J. Fluids Eng.* **1993**, *115*, 778–781. [CrossRef]
54. Alvi, S.H.; Sridharan, K.; Lakshmana Rao, N.S. Loss characteristics of orifices and nozzles. *J. Fluids Eng.* **1978**, *100*, 299–307. [CrossRef]
55. Martins, N.M.C.; Covas, D.I.C.; Meniconi, S.; Capponi, C.; Brunone, B. Characterisation of low-Reynolds number flow through an orifice: CFD results vs. laboratory data. *J. Hydroinform.* **2021**, *23*, 709–723. [CrossRef]
56. Martins, N.M.C.; Brunone, B.; Meniconi, S.; Ramos, H.M.; Covas, D.I.C. CFD and 1D approaches for the unsteady friction analysis of low Reynolds number turbulent flows. *J. Hydraul. Eng.* **2017**, *143*, 04017050. [CrossRef]
57. Brunone, B.; Maietta, F.; Capponi, C.; Duan, H.F.; Meniconi, S. Detection of partial blockages in pressurized pipes by transient tests: A review of the physical experiments. *Fluids* **2023**, *8*, 19. [CrossRef]

*fluids*

MDPI

*Article*

# Spatial and Energetic Organization of Coherent Structures in Couette–Poiseuille Turbulent Channels

**Sergio Gandía-Barberá and Sergio Hoyas ***

Instituto Universitario de Matemática Pura y Aplicada, Universitat Politècnica de València, 46022 València, Spain; serganba@upv.es

* Correspondence: serhocal@mot.upv.es

**Abstract**

Coherent structures play a pivotal role in wall-bounded turbulence, serving as primary carriers of momentum, energy, and scalar quantities across the flow. This study examines coherent structures, specifically streamwise streaks and intense Reynolds stress regions (Q structures), within a novel DNS dataset capturing a stepped transition from pure Poiseuille flow to pure Couette flow at $Re_\tau \approx 250$, based on the stationary wall. Structures are identified using a percolation algorithm to ensure well-defined boundaries, followed by three-dimensional clustering in Cartesian coordinates. They are further classified as wall-attached or wall-detached based on their proximity to the domain walls. Intense Reynolds stress structures are categorized into quadrants according to the signs of their averaged velocity components. The statistical properties of these structures—encompassing geometric characteristics, energy content, and spatial distribution—are thoroughly analyzed. Particular emphasis is placed on how these properties evolve across the transition from Poiseuille to Couette flow. The results reveal that increasing mean shear in Couette-like cases significantly influences the energy content and spatial distribution of the structures while their geometric characteristics remain relatively consistent across the dataset. This spatial distribution is closely linked to the large-scale structures of the streamwise velocity component in Couette flow, confirming that these structures are genuine physical features rather than artificial artifacts of the flow.

**Keywords:** DNS; Couette flows; Poiseuille flows

## 1. Introduction

Understanding the momentum transfer mechanism in wall turbulence is crucial to predicting drag in wall-bounded turbulent flows [1]. This mechanism is often explained by analyzing artifacts that carry most of the wall-normal momentum flux from the buffer to the logarithmic and outer layers. These artifacts, also known as coherent structures, were first described experimentally as streamwise streaks [2] and Reynolds-stress events [3].

The study of the geometry and interaction among coherent structures is extensive in boundary layers and turbulent channel flows [4–11]. In this study, we focus on the latter. Turbulent channel flows are defined by two parallel walls separated vertically by a distance of $2h$. Within the domain, the flow progresses in $x$-direction, generally being the longest dimension of the domain. Periodic boundary conditions are established in the streamwise and spanwise direction, while the no-slip condition is used in the wall-normal direction. The main control parameter is the friction Reynolds number, $Re_\tau = u_\tau h/\nu$, where $u_\tau$ is the friction velocity and $\nu$ is the kinematic viscosity [12]. We will use a superindex $^+$ to indicate any quantity adimensionalized with $u_\tau$ and $\nu$, as the wall-normal distance, $y^+ = yu_\tau/\nu$.

There are two large groups of channel flows: pressure-driven or Poiseuille [13–15] and wall-driven or Couette [16–18]. A combination of both is also possible [19–21], with surprising applications [22]. Even if the geometry is the same, and the mean flow and intensities are similar, Couette flows have a different particularity. They present long and wide structures that appear as streamwise rolls. Such structures are coherent regions of either positive or negative streamwise velocity fluctuations. They have been observed both experimentally [23,24] and numerically [16,17,19,25–29], persisting even if the flow is influenced by a cross-flow [30] or a streamwise pressure gradient [20]. Lee and Moser [17] obtained structures at least as long as $310h$ for $Re_\tau = 500$. On the other hand, Gandía-Barbera et al. [11] have shown that stratification processes in thermoconvective flows can effectively remove them. Hoyas & Oberlack [18] found that these rolls greatly affect flow statistics and may be finite for large Reynolds numbers. One question we aim to address in this paper is how these rolls impact the characteristics of the different structures within the flow.

This line of inquiry is connected with previous studies on the dynamics of coherent structures in turbulent flows. Hamilton et al. [31] studied the dynamics of the coherent structures present in the near-wall region of turbulent flows using a minimal Couette channel flow. The authors related the origin and disappearance of structures to a temporary quasi-cyclic process of regeneration consisting of three sequential subprocesses: streak formation, streak breakdown, and vortex regeneration. This process was proven to be conditioned to the spanwise dimension of the domain. By reducing the spanwise dimension, a mismatch in the subprocess's timing interrupts the regeneration cycle, which cancels both streaks and rolls. By merging the insights from these two areas, we aim to examine how streamwise rolls influence the overall behavior of coherent structures in wall-bounded turbulent flows. Later, Álcantara-Ávila et al. [32] showed that the flow statistics for domains shorter than approximately $2\pi h$ in the spanwise direction are unreliable.

However, these are not the only structures studied. Del Álamo et al. [4] studied the organization of vortex clusters defined by the discriminant of the velocity gradient tensor [33]. The authors observed that two classes of clusters populate the logarithmic region. One consists of small vortex packets detached from the wall, roughly homogeneous and isotropic. The other is formed by tall clusters rooted in the near-wall region below $y^+ \approx 20$, being self-similar in their Cartesian dimensions. The tall attached clusters are linked to intense velocity structures, which consist of a wall-normal ejection (i.e., burst event) surrounded by two inclined counter-rotating vortexes (i.e., rolls).

Lózano-Duran et al. [5] examined momentum transfer in turbulent channel flows through intense burst events. The authors found that these events can be separated into attached and detached families using the same threshold on $y^+$ as in [4]. Although the attached structures contain around 60% of the total Reynolds stress at various $Re_\tau$ values, the detached structures are dissipative objects of the order of a few local Kolmogorov scales. In fact, along the channel height, detached events with positive Reynolds stress contribution ($Q^+$) tend to be canceled by detached events with negative contribution ($Q^-$). The authors concluded that most of the Reynolds stress is carried by attached negative events extending up to the logarithmic layer and are spatially self-similar.

In a perspective paper, Jiménez [1] extensively studied eddies and coherent structures in wall-bounded turbulent flows. Eddies are statistical artifacts defined mainly through two-point correlations, while coherent structures include dynamics and are generally defined by filtering a given flow parameter, such as, for example, velocity fluctuations or Reynolds stress. The author describes two kinds of correlations for eddies: short ones for

https://doi.org/10.3390/fluids11010018

the transverse velocities (i.e., of the order of the distance from the wall) and a long one for the streamwise velocity (i.e., at least ten times longer than those of the transverse velocities). Coherent structures can generally be related to eddies. In particular, the transverse-velocity eddies correspond to structures defined by strong wall-normal or spanwise velocities or by the Reynolds stress. However, streamwise eddies do not correspond to particularly intense structures but to a concatenation of smaller coherent structures. Interestingly, the conditional flow field around structures of intense Reynolds stress is an inclined roll located between a high- and a low-speed streak of the streamwise velocity for either channels or turbulent homogeneous shear flows. Note that in the former, the impermeability condition of the wall deforms the structures, acting against their symmetry and alignment with the direction of the shear at 45 degrees.

Using a new paradigm, Cremades et al. [34] investigate coherent structures through an explainable deep-learning approach. The instantaneous velocity field derived from a turbulent channel flow simulation is utilized to forecast the velocity field over time using a U-net architecture [35]. The game-theoretic SHapley Additive exPlanations (SHAP) [36] algorithm is employed to evaluate the significance of each structure in the prediction process based on the predicted flow. The findings corroborate previous observations in the literature and extend them by revealing that the essential structures for flow reconstruction do not necessarily contribute the most to the Reynolds shear stress. Continuing this work, the authors state that the classic structures only paint a partial picture of wall-bounded turbulence. A new background to define coherent regions of great importance [37] is introduced using SHAP, obtaining that the most important structures do not coincide with the classical ones.

As noted in previous studies, the coherent structures involved in momentum transfer are sustained through a complex process involving not only linear transient mechanisms (i.e., Orr [38] and lift-up mechanisms) but also non-linearity [1]. Another approach to investigate the self-sustaining mechanism was to use scale decomposition and filtering by Motoori & Goto [39]. The authors examined two sustaining mechanisms: vortex stretching and real-space energy transfer. They characterized the contribution from mean shear to a given scale and among scales. The authors observed that the mean flow transfers energy to the largest scale at each height. For smaller scales, kinetic energy production from the mean flow becomes less important, and the energy transport term between scales plays a major role. Indeed, the threshold defining which mechanism (either production from mean shear or interscale transport) inserts energy on a given scale is the Corrsin scale, where the mean shearing and cascade time scales (i.e., the eddy turnover time, ETT) are balanced. Thus, structures larger than the Corrsin scale extract energy from the mean shear; conversely, the main energy source for structures smaller than the Corrsin scale is the energy cascade, being the donor scale, generally twice the size of the recipient scale.

Furthermore, the increase in computational power in the last decade has allowed the examination of momentum transfer through the spectral analysis of interscale transport ([40–45]). Kawata and Alfredsson [40] decomposed the Reynolds stress equation into a large- and a small-scale part by spatial filtering based on a spanwise cutoff wave number $k_z$. Filtering based on Fourier modes allows proper orthogonal decomposition, in which terms involving both scales are nullified ($\langle u_i^S u_j^L \rangle = \langle u_i^L u_j^S \rangle = 0$), where $u_i^L$ and $u_i^S$ are the large- and small-scale part of the perturbation velocity $u_i$ as $u_i = u_i^L + u_i^S$. In this way, the decomposition of the full-scale equation becomes less complex. For diverse Couette experiments with Reynolds numbers up to $Re_\tau = 108$, both the energy spectra $E_{kt}$ and the cospectra $E_{-uv}$ have their peaks around $\lambda_z/h = 3$. Similar spanwise wavelengths were associated with streaks in the center of Couette flows through one-point correlations of fluctuating streamwise velocity $u$ [20]. Turbulent energy transport $tr_{kt}$ is shown to bring

https://doi.org/10.3390/fluids11010018

energy from the large-scale energy-containing range to smaller scales; however, shear stress transport $tr_{-uv}$ indicates the opposite tendency; that is, an inverse energy cascade.

Kawata and Tsukahara [43] further investigated the mechanism of interscale transport in streamwise and spanwise reduced Couette channel flows. The influence of simulation domain size on very large-scale structures was examined, focusing on why streamwise length reduction is less critical than spanwise length reduction. While the streamwise length reduction affects the Reynolds stresses by suppressing energy redistribution among them, the spanwise length reduction tends to suppress the intensity of all velocity components. Also, in streamwise-minimal channels, the $x-$independent mode ($k_x = 0$) undertakes most of the streamwise turbulent energy ($k_{uu}$), production ($\mathcal{P}_{uu}$) and interscale transport ($T_{uu}$) from the longest wavelengths in the reference case. This fact may explain why streamwise length reduction does not eliminate the very large-scale structures, as the $x-$independent mode assumes its role. On the other hand, the authors observed that the spanwise-minimal channels could reproduce the same tendency on Reynolds shear stress as the reference case despite the obvious differences in the instantaneous flow fields of each case. This suggests that the interscale transfers observed through spanwise Fourier mode analysis may not represent the effect of the interaction between the near-wall and very-large-scale structures. It is very interesting how Kawata and Tsukahara [43] could link the Reynolds stresses' closed-loop transport to the subprocesses of the self-sustaining cycle that likely correspond. Such linkage may help explain how external influences (e.g., stratification) can affect specific production or transference terms and consequently break the self-sustaining cycle. For instance, Gandía-Barberá et al. [11] observed that stable stratification at $Ri_\tau \approx 3$ can suppress the outer streamwise streaks and large-scale counter-rotating rolls, which are common in neutral turbulent Couette channel flows.

The use of coherent structures to study turbulence organization is more frequent in turbulent Poiseuille flows than in Couette ones. First, the inversion of tangential Reynolds stress in Poiseuille flows produces a natural barrier for the growth of coherent structures that extract energy from $\langle uv \rangle$ [43]. This fact facilitates the identification of points and the clustering of structures. Second, while turbulent Couette flows are currently limited to friction Reynolds numbers around $Re_\tau = 10^3$ [18], turbulent Poiseuille flows are simulated at much greater values, being the limit currently at $Re_\tau = 10^4$ [15,46]. Considering that the friction Reynolds number defines the range of scales between the large shear-driven structures and the small structures affected by dissipation, one could argue that Poiseuille flows are preferred as they can provide greater separation between scales and a clear logarithmic region [15]. However, to observe large-scale streamwise elongated modes in Poiseuille flows, the friction Reynolds number must be above $Re_\tau = 2 \cdot 10^3$ [41]. This fact introduces obvious restrictions on computational cost and domain size.

On the other hand, turbulent plane Couette flows are well known to involve very large-scale streamwise elongated structures (i.e., streaks) in the channel center region, filling the entire channel wall-normal and streamwise directions, even at low $Re_\tau$ [17,29,31]. Ref. [20] measured the streamwise and spanwise wavelength of these structures at $Re_\tau = 125$ obtaining $(\lambda_x, \lambda_z) = (50h, 2.5h)$. Thus, a clear scale separation between outer and smaller-scale structures near the wall can be achieved at relatively low Reynolds numbers compared to other wall turbulence configurations, such as turbulent Poiseuille flows in channels, pipes, and boundary layers.

Large-scale phenomena in channel flows can be described from two different perspectives: either dimensional- and time-averaged or conditionally-averaged. In the former, the phenomena are depicted in two-dimensional fields of velocity fluctuations and streamwise vorticity $\omega_x$, showing alternating patterns along the streamwise and spanwise dimensions [17,20]. In the latter, large-scale structures are depicted in a three-dimensional

https://doi.org/10.3390/fluids11010018

space showing interaction among them [1,43]. Here, the conditional average presents a sweep-ejection pair, in which the ejection sits in a low-velocity streak and the sweep in a high-velocity one [1]. The interaction reveals an approximately streamwise roll, rising out of the ejection and sinking into the sweep.

Based on the interaction mentioned above, our study analyzes two types of coherent structures: streamwise streaks and bursts. The streamwise streaks are defined as regions of slowly moving fluid elongated in the direction of the mean flow. Since the seminal paper of Kline [2], low-speed streaks in wall-bounded flows play an important role in turbulence production near the wall and transport to the outer region of the flow. Streamwise vortices form low-speed streaks through a linear advection mechanism, in which low-speed fluid is lifted away from the wall by the vortex into a region of higher-speed fluid, producing a low-speed streak. On the other side of the vortex, high-speed fluid is pushed toward the wall, creating a high-speed streak [31]. This is the only mechanism by which energy from the mean flow can be transferred directly to the streaks. Conversely, streamwise streaks break in a wavy pattern transporting turbulent energy from $\langle u^2 \rangle$ to the velocity components $\langle v^2 \rangle$ and $\langle w^2 \rangle$ through pressure-strain correlation [43]. Bae and Lee [10] found that streaks tend to detach from the wall with a strong wall-normal velocity during the breakdown process. These detached streaks resemble ejections (i.e., Q2-events) when observed through conditional averages.

Bursting, in the form of ejections and sweeps (i.e., Q4-events), comprises structures of intense Reynolds stress. They play important roles in most structural models that explain how turbulent kinetic energy and momentum are redistributed in wall-bounded turbulence. The study of bursting in the form of coherent structures is limited to the past decade. In two separate works, Lozano-Durán [5,47] studied the characteristics of bursts from two different perspectives: first statistically in individual flow realizations and later time-resolved. Osawa and Jimenez [6] studied the statistical attributes of intense Reynolds stress structures and compared them to structures of a minimal norm on the stress tensor. Finally, Gandía-Barberá et al. [11] analyzed the effect of stable stratification on the statistical characteristics and intensity of bursts.

In the present article, the topology (e.g., length, width, height, volume), the intensity of the above-mentioned coherent structures, and the spatial interaction among them are studied in turbulent channel flows with diverse Couette–Poiseuille configurations at $Re_\tau = 250$. The simulation dataset goes from a pure Poiseuille flow to a pure Couette flow, covering intermediate configurations, which are achieved by controlling the velocity of the moving wall. As it will be observed later, introducing an additional degree of freedom in the study of channel flows allows the control of the Reynolds number on the moving wall. This feature makes C-P flows a good candidate for studying the interaction of the large-scale structures in the channel core with the near-wall turbulence [19] or to elucidate the minimum shear ratio necessary to develop very large-scale structures in the channel center. In particular, we are interested in the long and wide structures of the streamwise velocity present in Couette flows and absent in Poiseuille ones. These structures strongly affect the topology and distribution of coherent structures, creating a new self-sustaining cycle away from the viscous region and thus confirming that they are genuine physical features rather than artificial artifacts of the flow.

While previous studies have independently examined coherent structures in Poiseuille and Couette flows, as well as the mechanisms sustaining turbulence in each configuration, a unified understanding of how these structures change across a continuous transition between both flow types is still missing. The present work addresses this gap by analyzing coherent structures, namely streaks and Reynolds-stress–carrying Q events, across a controlled Couette–Poiseuille family at fixed Reynolds number. This approach allows us to

https://doi.org/10.3390/fluids11010018

isolate the effect of mean shear and, in particular, to investigate how the emergence of large streamwise rolls in Couette-like flows reorganizes the geometry, topology, and energetic contribution of coherent structures throughout the channel. By systematically quantifying how streaks and Q events evolve with shear ratio, our study reveals structural changes that are not captured in classical descriptions of wall turbulence and provides new evidence of the interplay between streamwise rolls and the spatial organization of momentum-carrying structures. In this way, the paper extends existing results and offers a new framework to interpret coherent-structure dynamics in mixed Couette–Poiseuille configurations.

The paper is organized as follows. The numerical methodology, the Couette–Poiseuille dataset, and the identification of coherent structures are explained in Section 2. The organization and characteristics of streaks and bursts on each C-P case are described in Sections 3 and 4, respectively. Finally, the conclusions, in Section 5, summarize the findings of this work. This work aligns with the United Nations Sustainable Development Goals 7 and 9, as a better understanding of turbulence can positively influence better energy use and boost innovation [48].

## 2. Methodology and Numerical Simulations

### 2.1. Couette–Poiseuille Simulation Dataset

In this study, a new dataset of direct numerical simulations of Couette–Poiseuille turbulent channel flows (C-P) at $Re_\tau \approx 250$ is calculated. The dataset comprises six C-P cases from pure Couette flow (i.e., C10P00) to pure Poiseuille flow (i.e., C00P10). Viscosity and bulk velocity were kept constant for all cases. See Table 1. The controlling parameter (e.g., degree of freedom) of the C-P dataset is the ratio of the shear stress between the two walls (i.e., moving and static wall) $\gamma = \tau_M/\tau_S$, which is restricted by definition to $-1 \leq \gamma \leq 1$ [19]. Based on the shape of the mean streamwise velocity profile, C-P cases with $\gamma > 0$ are considered Couette-like, while cases with $\gamma < 0$ are considered Poiseuille-like. Following this description, pure Couette and pure Poiseuille flows are ruled by $\gamma = 1$ and $\gamma = -1$, respectively. Special mention must be made of case C06P04 where $\gamma \approx 0$. This case exhibits nearly zero mean shear at the moving wall and can thus be used as a model for understanding turbulent flows near separation [19]. Under such conditions, the wall affects the core flow only through the impermeability condition; that is, wall-parallel velocity components are unaffected by the upper wall.

**Table 1.** Description of the C-P dataset. Two different Reynolds numbers are given depending on the local $Re_\tau$ at the moving ( fourth column) over the stationary (third) wall. $\gamma = \tau_M/\tau_S$ is the ratio of shear stress at the two walls [19]. The next two columns denote the computational time span while statistics were taken in wash-outs $(U_b/L_x)$ and eddy turn-overs $(u_\tau/h)$. $T$ is the computational time spanned by those fields. The last two columns list the percolation indexes used for streaks and Q-events identification acc. to Equations (4) and (5). Line shapes given in the second column are used to identify the cases throughout all the figures of the present paper.

| Case | Line | $Re_\tau^s$ | $Re_\tau^m$ | $\gamma$ | $TU_b/L_x$ | $Tu_\tau/h$ | $\alpha$ | $\beta$ |
|---|---|---|---|---|---|---|---|---|
| C10P00 | —○— | 250 | 250 | 1.00 | 28.2 | 38.6 | 5.00 | 2.25 |
| C08P02 | ——— | 252 | 97 | 0.38 | 32.1 | 44.3 | 4.50 | 2.25 |
| C06P04 | ——— | 254 | 6 | 0.02 | 24.0 | 33.4 | 4.00 | 2.00 |
| C04P06 | ——— | 256 | 54 | −0.21 | 20.4 | 28.7 | 4.00 | 1.70 |
| C02P08 | ——— | 263 | 139 | −0.53 | 25.9 | 37.3 | 4.00 | 1.40 |
| C00P10 | —□— | 275 | 275 | −1.00 | 36.4 | 54.9 | 4.00 | 1.40 |

The simulations were performed in a computational box of sizes $L_x = 8\pi h$, $L_y = 2h$ and $L_z = 6\pi h$. The streamwise, wall-normal, and spanwise coordinates are denoted by

https://doi.org/10.3390/fluids11010018

$x$, $y$, and $z$, and the corresponding velocity components are $U$, $V$, and $W$. Statistically averaged quantities are denoted by an overbar, whereas fluctuating quantities are denoted by lowercase letters, i.e., $U = \overline{U} + u$. Primes are reserved for intensities, $u' = (\overline{uu})^{1/2}$. Quantities averaged in time or along any particular direction are denoted using angle brackects; that is, $\langle U \rangle_{xzt}$ is the streamwise velocity averaged in the homogeneous directions and time.

The C-P flows can be described through the mass balance and momentum equations:

$$\partial_j U_j = 0, \tag{1}$$

$$\partial_t U_i + U_j \partial_j U_i = -\partial_i P + \frac{1}{\mathrm{Re}_\tau} \partial_{jj} U_i, \tag{2}$$

where repeated subscripts indicate summation over 1, 2, and 3. The pressure term includes the constant density $\rho$. The previous equations are transformed into an equation for the wall-normal vorticity $\omega_y$ and for the Laplacian of the wall-normal velocity $\phi = \nabla^2 v$. The spatial discretization uses dealiased Fourier expansions in $x$ and $z$, and seven-point compact finite differences in $y$, with fourth-order consistency and extended spectral-like resolution [49]. The temporal discretization is a third-order semi-implicit Runge-Kutta scheme [50]. The algorithm is described in [51]. This code exhibits significant parallelizability, encompassing I/O operations, facilitating computations demanding substantial computational power [18]. Results from this code have been used largely by the community [52–55], so we consider the code and its results validated.

The mesh has a size of $(768, 251, 1152)$ points, which gives a resolution of 8.2 and 4.1 wall units in $x$ and $z$, respectively. The wall-normal grid spacing is adjusted to keep the resolution at $\Delta y = 1.5\eta$, i.e., approximately constant in terms of the local isotropic Kolmogorov scale $\eta = (\nu^3/\varepsilon)^{1/4}$. In wall units, $\Delta y^+$ varies from 0.37 at the wall to $\Delta y^+ \simeq 2.68$ at the centreline. This grid size is similar to the one typically used in channel flows for Poiseuille and Couette flows [15,18]. The simulations ran for at least 25 ETT to obtain accurate statistical information [56].

### 2.2. The Identification of Coherent Structures

This study defines coherent structures as three-dimensional regions of the flow in which given conditions are fulfilled. The streamwise streaks are identified using the method used in [10]. In a streak, the streamwise velocity fluctuation shall be negative, and the wall-parallel velocity shall surpass a set threshold,

$$u(x,y,z,t) < 0, \tag{3}$$

$$\sqrt{u^2(x,y,z,t) + w^2(x,y,z,t)} > \alpha u_\tau^s, \tag{4}$$

where $u_\tau^s$ is the friction velocity measured on the static wall, and the parameter $\alpha$ is the percolation index for streak identification. It is also possible to define high-velocity streaks where $u > 0$. However, as low-velocity streaks are thought to be more important close to the wall [10], we will only focus on this second, using only streaks to denote them.

The structures of intense Reynolds stress (i.e., bursts) are identified through the same approach as in [5]. Intense Reynolds stress structures are defined by

$$|\tau(x,y,z,t)| > \beta\, u'(y) v'(y) \tag{5}$$

where $\tau(x,y,z,t) = -u(x,y,z,t)v(x,y,z,t)$ is the instantaneous point-wise tangential Reynolds stress and $\beta$ is the percolation index for $uv$-structure identification.

https://doi.org/10.3390/fluids11010018

Both coherent structures are defined by connecting identified points in the six orthogonal directions of the direct numerical simulation Cartesian mesh. This algorithm is detailed in [57]. Once the clusters satisfying the threshold are identified, coherent structures with a volume less than $30^3$ wall units are discarded to reduce noise in identifying structure interactions. A coherent structure's volume is measured here as the sum of the volume of individual structure voxels, defined as in [58].

The percolation indexes for streaks (4) and $uv$-structures (5) are estimated through percolation analysis [4,5]. Figure 1a,b show the results of the percolation analysis for streaks and $uv$-structures, respectively. In both figures, the volume ratio $V_{\max}/V_{\text{tot}}$, composed by the maximum volume of a single structure ($V_{\max}$) and the sum of the volume of all identified structures ($V_{\text{tot}}$), tends to unity at low percolation values. Here, the scenario comprises a very large coherent structure that merges most identified points and several small structures without a relevant volume proportion. On the other hand, at high percolation ratios, both the volume ratio and the ratio of identified objects $N/\max(N)$ become small, resulting in very few but very intense structures.

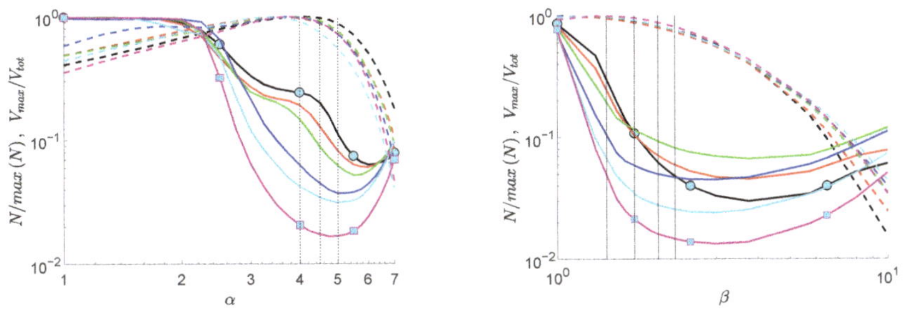


(**a**) (**b**)

**Figure 1.** Lines as in Table 1. Percolation study for identifying streaks (**a**) and $uv$-structures (**b**). At each percolation index, the curves depict the ratio of the volume of the largest identified object against the volume of all identified objects $V_{\max}/V_{\text{tot}}$ (solid lines) and the ratio of the number of identified objects against the maximum number of objects $N/\max(N)$ (dashed lines). To avoid many tiny structures, those with a volume less than $V^+ = 30^3$ have been removed before the study. The vertical dotted lines denote the selected percolation indexes according to Table 1.

In Figure 1a, the evolution of the volume ratio has two different trends. The ratio decreases steadily in cases from C00P10 to C04P06, whilst in cases from C06P04 to C10P00, the ratio shows a saddle at $\alpha \approx 3.75$. This fact provides the first evidence that the intensity of streamwise streaks is organized differently in the cases with a predominant Couette proportion. Pirozzoli et al. [19] showed that in a case with shear stress ratio $\gamma \approx 0$ (similar to our C06P04), the streamwise streaks can grow up to the channel center as in pure Couette flows. Similarly, Gandía-Barberá et al. [20] determined that the presence of alternating very-large streamwise streaks in the channel center starts in case C06P04.

In Figure 1b, the volume ratio decreases faster at $1 \leq \beta \leq 2$ when comparing cases C00P10-C04P06 against C06P04-C10P00. However, it is worth noting how the volume ratio of cases C04P06, C06P04, and C08P02 reach their minimum in greater values than C10P00. This can be explained by the fact that in cases C04P06 to C08P02, the shear Reynolds number at the moving wall (Table 1) is too low to generate relevant structures in the upper region of the domain. Therefore, the overall identified volume $V_{tot}$ in these cases is less than in case C10P00 due to the missing contribution of the upper area, increasing the volume ratio.

https://doi.org/10.3390/fluids11010018

The requirements for selecting a percolation index are to reach a volume ratio below 10% while keeping the ratio of identified objects near unity [4]. In this way, it is ensured that a representative population structure is identified and structures are separated enough to provide valid statistical results. The selected percolation indexes are gathered in Table 1. Similar percolation indexes were selected by Bae et al. [10] for streaks ($\alpha$) in C00P10. Furthermore, the percolation indexes for $uv$-structures are similar to the selected values in [5] and [11] for C00P10 and C10P00 cases, respectively. Small variations in the indices do not alter the qualitative or quantitative results of the analysis. In our case, all reported ratios remain essentially unchanged under reasonable perturbations of $\alpha$ and $\beta$. Moreover, this same methodology has been used consistently in our previous works ([34,37]) with comparable robustness.

## 3. Streaks

### *3.1. The Geometry and Shape of Individual Streaks*

Streaks and other coherent structures are generally classified first based on their distance to the nearest domain wall [5]. Figure 2 shows the joint probability density function (PDF) of the wall-normal distance to the closest wall from streak bottom $(y^+_{min})$ and top $(y^+_{max})$ on the cases listed in Table 1. Two areas gather at least 50% of the plotted data in all cases. These areas are separated by a transition zone with less data density at $y^+_{min} = 20$. Thus, this value defines two kinds of streaks: wall-attached and wall-detached [5]. The former, from now on referred to simply as attached, are defined by $y^+_{min} \leq 20$ while the latter, detached, are defined by $y^+_{min} > 20$. A similar streak classification is employed in [10]. Both attached and detached regions in Figure 2 exhibit a power-law relation between $y^+_{min}$ and $y^+_{max}$. The attached region presents $y^+_{max} \propto (y^+_{min})^{-4.9}$ for cases C00P10, C02P08 and C04P06. On the contrary, cases C06P04, C08P02 and C10P00 have a distribution $y^+_{max} \propto (y^+_{min})^{-7.6}$. This fact denotes that the vertical growth of the attached streaks behaves differently depending on the sign of the shear stress ratio, $\gamma$. Moreover, the change in the proportionality is not gradual with the shear stress ratio, as inclination only changes once the shear ratio reaches positive values. That is, Couette-like cases (i.e., from C06P04 to C10P00) follow the same distribution independently of the value of the shear stress ratio. The same occurs in Poiseuille-like cases. Therefore, Figure 2 indicates that the mechanism of streak generation changes after case C04P06. This fact is aligned with the observations in [20], in which the presence of alternating very-large streamwise streaks in the channel center starts in case C06P04.

On the other hand, the distribution of detached streaks follows $y^+_{max} \propto (y^+_{min})^{0.76}$. This distribution does not change among the different cases, suggesting that the height of these streaks is less affected by the underlying mechanism in each case.

The organisation of streaks in streamwise- and time-averaged planes is shown in Figure 3. In cases with a negative shear ratio $(\gamma)$, no spanwise pattern in the arrangement of streaks is observed. However, an alternating pattern appears for positive shear ratios, with a spacing of approximately $3.75h$ between streak regions. The alternating spanwise arrangement of streaks emerges first at the bottom static wall, as the friction Reynolds number in this region is always in the turbulent regime. See Table 1. The lack of streaks in the upper moving wall in cases C04P06 to C08P02, independently of the shear ratio, is due to the insufficient friction Reynolds number in this region to sustain turbulence.

Furthermore, this organization remains consistent when the shear ratio becomes positive, regardless of the Couette proportion in each case. This behavior aligns with the results for the attached streaks seen in Figure 2.

Following the streak classification in [10], the authors explored the potential identification of tall attached structures in the simulations. These structures would not only

https://doi.org/10.3390/fluids11010018

extend down to $y_{min} \leq 20$, but would also be taller in $y_{max}$ or $y_{cm}$ (i.e., the height of the structure's center of mass) than the rest of the attached structures. Figure 4 shows the PDF of the $y_{cm}$ associated only with attached streaks. The figure does not present any distinction among attached streaks. Nevertheless, after defining structures with $y^+_{cm} > 50$ as potential tall-attached streaks and studying their characteristics, these structures were found to be conglomerates of attached streaks. The reason is that potential tall-attached structures exhibited dimensional ratios different from those of attached streaks in our study, which deviates from the self-similarity observed in all types of attached structures, whether tall-attached or just attached [10]. Thus, the streaks are classified in this study as either attached ($y_{min} \leq 20$) or detached ($y_{min} > 20$).

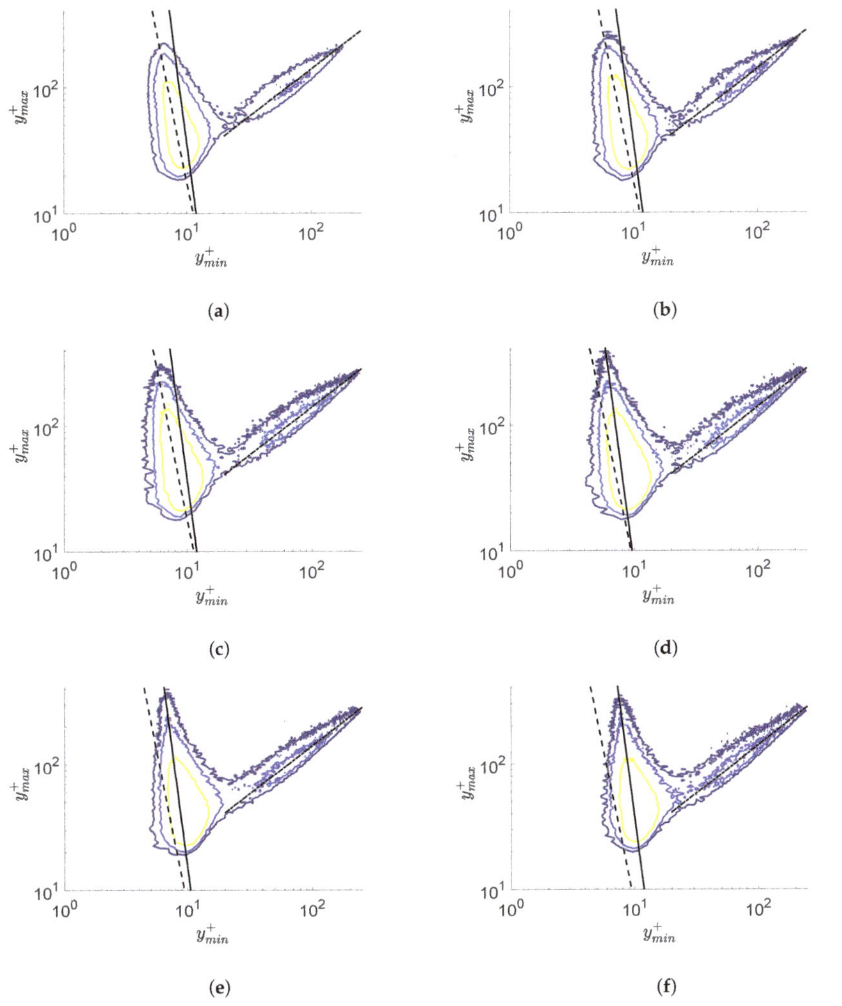

**Figure 2.** Joint PDF of the wall-normal distance to the closest domain wall from streak bottom $(y^+_{\text{min}})$ and top $(y^+_{\text{max}})$. **Top row**: cases C00P10 (**a**) and C02P08 (**b**). **Middle row**: cases C04P06 (**c**) and C06P04 (**d**). **Bottom row**: cases C08P02 (**e**) and C10P00 (**f**). Contours contain 50%, 70%, and 90% of the total data (Dark blue to yellow). The dashed line represents $y^+_{\text{max}} \propto (y^+_{\text{min}})^{-4.9}$, the solid line $y^+_{\text{max}} \propto (y^+_{\text{min}})^{-7.6}$ for attached streaks in Poiseuille and Couette cases, respectively, and the dash-dotted line $y^+_{\text{max}} \propto (y^+_{\text{min}})^{0.76}$ for detached streaks.

https://doi.org/10.3390/fluids11010018

The average ratio of attached and detached streaks and their volume is listed in Table 2. Attached streaks are predominant among them in all cases, making up, on average, 70% of all streaks. These results indicate that the use of different percolation indexes, $\alpha$, across cases (See Table 1) influences neither the number of streaks of each type nor their overall volume. Indeed, the results highlight a similar scenario in all cases despite the clear reallocation observed in Figure 3.

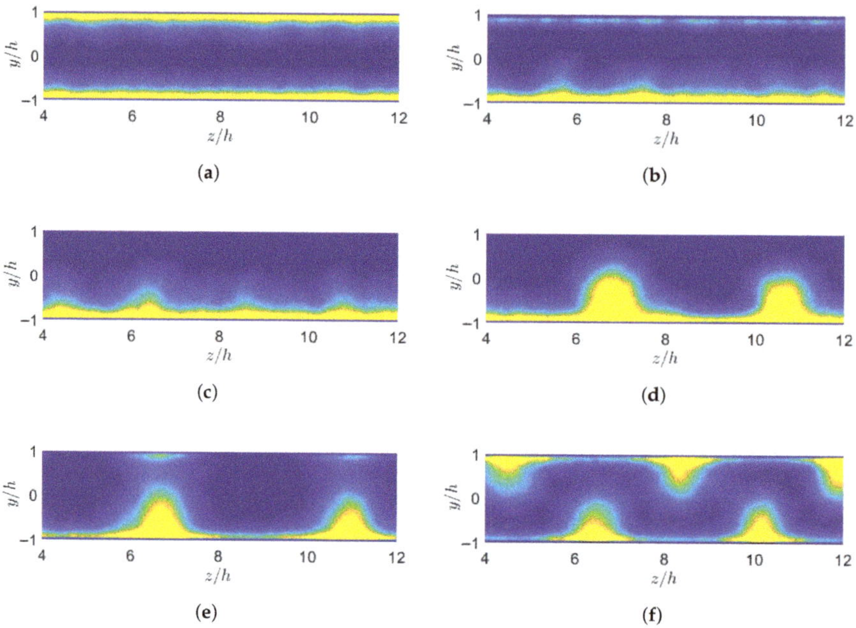


**Figure 3.** Streak points position averaged in time and $x$, normalized by the local maximum. Blue indicates a lack of streak points, while yellow marks the regions of highest density. **Top row**: cases C00P10 (**a**) and C02P08 (**b**). **Middle row**: cases C04P06 (**c**) and C06P04 (**d**). **Bottom row**: cases C08P02 (**e**) and C10P00 (**f**). To improve visibility, the values of $z/h$ are limited to the two central rolls.

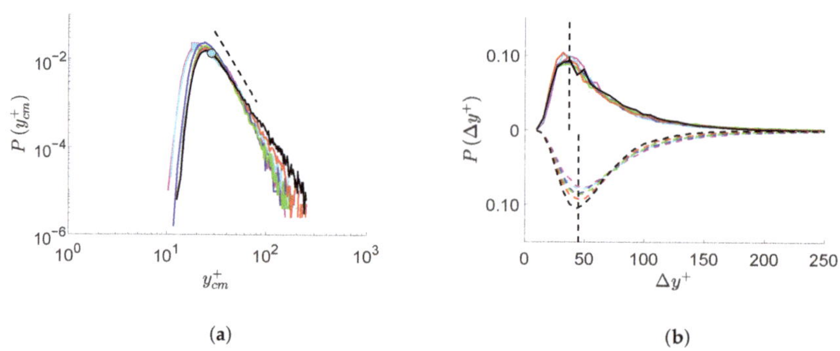


**Figure 4.** Lines as in Table 1. (**a**) PDF of the center of mass $(y^+_{cm})$ of attached streaks. The dashed line represents a decay proportional to $(y^+_{cm})^{-3.75}$. (**b**) PDF of the height of attached (solid lines) and detached (dashed lines) streaks. Vertical dashed lines are placed at $\Delta y^+ = 37$ and $\Delta y^+ = 45$ for reference.

As streaks are typically meandering structures, obtaining a simpler geometric configuration is necessary to study their spatial orientation [59]. Each identified streak is contained within a Cartesian box aligned with the domain axes to achieve this. The box's

https://doi.org/10.3390/fluids11010018

length, width, and height are defined by the parameters $\Delta x$, $\Delta z$, and $\Delta y$, respectively (see Figure 5). A spline is defined for each identified streak by calculating the streak's center of mass at each $\Delta y - \Delta z$ plane. The spline points $(x_s, y_s, z_s)$ are then used to generate a linear approximation $(x_s, A_2x_s + B_2, A_3x_s + B_3)$ based on least squares fitting, where:

$$A_k = \frac{N\sum_{i=1}^{N} x_{s_i}x_{k_i} - \sum_{i=1}^{N} x_{s_i}\sum_{i=1}^{N} x_{k_i}}{N\sum_{i=1}^{N} x_{s_i}^2 - \left(\sum_{i=1}^{N} x_{s_i}\right)^2},$$
$$B_k = \frac{\sum_{i=1}^{N} x_{k_i}\sum_{i=1}^{N} x_{s_i}^2 - \sum_{i=1}^{N} x_{s_i}\sum_{i=1}^{N} x_{s_i}x_{k_i}}{N\sum_{i=1}^{N} x_{s_i}^2 - \left(\sum_{i=1}^{N} x_{s_i}\right)^2}, \tag{6}$$

being $N$ the amount of spline points and $k = \{2, 3\}$, where $x_2 = y_s$ and $x_3 = z_s$.

**Table 2.** Cases as in Table 1. The ratio of attached ($N_{att}$) and detached ($N_{det}$) streaks per case, normalized with the total amount of identified streaks ($N_{all}$). The ratio of attached ($V_{att}$) and detached ($V_{det}$) streak volumes per case normalized with the simulation domain volume ($V_{dom}$).

| Case | $N_{att}/N_{all}$ | $N_{det}/N_{all}$ | $V_{att}/V_{dom}$ | $V_{det}/V_{dom}$ |
|---|---|---|---|---|
| C00P10 | 0.74 | 0.26 | 0.0145 | 0.0011 |
| C02P08 | 0.72 | 0.28 | 0.0097 | 0.0009 |
| C04P06 | 0.66 | 0.34 | 0.0101 | 0.0012 |
| C06P04 | 0.65 | 0.35 | 0.0142 | 0.0012 |
| C08P02 | 0.67 | 0.33 | 0.0113 | 0.0014 |
| C10P00 | 0.66 | 0.34 | 0.0131 | 0.0021 |

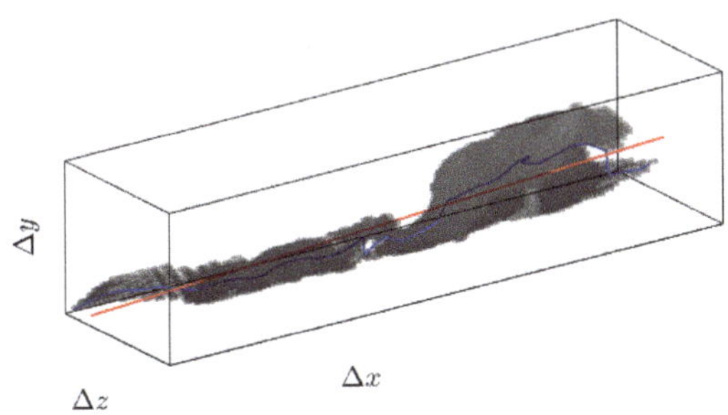


**Figure 5.** A single identified attached streak (gray) is contained in a Cartesian box. The blue line is the streak spline defined by the streak center of mass at each $\Delta y - \Delta z$ plane. The red line is the linear approximation of the spline following the least squares fitting (Equation (6)). The flow goes from bottom left to top right along the x-axis.

The streak elevation, heading angles, and meandering are calculated from the linear approximation in reference to the Cartesian box. The elevation angle ($\theta$) and heading angle ($\psi$) in radians are defined as follows:

$$\theta = \arctan(A_2),$$
$$\psi = \arctan(A_3). \tag{7}$$

The streak meandering, $\epsilon$ is the root mean square of the distance between the linear approximation and the spline at each $x_s$ location, as in [59].

The study of the sizes of the circumscribing boxes for the attached $Q^-$ structures shows how the structures tend to grow in each Cartesian direction. Self-similarity, as well as growth patterns, can be extracted from probabilistic plots of the box dimensions

https://doi.org/10.3390/fluids11010018

and ratios. The PDFs of streak dimensional ratios are depicted in Figure 6 based on box dimensions (Figure 5). A good collapse of the dimensional ratios is observed independent of the case, except for $\Delta x/\Delta y$ on attached streaks. Poiseuille-like attached streaks have a longitudinal ratio $\Delta x/\Delta y = 5.0$ greater than Couette-like streaks with $\Delta x/\Delta y = 4.0$. This fact can be traced to the different percolation ratios selected for Poiseuille- and Couette-like streaks. A 2-point correlation of $u'$ along the streamwise direction in Couette flows shows a very large correlation, generally exceeding the domain length [17,20,27]. Hence, a greater percolation index is needed to break the streaks (i.e., $u'$ structures) into coherent structures. See Figure 1. Bae et al. [10] reported $\Delta x/\Delta y = 7.0$ for attached streaks on a Poiseuille flow using $\alpha = 3.4$. Thus, the ratio $\Delta x/\Delta y$ on the attached streaks is sensitive to the percolation index selected. Conversely, detached streaks show a perfect agreement independent of the case, collapsing at $\Delta x/\Delta y = 1.0$. The ratio $\Delta z/\Delta y$ shows a good agreement for all streaks independent of the case, collapsing for attached streaks at $\Delta z/\Delta y = 0.9$ and for detached streaks at $\Delta z/\Delta y = 0.75$. These values match the findings in [10].

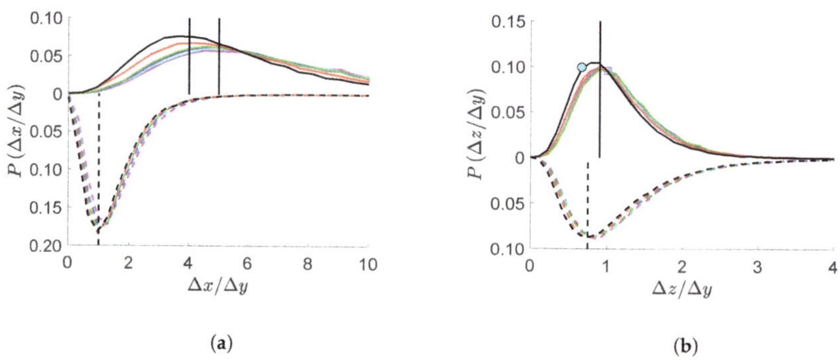


**Figure 6.** Lines as in Table 1. Solid lines correspond to attached streaks, dashed lines to detached streaks. (**a**) PDF of the aspect ratio $\Delta x/\Delta y$ of identified streaks. (**b**) PDF of the aspect ratio $\Delta z/\Delta y$ of identified streaks. Vertical lines in (**a**) mark $\Delta x/\Delta y = \{1.0, 4.0, 5.0\}$; vertical lines in (**b**) mark $\Delta z/\Delta y = \{0.75, 0.90\}$.

The streak dimensions are analyzed through 2-dimensional PDFs of the streak length, $\Delta x^+$, against the streak height, $\Delta y^+$ (Figure 7a), and the streak width, $\Delta z^+$ (Figure 7b), in wall units. Both figures reveal a strong collapse across cases within the region $\Delta x^+ \leq 400$ and $\Delta y^+ \leq 70$. The data are observed to follow the trend $\Delta x \sim \Delta y^2$ in the former and $\Delta x \sim \Delta z$ in the latter. These trends are consistent with those reported by [10] for streaks in Poiseuille channel flows. However, the analysis presented here demonstrates that the trends are also applicable to streaks identified in pure Couette flows as well as in Couette–Poiseuille combinations. Furthermore, Poiseuille-like cases are found to exhibit longer streaks, extending up to $\Delta x^+ = 600$. The deviation of streaks with $\Delta x^+ > 400$ from the observed data trends may suggest that such streaks are conglomerates of smaller ones.

Streaks are further evaluated based on their elevation and heading angles as described in Equation (7). The PDF of heading and elevation angles is depicted in Figure 8a and Figure 8b, respectively. The PDF of the streaks' heading angle is centered at $\psi = 0$ degrees and symmetric along this axis for either attached or detached streaks. The predominant heading angle aligned with the streamwise direction is also observable in the stream-spanwise plane representation of streaks in the literature [60,61]. Symmetry may be a consequence of the periodic boundary conditions in the X-Z plane of the simulation domain. While the heading angle range for attached streaks is mainly within $\pm 10$ degrees, detached streaks have a broader range up to $\pm 40$ degrees. On the other hand, the PDF for streaks elevation angle is centered at $\theta = 1.25$ degrees for attached streaks and is not

https://doi.org/10.3390/fluids11010018

symmetric. The non-uniformity of the mean shear caused by the impermeability of the wall reduces the inclination of attached structures with respect to the free stream [1]. Detached structures in Figure 8b have a greater average elevation angle as they are located further from the domain walls. Curiously enough, Figure 8a,b show similar average values for attached and detached streaks independently of the Couette–Poiseuille case.

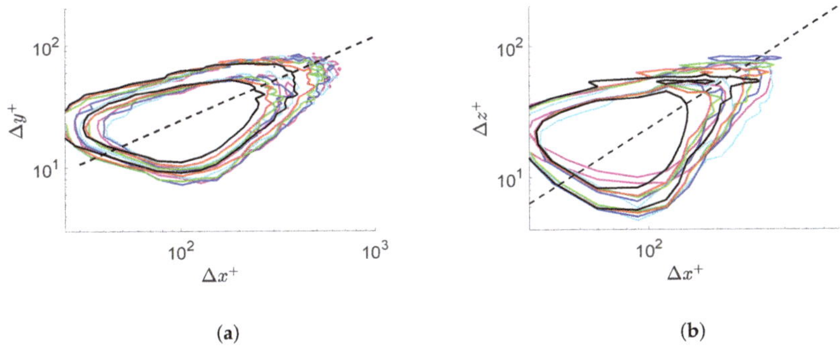


**Figure 7.** Lines as in Table 1. (**a**) Two-dimensional PDF of attached streaks' length in wall units $\Delta x^+$ vs. height $\Delta y^+$, with dashed line $\Delta x \sim \Delta y^2$. (**b**) Two-dimensional PDF of attached streaks' width $\Delta z^+$ vs. length $\Delta x^+$, with dashed line $\Delta x \sim \Delta z$. Contours represent 50% and 80% of the data.

Figure 9a shows the relation between the streak meandering and streak height, both in wall units. In all cases, the meandering behaviour increases with distance from the wall. Such a trend is also observed in canonical smooth-wall boundary layers [59]. The meandering scales exponentially with the streak height with exponent 1.33, showing an almost linear increase of meandering with streak height. Finally, the collapse among cases in Figure 9a shows that streaks have similar meandering behaviour independently of the shear ratio (i.e., Couette–Poiseuille contributions).

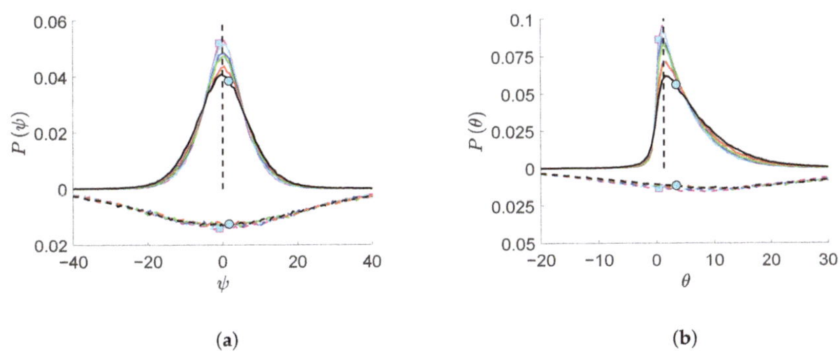


**Figure 8.** Lines as in Table 1. Solid lines correspond to attached streaks and dashed lines to detached streaks. (**a**) PDF of the heading angle $\psi$ (in degrees) of the identified streaks. The vertical dashed line is located at $\psi = 0°$. (**b**) PDF of the elevation angle $\theta$ (in degrees) of the identified streaks. The vertical dashed line is located at $\theta = 1.25°$.

Figure 9b depicts the relation between the streak volume and streak height, both in wall units. The streak volume is the sum of the volume of all the voxels defining the streak. The streak volume increases with streak height and scales exponentially as per $(\Delta y^+)^{2.60}$. This exponent is considered a crude fractal dimension of the object provided that the object is self-similar [4], being characteristic of full shell- or flake-like objects when its value is around 2.5 [6]. Here, the collapse among cases in Figure 9b shows that streaks have a

https://doi.org/10.3390/fluids11010018

similar increase in volume independently of the shear ratio, and in all cases, streaks can be portrayed as full shell- or flake-like objects.

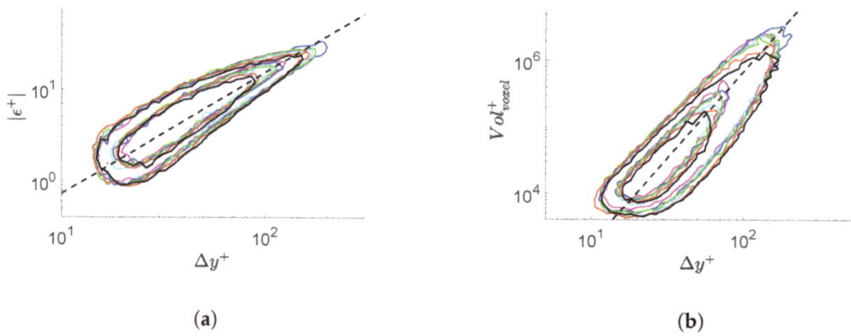


**Figure 9.** Lines as in Table 1. (**a**) Two-dimensional PDF of the meandering $|\epsilon^+|$ (in wall units) versus streak height $\Delta y^+$ (in wall units). The black dashed line indicates $|\epsilon^+| \propto (\Delta y^+)^{1.33}$. (**b**) Two-dimensional PDF of the streak volume in voxels $V^+_{\text{voxel}}$ versus streak height $\Delta y^+$. The black dashed line indicates $V^+_{\text{voxel}} \propto (\Delta y^+)^{2.60}$. In both cases, only attached streaks are used. Contours represent 50% and 99% of the total data.

### *3.2. Relative Position and Absolute Distance Between Streaks*

Figure 3 shows a global redistribution of the streaks along the domain spanwise dimension. Therefore, it is interesting to see how this redistribution may affect the distances between streak pairs. The relative position between the structures $i$ and $j$ is defined by the vector:

$$\vec{\delta}^{(i,j)} = \frac{\vec{x}^{(j)}_{mc} - \vec{x}^{(i)}_{mc}}{\overline{\Delta y}^{(i,j)}} \tag{8}$$

where $x^{(i)}_{mc}$ is the position vector of the structure $i$ mass center in Cartesian coordinates. Following the work of [5,6], the relative position $\delta^{(i,j)}$ is normalized with the mean wall-normal height of the selected structures, $\overline{\Delta y}^{(i,j)} = \left(\Delta y^{(i)} + \Delta y^{(j)}\right)/2$.

The relative position is measured against structures of similar size [5,62,63]. The filter to define if two structures can be considered as a pair is:

$$1/2 < \Delta y^{(j)}/\Delta y^{(i)} < 2. \tag{9}$$

Consecutive to the definition of relative distance in Equation (8), the absolute distance between structure $i$ and structure $j$ is defined as

$$R_{(i,j)} = \left\|\left(\vec{x}^{(j)}_{mc} - \vec{x}^{(i)}_{mc}\right)\right\| = \sqrt{\left(x^{(j)}_{mc} - x^{(i)}_{mc}\right)^2 + \left(y^{(j)}_{mc} - y^{(i)}_{mc}\right)^2 + \left(z^{(j)}_{mc} - z^{(i)}_{mc}\right)^2} \tag{10}$$

In this work, the figures related to structure distances are compiled using only the closest structure $(j)$ to the reference one $(i)$ as in [6].

Figure 10 shows the two-dimensional PDF of the relative position between two streaks of diverse types along the streamwise $(\delta x)$ and spanwise $(\delta z)$ directions. Figure 10a,b show the spatial distribution between two attached streaks in pure Couette and pure Poiseuille cases, respectively. In both cases, attached streaks tend to form pairs aligned spanwise with a relative spanwise distance $(\delta_z)$ between 0.8 and 3 units. However, in the pure Couette case, attached streaks can be paired streamwise, too, with an average relative distance $(\delta_x)$ of 2.8 units.

Figure 10c,d define the pair formed by an attached and detached streak. The spatial distribution of the pair is isotropic along the streamwise and spanwise directions, with

https://doi.org/10.3390/fluids11010018

50% of the structures at a distance below 4 units. This distribution applies to either pure Poiseuille or pure Couette flow. Generally, detached streaks tend to be located at a positive streamwise position regarding their original attached streak after detaching [64]. However, in our study, the closest distance between structures is calculated without considering the origin of each structure.

The spatial distribution formed by a pair of detached streaks is depicted in Figure 10. Detached streaks tend to form pairs aligned in the streamwise direction with an average relative distance $\delta_x$ below 2.8 units. Bae et al. [64] show that detached streaks have a greater averaged wall-normal velocity than attached streaks in the buffer layer. The authors consider the generation of detached streaks through a streak breakup as a strong ejection or bursting effect, which is conditioned on the presence of a Q2 structure. Curiously enough, ref. [5] observed that Q2 structures tend to form pairs aligned in the streamwise direction. This fact agrees with our observation on the organization of detached streaks.

The closest absolute distance on average is depicted in Figure 11 for pairs of attached streaks (a) and pairs of attached and detached streaks (b). Attached streak pairs decrease their absolute distance in all cases, reaching a minimum at $\Delta y^+_{(i)} = 40$, which coincides with the most frequent streak height in wall-units, as depicted in Figure 4b. Thus, the absolute distance between attached streaks reduces as more streaks of a given height are present. Note that only streaks of similar height are compared against each other by using the filter in Equation (9). After reaching the minimum, the absolute distance increases proportionally to $(\Delta y^+)^{0.5}$ for streaks reaching $\Delta y^+_{(i)} \approx 150$ in all cases. After this threshold, the absolute distance tends to increase even further, becoming proportional to $\Delta y^+_{(i)}$. However, the latter must be considered with precaution as the number of streaks employed for the data in this range is much smaller.

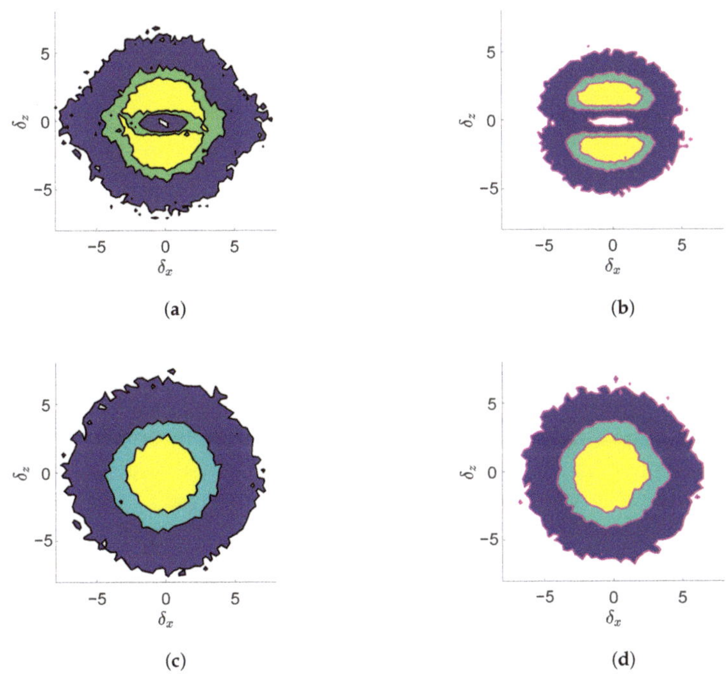


**Figure 10.** *Cont.*

https://doi.org/10.3390/fluids11010018

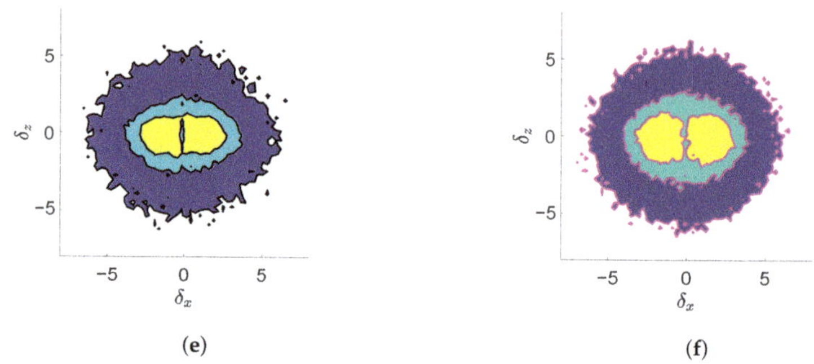


**Figure 10.** Two-dimensional PDFs of the closest relative position of (**a**,**b**) two attached streaks, (**c**,**d**) one attached and one detached streak, and (**e**,**f**) two detached streaks. The left column corresponds to case C10P00 and the right column to case C00P10. Contours contain 30%, 50%, and 80% of the total data.

Similar behavior is observed in pairs of attached and detached streaks in Figure 11b. The closest absolute distance is observed in streaks with a height of around 40 wall units, as the number of detached streaks with this height also peaks in Figure 4b. After the minimum distance, the distributions no longer collapse among cases. For a given $\Delta^+_{(i)}$, the absolute distance is greater in Poiseuille-like cases (i.e., negative shear ratio) than in Couette-like cases (i.e., positive shear ratio). Since attached-attached pairs show a good collapse in Figure 11b, the lack of agreement among attached-detached streaks must stem from changes in the distribution of detached streaks. Indeed, the number of detached streaks and their volume increase in Couette-like cases, with a maximum in case C10P00. See Table 2. Therefore, as the number and volume of detached streaks are highest at C10P00, the closest distance between attached and detached streaks reduces.

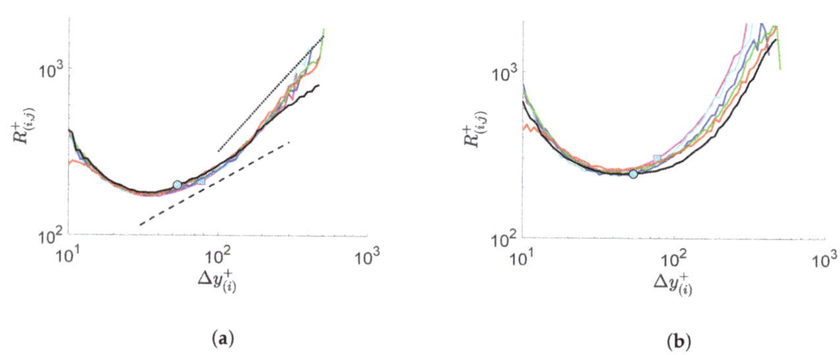


**Figure 11.** Lines as in Table 1. Closest absolute distance, on average, between streaks of types $i$ and $j$ ($R^+_{i,j}$) versus the height of streak $i$, $\Delta y^+_i$, both in wall units. (**a**) Streaks $i$ and $j$ are attached and attached, respectively. The black dashed line indicates $R^+_{i,j} \propto (\Delta y^+)^{0.5}$, and the dotted line $R^+_{i,j} \propto \Delta y^+$. (**b**) Streaks $i$ and $j$ are attached and detached, respectively.

## 4. Intense Reynolds Stress Structures

The points belonging to intense Reynolds stress structures are identified using the threshold defined in Equation (5). Each structure is engulfed in a 3-dimensional box aligned with the Cartesian coordinate system following the same procedure as for streaks, see Figure 5. The ratio of Q structures belonging to a given quadrant and their volume ratio is listed in Table 3. Here, the data is presented for either all structures or attached

https://doi.org/10.3390/fluids11010018

structures. Data relative to detached structures can be extracted by subtracting the attached data from the overall data.

**Table 3.** Cases as in Table 1. Time-averaged data of the identified Q events. $N_i$ is the numerical fraction of structures classified in each $uv$ quadrant among all identified structures. $V_i$ is the volume fraction of structures classified in each $uv$ quadrant among the overall identified volume. Ratios are classified in a given structure class, being all of them or *attached*. The attached structures have $y^+_{min} < 20$.

| Case | $N_{Q1}$ | $N_{Q2}$ | $N_{Q3}$ | $N_{Q4}$ | $V_{Q1}$ | $V_{Q2}$ | $V_{Q3}$ | $V_{Q4}$ |
|---|---|---|---|---|---|---|---|---|
| C00P10 | 0.181 | 0.266 | 0.204 | 0.349 | 0.039 | 0.671 | 0.035 | 0.254 |
| C00P10 (att) | 0.017 | 0.130 | 0.021 | 0.165 | 0.010 | 0.604 | 0.005 | 0.177 |
| C02P08 | 0.173 | 0.273 | 0.203 | 0.351 | 0.042 | 0.664 | 0.049 | 0.245 |
| C02P08 (att) | 0.020 | 0.150 | 0.026 | 0.192 | 0.012 | 0.618 | 0.011 | 0.185 |
| C04P06 | 0.145 | 0.326 | 0.179 | 0.350 | 0.054 | 0.555 | 0.150 | 0.241 |
| C04P06 (att) | 0.022 | 0.188 | 0.023 | 0.198 | 0.037 | 0.516 | 0.123 | 0.188 |
| C06P04 | 0.118 | 0.395 | 0.131 | 0.356 | 0.048 | 0.405 | 0.051 | 0.496 |
| C06P04 (att) | 0.017 | 0.231 | 0.013 | 0.216 | 0.033 | 0.351 | 0.032 | 0.441 |
| C08P02 | 0.105 | 0.447 | 0.095 | 0.353 | 0.039 | 0.500 | 0.046 | 0.415 |
| C08P02 (att) | 0.022 | 0.276 | 0.010 | 0.232 | 0.021 | 0.440 | 0.030 | 0.365 |
| C10P00 | 0.119 | 0.444 | 0.092 | 0.346 | 0.042 | 0.525 | 0.049 | 0.384 |
| C10P00 (att) | 0.022 | 0.260 | 0.010 | 0.221 | 0.018 | 0.443 | 0.030 | 0.323 |

### 4.1. The Geometry and Shape of Individual Structures

Figure 12 depicts the joint PDF of the structure's minimum distance to the nearest wall $(y^+_{min})$ against its maximum distance $(y^+_{max})$. A vertical line at $y^+_{min} = 20$ separates between attached and detached structures following the literature definition [5]. Among the attached structures, we can differentiate two zones: the first zone with $y^+_{min} < 1$, and the second zone with $1 \leq y^+_{min} \leq 20$. While the quadrant contributions to the first zone remain similar among cases, the second zone differs from case to case, being therefore affected by the shear ratio, $\gamma$. The first zone has a constant averaged contribution of 43% Q2 structures and 53% Q4 structures, among cases. It is obvious that Q4 structures will be more present in this zone as they tend to grow with negative wall-normal velocity, reaching closer positions to the wall. Conversely, the contribution of Q2 and Q4 to the second zone is not constant. The Poiseuille-like cases show a 50–41% contribution of Q2 and Q4, respectively, while Couette-like cases show a 75–20% contribution.

To examine whether attached structures exhibit different geometric characteristics based on their zone, we analyze their center-of-mass height relative to the nearest wall $\left(y^+_{cg}\right)$ and their maximum height $(y^+_{max})$, as shown in Figure 13a. The joint probability density function (PDF) reveals a consistent linear trend, $\left(y^+_{max} \propto y^+_{cg}\right)$, across all attached structures, regardless of their zone or quadrant (e.g., Q2 or Q4). However, this linear relationship is modified for structures with $y^+_{cg} \leq 15$ due to the influence of the wall. This region is predominantly occupied by Q4 structures, whose negative wall-normal velocity flattens them against the wall, as previously reported by [11].

To compare these characteristics with detached structures, Figure 13b presents the same parameters for detached structures. The results show excellent agreement between pure Couette and pure Poiseuille cases, following a trend of $\left(y^+_{max} \propto y^+_{cg}\right)^{0.82}$, which differs from the trend observed in attached structures. Based on these findings, we classify all structures with $y_{min} \leq 20$ as attached and those with $y_{min} > 20$ as detached.

Figure 14 shows the self-similar behavior of the attached and detached Q events in all simulation cases. The ratio $\Delta x/\Delta y$ has the highest probability at 2.0 and 1.0 for attached and detached structures, respectively. On the other hand, the ratio $\Delta z/\Delta y$ has the highest probability at 0.7 and 0.65 for attached and detached structures, respectively.

https://doi.org/10.3390/fluids11010018

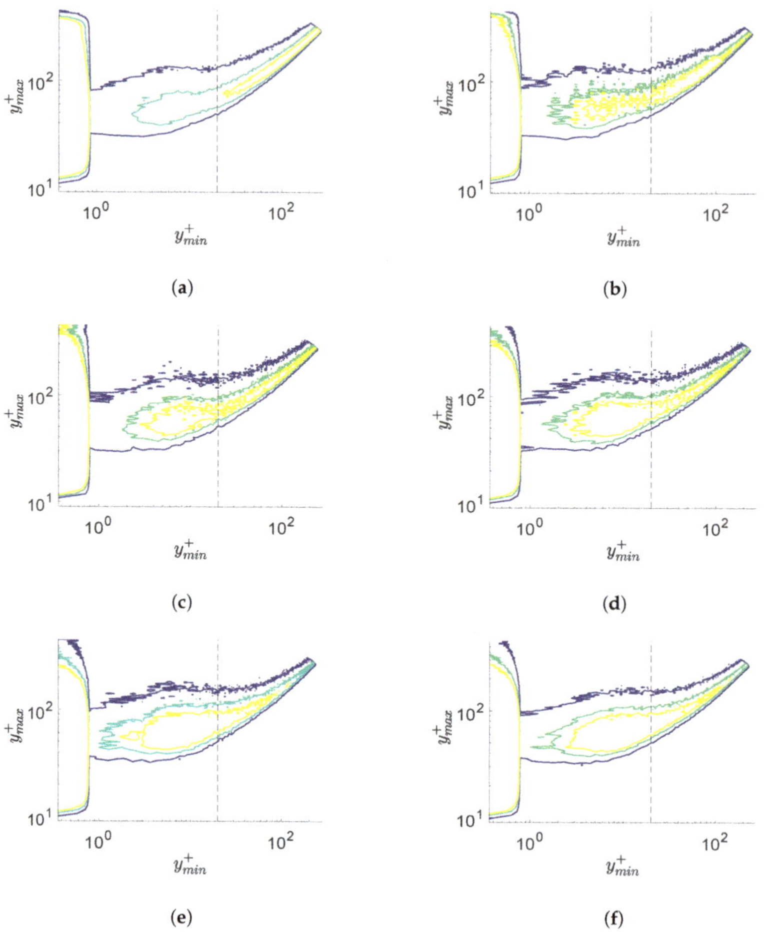

Figure 12. Joint PDF of the wall-normal distance to the closest domain wall from the structure bottom $(y^+_{min})$ and top $(y^+_{max})$. Cases are C00P10 (a), C02P08 (b), C04P06 (c), C06P04 (d), C08P02 (e) and C10P00 (f). For each case, the plotted contours contain 50%, 70%, and 90% of the total data. The dashed line represents $y^+_{min} = 20$.

Further developing the above-mentioned ratios, the attached structures in all cases (Table 1) follow

$$\Delta x \approx 2\Delta y \approx 2.86\Delta z, \quad (11)$$

whilst detached structures in all cases follow

$$\Delta x \approx \Delta y \approx 1.54\Delta z. \quad (12)$$

A similar ratio as described in Equation (11) is reported by [6] in turbulent Poiseuille flow. The authors demonstrated the self-similarity for $Q^-$ (i.e., both Q2 and Q4 together) objects at $Re_\tau = 934$ and $Re_\tau = 2003$, with $\Delta x \approx 2\Delta y \approx 2.5\Delta z$. Although we include all types of attached Q structures in this analysis, the majority are $Q^-$ structures (see Table 3), allowing for a meaningful comparison. This indicates that the self-similarity trend of attached Q structures is independent of the shear ratio and Reynolds number, at least within the range $Re_\tau \leq 2003$.

https://doi.org/10.3390/fluids11010018

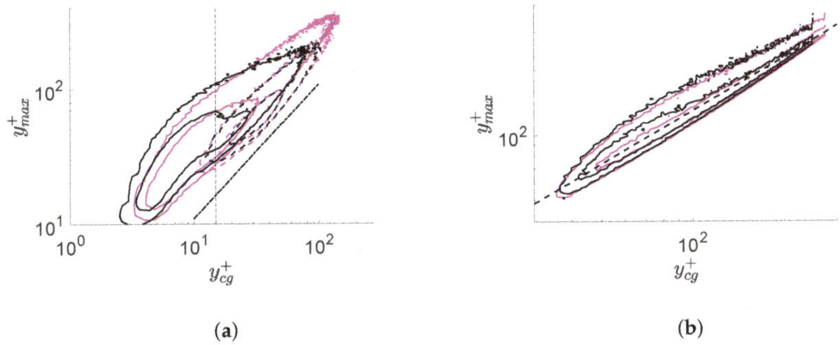


**Figure 13.** (**a**) Two-dimensional PDF of the Q-structure center-of-mass height $y^+_{cg}$ versus the structure maximum height $y^+_{max}$ in wall units for attached Q structures. Solid and dashed lines correspond to attached Q structures with $y^+_{min} \leq 1$ and $1 < y^+_{min} \leq 20$, respectively. The vertical dashed line is at $y^+_{cg} = 15$. The dash-dotted line indicates $y^+_{max} \propto y^+_{cg}$. (**b**) Two-dimensional PDF for detached Q structures. The dashed line indicates $y^+_{max} \propto (y^+_{cg})^{0.82}$. Contours represent 50% and 90% of the total data.

As a result of this self-similarity, the volume of the bounding boxes containing Q structures (Figure 5) can be expressed as $V^+_{\text{box}} = \Delta x^+ \times \Delta y^+ \times \Delta z^+ \propto (\Delta y^+)^3$ [11]. However, coherent structures often exhibit elongated shapes resembling filaments or ribbons [65], which do not entirely fill the bounding box. To better characterize the intrinsic shape of these structures, we employ parameters such as volume in voxels and meandering. Other studies use different methodologies, such as a box-counting algorithm [65], the structure's genus [58], or shape definitions based on Minkowski functionals [66].

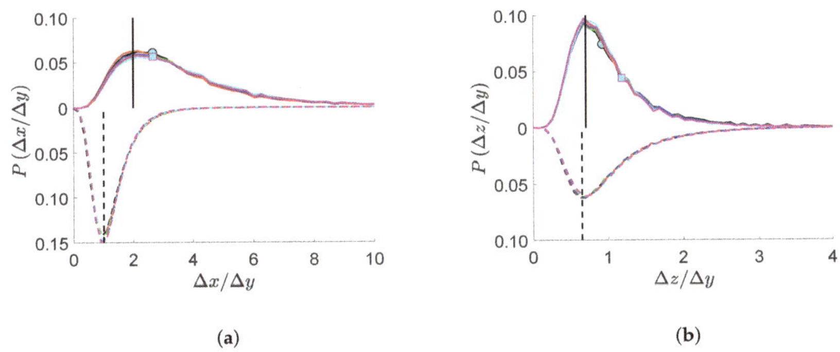


**Figure 14.** Lines as in Table 1. Solid lines are for attached Q-structures, and dashed lines are for detached Q-structures. (**a**) PDF of the structure aspect ratio $\Delta x/\Delta y$. Vertical lines are located at $\Delta x/\Delta y = \{1.0, 2.0\}$. (**b**) PDF of the structure aspect ratio $\Delta z/\Delta y$. Vertical lines are located at $\Delta z/\Delta y = \{0.65, 0.70\}$.

Figure 15a shows that the volume of attached structures scales with $(\Delta y^+)^D$, where $D = 2.32$. This scaling aligns with results from similar studies [5,11]. In these studies, the authors interpreted the scaling of object volume with height as an estimate of fractal dimension. They concluded that an exponent $D$ between 2.25 and 2.45 corresponds to structures resembling spongelike objects composed of flakes. Notably, in Figure 15a, attached structures with $\Delta y^+ \leq 20$ deviate from this trend. These structures are typically Q4 structures flattened against the wall due to their negative wall-normal velocity component.

The meandering of the structures is evaluated using the same definition as for streaks in Section 3. Figure 15b compares the meandering of attached Q structures against their

https://doi.org/10.3390/fluids11010018

height. The results show good agreement across simulation cases, indicating that the meandering scales as $(\Delta y^+)^{1.2}$. This scaling is similar to that observed for streaks in Figure 9. We conclude that coherent structures tend to become increasingly wavy with wall distance, leading to higher meandering [59]. Interestingly, attached structures with $\Delta y^+ \leq 20$, previously associated with wall flattening, still follow the same meandering trend as other structures. This observation suggests that meandering is governed primarily by the growth of the structure in the streamwise and spanwise directions.

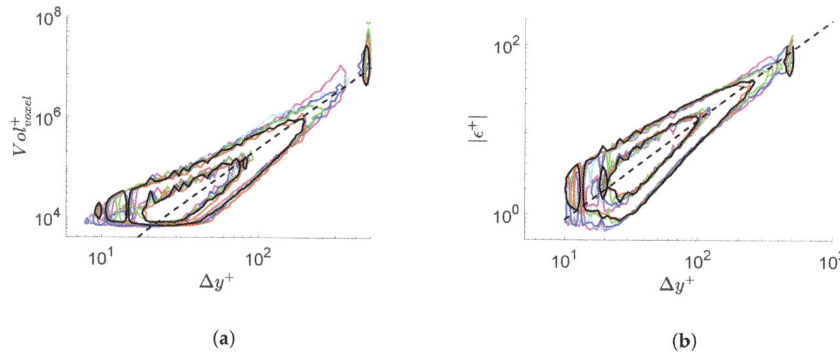


**Figure 15.** Lines as in Table 1. (**a**) Joint PDF of the Q structure height against structure volume measured in voxels. (**b**) Joint PDF of the Q structure height against structure meandering. All quantities are expressed in wall units. In both panels, only attached structures are considered. Contours contain 50% and 90% of total data. The lines are $(\Delta y^+)^{2.32}$.

Figure 16 presents the PDFs of these angles as defined in Equation (7). A perfect agreement is observed between cases in either attached or detached structures. Like streaks, the Q structures tend to be aligned on average with the streamwise direction. Structures whose characteristic length is greater than the Corrsin length [67] interact with the mean shear; thus, they grow in the direction of the mean shear, $\psi = 0°$. Attached Q structures comply with this condition since the Corrsin scale ($L_c$) in channel flows is $L_c \approx y$ and all attached structures are larger than $L_c$ [1]. For this reason, the maximum of the heading angle PDF is located at 0 degrees in all cases for attached structures. Finally, detached $Q^-$ in channel flows are shorter than the attached ones, but not wider [1]. This results in several detached structures being smaller than the Corrsin scale and, therefore, a greater dispersion on their heading angle. Nevertheless, detached structures also have a predominant alignment with the streamwise direction. This agreement between both structures may stem from the fact that detached structures can be linked to attached structures during their lifetime. Attached Q2 structures tend to grow steadily away from the wall and end up detaching from it. Conversely, detached Q4 structures can be born away from the wall and move down relatively quickly, attaching to the wall and staying attached thereafter [47].

On the other hand, the elevation angle for attached structures peaks at $\theta = 3°$ and covers a wide range until $\theta = 20°$. The range in attached structures is narrow in comparison to detached structures. The latter peaks at $\theta = 12°$ and covers a range of $-20° \leq \theta \leq 40°$. This trend matches the observations of Jiménez [1]. The author reports that the elevation angle of the flow field conditioned to a Q2–Q4 pair is truncated by the presence of the domain wall. Consequently, the flow field conditioned to the presence of attached Q structures is less symmetric and less inclined in the streamwise direction than a case without walls (i.e., stationary homogeneous shear turbulence) [62]. When conditioning the flow field to the presence of detached Q structures, the elevation angle is still truncated by the wall, but to a lesser magnitude. Thus, the elevation angle is higher than for attached structures.

https://doi.org/10.3390/fluids11010018

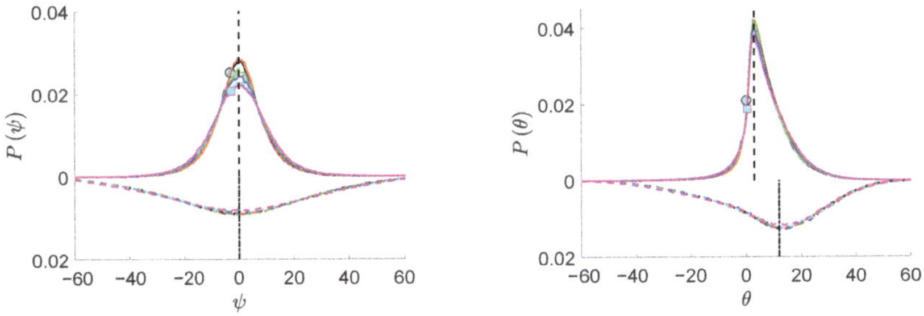


(**a**) (**b**)

**Figure 16.** Lines as in Table 1. Solid lines are for attached Qs, and dashed lines are for detached Qs. (**a**) PDF of the heading angle ($\psi$) in degrees. The vertical lines are defined at $\psi = 0$ deg. (**b**) PDF of the elevation angle ($\theta$) in degrees. The vertical dashed line is defined at $\theta = 3$ deg. The vertical dash-dotted line is defined at $\theta = 12$ deg.

### *4.2. The Intensity of Individual Structures*

The intensity of individual structures is analyzed in terms of Reynolds stress and turbulent kinetic energy, with a focus on their contributions to the averaged quantities $\langle uv \rangle$ and $\langle k \rangle$. For attached structures, we only take into account the ones of type Q2 ($u_m < 0$, $v_m > 0$) and Q4 ($u_m > 0$, $v_m < 0$), while for detached structures we group them into $Q^-$ ($u_m v_m < 0$) and $Q^+$ ($u_m v_m > 0$). Both the neglect of attached $Q^+$ structures and the grouping of detached structures are based on the results presented in Table 3. The structure-averaged streamwise and wall-normal fluctuation velocities, $u_m$ and $v_m$, are defined as:

$$u_m = \frac{\int_\Omega u\,dV}{\int_\Omega dV}, \quad v_m = \frac{\int_\Omega v\,dV}{\int_\Omega dV} \tag{13}$$

where the domain $\Omega$ is determined by the connected points that define the structure.

Here, we observe that the numerical fraction and volume ratio of $Q^+$ events are significantly lower than those of $Q^-$ events. As described by [1], an explanation for this imbalance of structure types is that $Q^-$ events extract energy from the shear, while $Q^+$ events lose it. Consequently, the former tend to grow, with some individuals reaching the domain walls and developing attached structures, while the latter tend to vanish before reaching a domain wall.

The contribution of the attached and detached Q structures to the overall Reynolds stress along the wall-normal direction is depicted in Figure 17a,b. Note that only the lower part of the simulation domain is depicted as in this region, the friction Reynolds number (Table 1) is similar among cases. In Figure 17a, attached structures reach, in most cases, constant quadrant contributions to the momentum flux at $y^+ > 50$. This fact stems from using an identification threshold dependent on the wall distance as reported in [11].

Independent from the simulation case, the attached Q4 events are less energetic than the attached Q2 events at $y^+ > 20$, approximately ([5,11]). This characteristic is related to the fact that attached Q4 structures tend to come nearer to the wall due to their negative $v_m$ component. Thus, the Reynolds stress near the wall ($y^+ \leq 20$) is mainly dominated by attached Q4 structures, covering up to 50% of the averaged Reynolds stress. Curiously enough, in this region, the contribution of the attached Q4 structures collapses perfectly among cases, and even along the wall-normal direction, their contribution is just slightly affected by the simulation case. Conversely, the contribution of attached Q2 structures near

https://doi.org/10.3390/fluids11010018

the wall is small and dependent on the simulation case. The former is due to the reduced presence of this kind of structure. See the discussion related to Figure 13a.

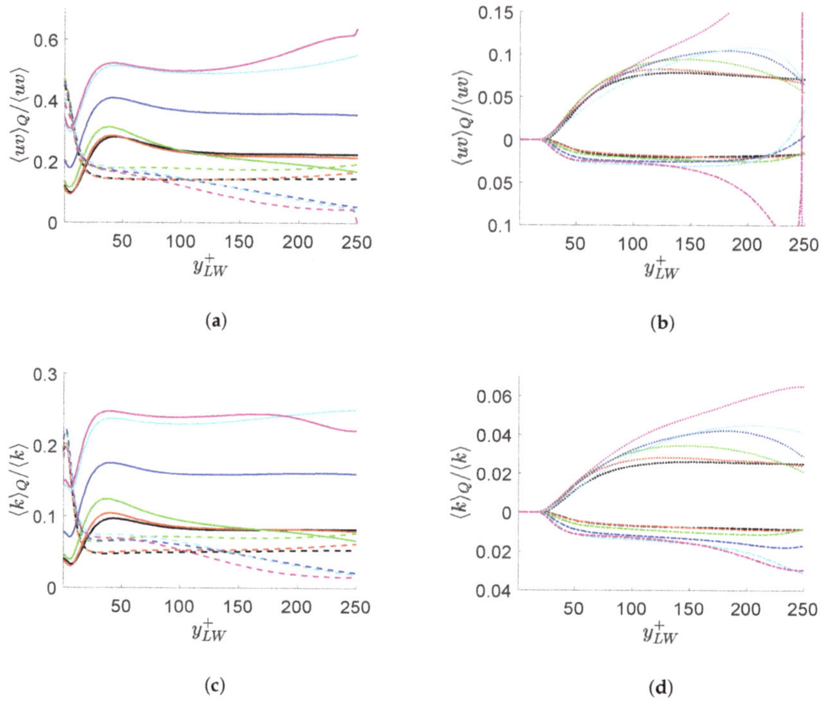

**Figure 17.** Lines as in Table 1. Contribution of Q structures to the overall Reynolds stress (attached structures in (**a**), detached structures in (**b**)) and turbulent kinetic energy (attached structures in (**c**), detached structures in (**d**)) along the wall-normal direction near the lower domain wall, $y^+_{LW}$. Lines: (**a**,**c**) —, attached Q2; – – –, attached Q4; (**b**,**d**) ···, detached $Q^-$; –·–, detached $Q^+$.

Figure 17b illustrates the contribution of detached structures to the momentum flux. Due to their definition, detached structures cannot exist at $y^+ \leq 20$, leading to a 0% contribution in this region. In the range $20 < y^+ < 100$, detached $Q^-$ structures show an increasing Reynolds stress contribution, reaching up to 8%, regardless of the case. Beyond this range, most cases maintain a constant contribution below 10% in the wall-normal direction. An exception occurs in the pure Poiseuille case, where the contribution rises toward the channel center, as previously observed in [11]. This trend in the pure Poiseuille and other Poiseuille-like cases may be attributed to the reduction of momentum flux near the channel center, as shown in Figure 18a, while the kinetic energy remains similar Figure 18b. The decreasing averaged Reynolds stress limits the growth of structures originating above the buffer layer, resulting in predominantly detached and less energetic structures. In contrast, Couette-like cases sustain a nearly constant averaged Reynolds stress in the channel center, producing stable quadrant contributions from both attached and detached structures. Detached $Q^+$ structures, however, consistently contribute approximately 3% to Reynolds stress across all cases. Overall, detached structures are only weakly influenced by the mean shear and exhibit a nearly constant Reynolds stress contribution along the wall-normal direction.

The contributions of attached and detached $Q$ structures to the averaged turbulent kinetic energy are depicted in Figure 17c,d. Although the contributions of attached structures resemble their trends in Reynolds stress, their impact on turbulent kinetic energy is

https://doi.org/10.3390/fluids11010018

considerably smaller across all wall-normal regions. This difference arises because $Q$ structures are identified for their intense Reynolds stress (via Equation (5)), not their turbulent kinetic energy. Attached $Q2$ structures contribute between 25% and 10% to the averaged turbulent kinetic energy, while attached $Q4$ structures remain below 10% throughout.

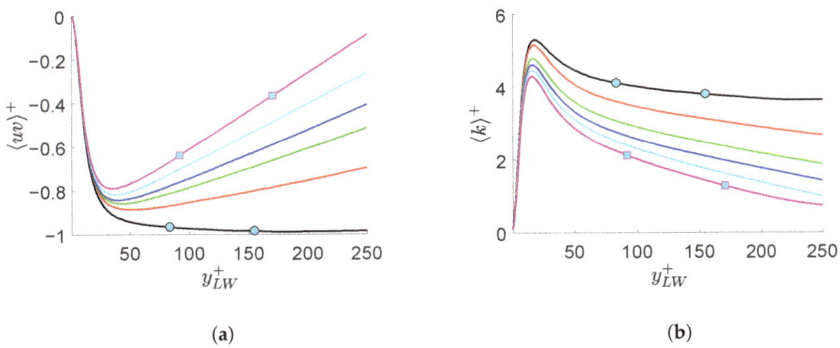


**Figure 18.** Lines as in Table 1. (**a**) Non-dimensional averaged Reynolds stress $\langle uv \rangle^+$ along the wall-normal direction. (**b**) Non-dimensional averaged turbulent kinetic energy $\langle k \rangle^+$ along the wall-normal direction. Both magnitudes are represented along the domain's lower wall $y^+_{LW}$ and averaged in time, streamwise and spanwise directions.

Detached $Q$ structures also contribute less to turbulent kinetic energy than to Reynolds stress, particularly in the region $y^+ < 100$. Detached $Q^-$ structures contribute less than 4%, with the pure Poiseuille case being the only exception. Contributions from detached $Q^+$ structures remain below 2%.

Therefore, attached $Q2$ structures are consistently more energetic than attached $Q4$ structures in the region $y^+ > 20$ across all C-P cases. In the viscous and buffer layers, attached $Q4$ structures dominate and account for a significant share of $\langle uv \rangle$. Detached structures are generally less energetic than attached ones. However, in Poiseuille-like cases, detached $Q^-$ structures in the outer region ($y^+ > 100$) show similar proportions of $\langle uv \rangle$ and $\langle k \rangle$ as attached $Q4$ structures.

The Reynolds stress and turbulent kinetic energy averaged within the structure domain, $\Omega$, are evaluated using the velocity components defined in Equation (13). To facilitate comparison across simulation cases, their one-dimensional probability density functions (PDFs) are shown in Figure 19, distinguishing between attached and detached structures. For attached structures, Q2 consistently exhibits higher values of $\langle uv \rangle_\Omega$ and $\langle k \rangle_\Omega$ compared to Q4 across all cases. Furthermore, Q2 structures occupy a larger domain volume than Q4 structures, as detailed in Table 3.

This combination of higher intensity and larger domain volume explains the dominant contribution of attached Q2 structures to the streamwise and spanwise-averaged Reynolds stress and turbulent kinetic energy (TKE), as illustrated in Figure 17a. These characteristics establish Q2 structures as the primary contributors to the spatially averaged values of these quantities.

An increase in the shear ratio further enhances the energy content of both attached structures. Consequently, attached structures in Couette-like cases contain more momentum flux and turbulent kinetic energy than those in Poiseuille-like cases [11]. This trend is also visible in Figure 20, where the contours of $u^+_m$ and $v^+_m$ display higher absolute values as the case transitions from C00P10 to C10P00. Interestingly, the attached Q4 structures exhibit two distinct regions of intense momentum flux ($\tau^+_m = u^+_m \cdot v^+_m$). These regions persist across simulation cases, with the less intense region being largely independent of the simulation case and, therefore, the shear ratio. In fact, a nearly perfect collapse is observed in the less

https://doi.org/10.3390/fluids11010018

intense region of Q4 among Couette-like cases, as shown in Figure 20b. This region has an averaged value of $\tau_m^+ \approx 2$, a pattern also evident in Figure 19a. In these PDFs, attached Q4 structures exhibit a prominent peak at higher values, which vary with the case, and a secondary, weaker peak around $\langle uv\rangle_\Omega^+ \approx 2.0$ across cases. Notably, this duality of intense regions is observed in UV graphs but not in TKE figures.

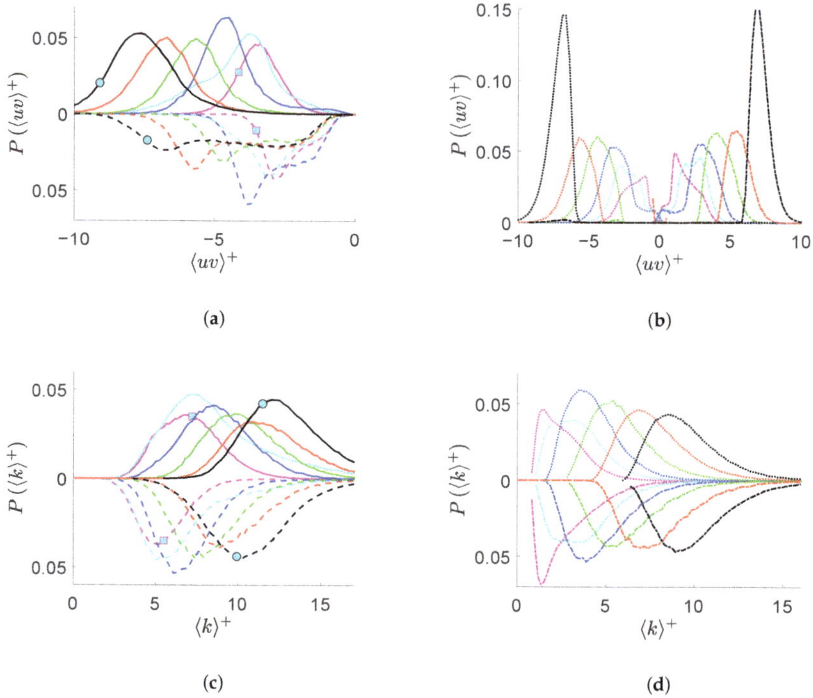


**Figure 19.** Lines as in Table 1. One-dimensional PDF of Reynolds stress (**a**,**b**) and turbulent kinetic energy (**c**,**d**) contained in attached (**a**,**c**) and detached (**b**,**d**) Q structures. Lines: (**a**,**c**) —, attached Q2; – – –, attached Q4; (**b**,**d**) ···, detached $Q^-$; –·–, detached $Q^+$.

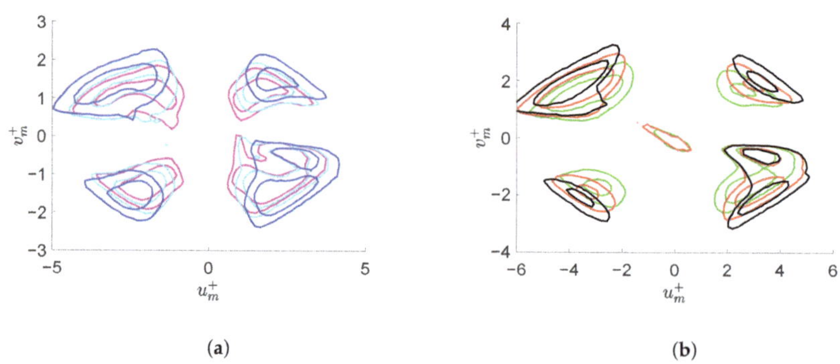


**Figure 20.** Lines as in Table 1. Two dimensional PDF of the average fluctuating streamwise velocity against the average fluctuating wall-normal velocity of all identified Qs. Both quantities are expressed in wall units. (**a**) Poiseuille-like cases. (**b**) Couette-like cases. Contours contain 50% and 85% of total data.

The analysis of detached structures in Figure 19 (right) shows that detached structures increase their intensity components following the same trend as attached structures. As

https://doi.org/10.3390/fluids11010018

observed by [47], structures are controlled by the shear, mainly generated by the wall, and can be born at any height, being attached or detached at the beginning of their lifetime. The author observed that Q2s are generally born in the buffer layer and rise, while Q4s are born away from the wall and tend to approach it. For this reason, a structure can evolve during its lifetime from detached to attached or vice versa.

Moreover, the increasing intensity of detached structures is somehow surprising, as we showed in Figure 17 that their contribution to the total Reynolds stress and TKE is never large [5]. The reason for their limited contribution is that generally detached structures represent a small volume fraction among all structures identified as observed in Table 3. We generally consider that detached structures may be either at the beginning of their lifetime or near their end, as a product of a split from a bigger attached structure. Finally, $Q^+$ structures also increase their intensity with the shear ratio of our simulations. We can then conclude that the shear increases the intensity, both TKE and UV, of Q structures independently of their height from the nearest wall and for all cases.

### *4.3. Relative Position and Absolute Distance Between Structures*

The relative distances between attached pairs in the streamwise and spanwise directions are shown in Figure 21. Attached pairs of the same type (i.e., Q2–Q2 or Q4–Q4) predominantly align in the streamwise direction, as illustrated in Figure 21a,b. This alignment was previously observed by [5] in pure Poiseuille channel flows, where it was noted that structures sufficiently tall to be influenced by the mean shear tend to exhibit organized patterns. In the present study, the pure Poiseuille case (C00P10) displays an average streamwise distance between attached Q2 structures of $\delta_x \approx 1.8$, which reduces to $\delta_x \approx 1.08$ in the pure Couette case. Notably, this reduction in distance is specific to Q2 pairs and does not appear in the other configurations presented in Figure 21.

Attached Q4 pairs also exhibit streamwise alignment with similar streamwise distances to Q2 pairs, though the alignment is less pronounced. In the pure Couette case, the organization of attached pairs of the same type becomes more isotropic, with the probability density function (PDF) tending toward a circular distribution. This trend is consistent with the behavior of attached streaks, as shown in Figure 10, when comparing pure Poiseuille and pure Couette flows.

Finally, the attached Q2–Q4 pairs exhibit a preferred alignment in the spanwise direction, consistent with findings from previous studies [5] and as shown in Figure 21e,f. These pairs are typically separated by $\delta_z \approx 0.8$, regardless of the simulation case. Figure 21 highlights that increasing the shear ratio (from Poiseuille to Couette) primarily affects the distance and organization of attached Q2 pairs.

The analysis of distances between structure pairs is extended by focusing on the minimum absolute distance between attached structures of similar height, as shown in Figure 22. The absolute distance is calculated using Equation (10). Unlike the approach in Figure 21, this analysis incorporates the wall-normal distance between pairs. Results presented in Figure 22a,b indicate that the minimum absolute distance in wall units scales consistently across simulation cases for attached Q2 and Q4 structure pairs. In both cases, the scaling follows $(\Delta y^+)^{0.55}$.

However, the minimum distance between attached Q2 and Q4 pairs does not collapse as neatly as it does for pairs of the same type. A gradual increase in distance is observed when transitioning from pure Poiseuille to pure Couette flow. This trend is attributed to the alignment of Q2–Q4 pairs in the spanwise direction, where Q2 structures typically align with low-speed streaks and Q4 structures with high-speed streaks. The spanwise size of streaks above the buffer layer increases with mean shear, reaching up to 2.5 times the channel's semi-height in pure Couette flows [20]. Consequently, the gradual increase in

https://doi.org/10.3390/fluids11010018

the absolute distance between Q2–Q4 pairs, as shown in Figure 22c, is likely related to the enhanced spanwise separation between streaks driven by the moving wall.

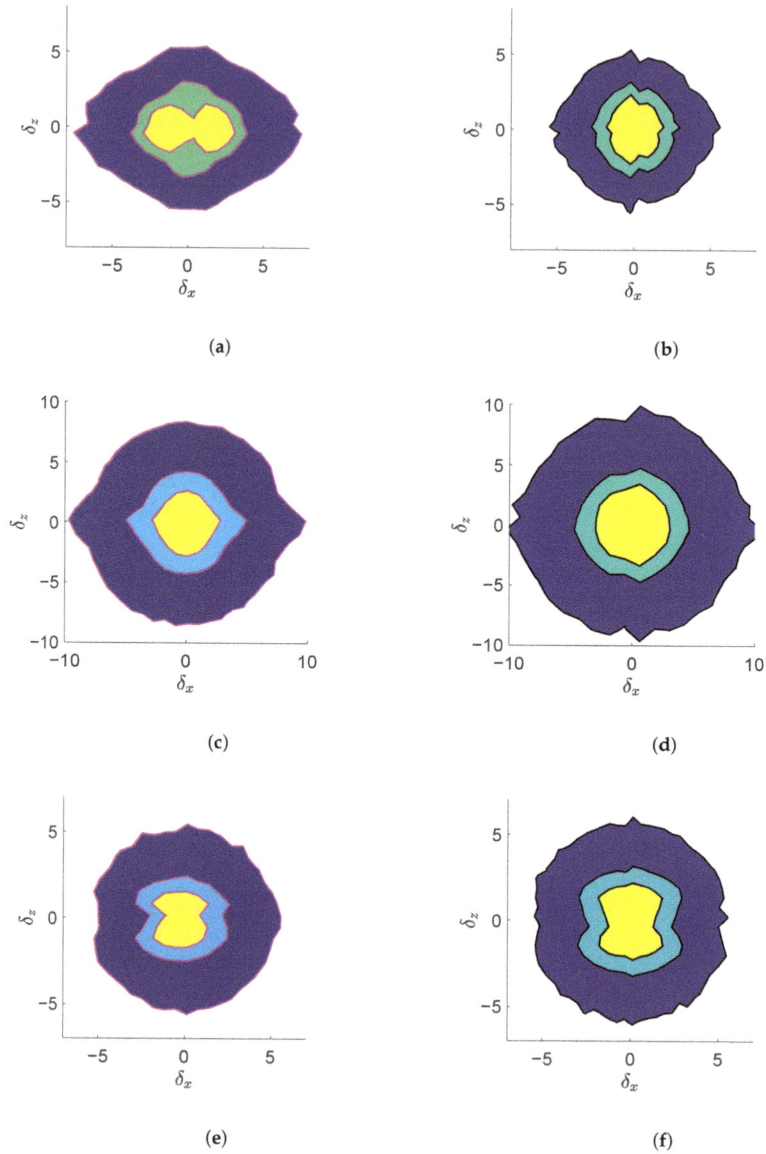


**Figure 21.** Two-dimensional PDF of the closest relative position in streamwise ($\delta_x$) and spanwise ($\delta_z$) directions of (**a**,**b**) two attached Q2 structures, (**c**,**d**) two attached Q4 structures, and (**e**,**f**) an attached Q2–Q4 pair. The left column is for case C00P10. The right column is for case C10P00. Contours contain 20%, 50%, and 90% of the total data.

Distances between structures exhibit a minimum value, a pattern similar to the behavior observed for streaks in Figure 11, particularly for structure heights below 100 wall units. This minimum corresponds to the peak probability of $\Delta y^+$, as illustrated in Figure 22d, where $\Delta y^+ = 40$ for attached Q2 structures and $\Delta y^+ = 30$ for attached Q4 structures. These specific $\Delta y^+$ values increase the likelihood of matching structures, thereby reducing their

https://doi.org/10.3390/fluids11010018

absolute distance and creating a minimum in the distribution. Matching of structures is based on similar height criteria, as defined by the filter in Equation (9).

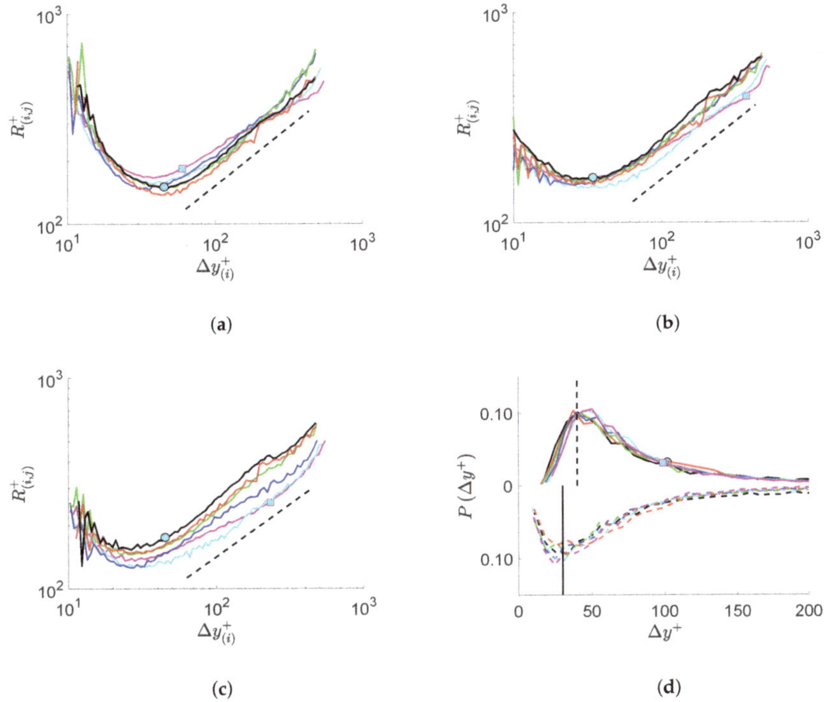

**Figure 22.** Lines as in Table 1. Closest absolute distance in average between Reynolds stress structures of types $i$ and $j$ $\left(R^+_{i,j}\right)$ against the height of structure $i$, $\Delta y^+_i$. Both quantities are in wall units. (**a**) Structure types $i$ and $j$ are Q2-attached and Q2-attached, respectively. The black dashed line indicates $R^+_{i,j} \propto \left(\Delta y^+\right)^{0.55}$. (**b**) Structure types $i$ and $j$ are Q4-attached and Q4-attached, respectively. The black dashed line indicates $R^+_{i,j} \propto \left(\Delta y^+\right)^{0.55}$. (**c**) Structure types $i$ and $j$ are Q2-attached and Q4-attached, respectively. The black dashed line indicates $R^+_{i,j} \propto \left(\Delta y^+\right)^{0.50}$. (**d**) PDF of structure height in wall-normal units $\left(\Delta y^+\right)$. Solid lines and dashed lines are related to the attached Q2 and Q4 structures, respectively. Vertical lines are located at $\Delta y^+ = 30$ and $\Delta y^+ = 40$.

The analysis is extended to detached structures, focusing on detached Q2 and Q4 structures, while excluding detached Q1 and Q3 structures due to their limited energetic contribution, as shown in Figure 17. Detached Q2–Q2 and Q4–Q4 pairs exhibit isotropic organization in the streamwise-spanwise plane, as evidenced by the circular probability contours in Figure 23, which indicate no preferential alignment. Regardless of the simulation case, 50% of the structure pairs are confined within the region defined by $\delta_x < 2$ and $\delta_z < 2$.

A more detailed understanding of structure distances is provided by incorporating Figure 24. Detached structures show a minimum distance at the peak probability of $\Delta y^+$, which occurs at $\Delta y^+ = 50$ for both detached Q2 and Q4 structures. Unlike attached structures, detached structures are far enough from the wall to avoid geometric constraints, allowing them to grow taller. In contrast, attached Q4 structures in Figure 22 display smaller $\Delta y^+$ values due to their tendency to flatten against the wall.

Once the minimum distance is exceeded, the scaling of absolute distance among detached structures diverges from that of attached structures. Depending on the simu-

https://doi.org/10.3390/fluids11010018

lation case, the absolute distance scales between $(\Delta y^+)^{0.63}$ and $(\Delta y^+)^{0.90}$, demonstrating variability across different conditions.

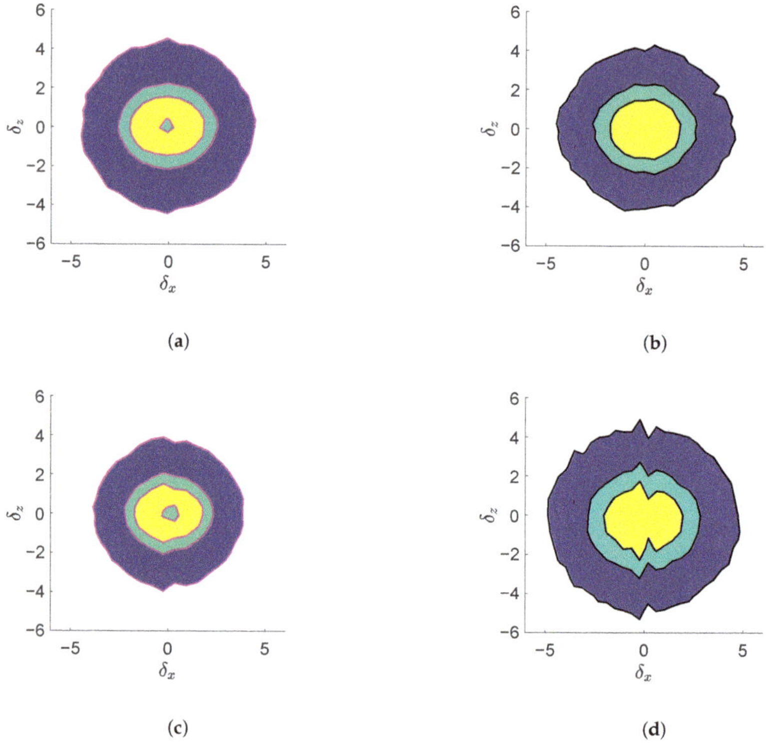

**Figure 23.** Two-dimensional PDF of the closest relative position in streamwise ($\delta_x$) and spanwise ($\delta_z$) directions of (**a**,**b**) two detached Q2 structures, (**c**,**d**) two detached Q4 structures. The left column is for case C00P10. The right column is for case C10P00. Contours contain 20%, 50%, and 90% of the total data.

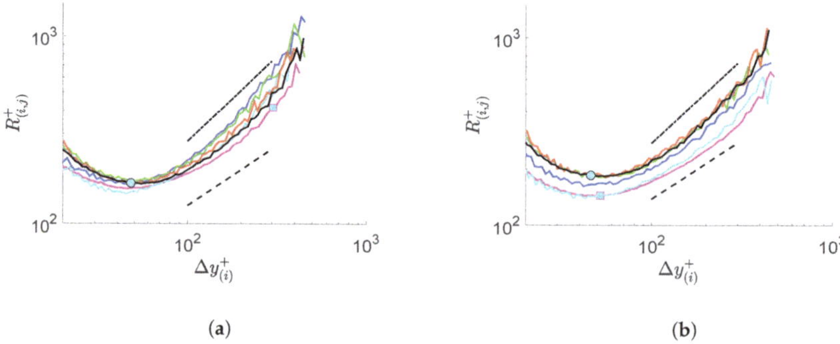

**Figure 24.** *Cont.*

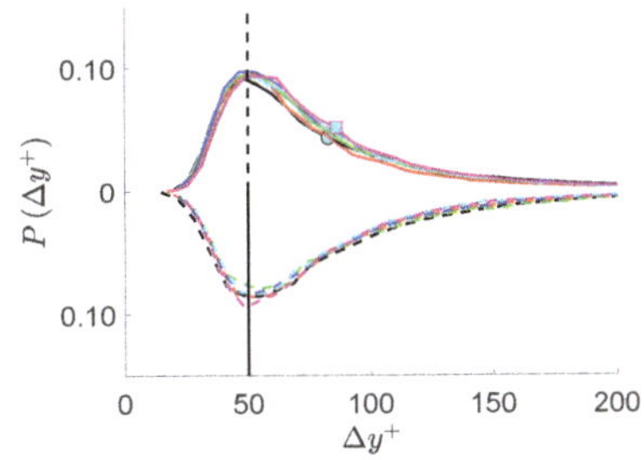


(c)

**Figure 24.** Lines as in Table 1. Closest absolute distance in average between Reynolds stress structures of types $i$ and $j$ $\left(R^+_{i,j}\right)$ against the height of structure $i$, $\Delta y^+_i$. Both quantities are in wall units. (**a**) Structure types $i$ and $j$ are Q2-detached and Q2-detached, respectively. (**b**) Structure types $i$ and $j$ are Q4-detached and Q4-detached, respectively. The black dash-dotted line indicates $R^+_{i,j} \propto \left(\Delta y^+\right)^{0.63}$. The black dashed line indicates $R^+_{i,j} \propto \left(\Delta y^+\right)^{0.90}$. (**c**) PDF of structure height in wall-normal units $\left(\Delta y^+\right)$. Solid lines and dashed lines are related to detached Q2 and detached Q4 structures, respectively. Both vertical lines are located at $\Delta y^+ = 50$.

## 5. Conclusions

This study contributes to the understanding of the characteristics of coherent structures in various channel flows. The analyzed flows span a stepped transition from pure Poiseuille to pure Couette configurations, providing a range of cases between these two extremes. The friction Reynolds number was chosen in all cases to sustain turbulence and ensure sufficient scale separation in the pure Couette case, allowing for the distinction between near-wall and large-scale structures. This dataset is thus well-suited for investigating coherent structures based on their momentum source, whether pressure-driven or shear-driven, despite the relatively low-friction Reynolds number.

The structures analyzed include low-speed streaks and intense Reynolds stress structures (Q structures), due to their significance in wall-bounded turbulence dynamics and energy transfer. Low-speed streaks, formed by the lift-up effect, redistribute momentum in the near-wall region and play a central role in turbulence production and maintenance through interaction with the mean shear. Conversely, Q events, such as bursts (Q2) and sweeps (Q4), are critical for driving Reynolds stresses and transferring energy from the mean flow to turbulent fluctuations.

The spanwise organization of streaks is heavily influenced by the shear ratio. In Couette-like cases (positive shear ratio), attached streaks reach the channel center and are separated spanwise by approximately $3.75h$. Remarkably, this organization emerges even in the first dataset case with a positive shear ratio. Additionally, the wall-normal growth of streaks varies across cases: attached streaks in Couette-like flows grow taller than those in Poiseuille-like flows, as shown by the probability distributions of their minimum and maximum distances from the wall. Detached streaks, however, exhibit consistent growth tendencies across all cases, as they are unaffected by the mean shear.

Relative distances between streaks reveal that the shear ratio primarily affects attached pairs. Couette-like cases exhibit higher streamwise organization of attached streak pairs compared to Poiseuille-like cases, though attached pairs in both flows are predominantly organized spanwise. Despite these spatial differences, streaks exhibit similar characteristics across cases, including angular inclination, meandering, volume, and dimensional ratios. This consistency suggests that while shear significantly impacts the spatial location and growth of streaks, their dimensional characteristics remain unaffected.

https://doi.org/10.3390/fluids11010018

Q structures, particularly bursts (Q2) and sweeps (Q4), play a crucial role in the redistribution of momentum and energy within turbulent flows. These structures sustain turbulence by driving Reynolds stresses and facilitating energy transfer.

The spatial distribution and intensity of Q structures vary with the shear configuration. In Poiseuille-like cases, Q2 structures are significantly more volumetric than Q4 structures, but this disparity decreases in Couette-like configurations. In the latter, Q2 structures reduce in volume while Q4 structures expand. Furthermore, Couette-like configurations favour attached Q structures, whereas Poiseuille-like cases exhibit a more balanced distribution of attached and detached structures.

The shear ratio has limited influence on other Q structure properties, such as dimensional ratios $(\Delta x/\Delta y, \Delta z/\Delta y)$, angular inclination, volume, and meandering. These properties show strong agreement across shear cases, indicating robust structural similarities.

The intensity of Q structures was also studied along the wall-normal direction. In the near-wall region ($y^+ < 20$), Q4 structures dominate the intensity contribution, which remains consistent across shear cases. At wall distances $y^+ \geq 20$, Q2 structures dominate, peaking at $y^+ \approx 40$. Beyond $y^+ > 50$, the intensity contribution of all Q structures stabilizes, reflecting the wall-normal percolation method used. Individual Q structures increase in intensity with shear, with the most intense structures observed in pure Couette flows. Interestingly, a common region of attached Q4 structures with $\tau_m^+ \approx -2$ is observed across all cases, driven by Q4 structures located in the near-wall region ($y^+ < 20$).

The study also examined the relative and absolute distances between Q structures. Pairs of attached Q structures of the same type (Q2–Q2 and Q4–Q4) predominantly align streamwise, whereas Q2–Q4 pairs align spanwise. Shear ratio primarily affects the streamwise distance between Q2–Q2 pairs, reducing it from $\delta_x \approx 1.8$ in Poiseuille-like cases to $\delta_x \approx 1$ in Couette-like cases. The absolute distance between attached pairs of the same type scales proportionally to $(\Delta y^+)^{0.55}$, showing good agreement across all shear cases. However, this agreement is not observed for attached pairs of different types, where Q2–Q4 pairs exhibit progressively larger distances with increasing shear. This spatial relocation of Q2 and Q4 structures is attributed to the influence of very large streamwise rolls.

To sum up, the changes in the mean shear due to the domain moving wall modify the characteristics of the coherent structures, such as streaks and Q structures. The geometry, topology, energetic intensity, and spatial organisation of these structures are affected by the gradual increase in shear ratio, although in different ways. While the geometry and topology are less influenced by the changes in shear, often showing a perfect collapse among simulation cases, the energetic intensity increases incrementally with the shear ratio. In contrast, the spatial organisation of coherent structures changes with the sign of the shear ratio, presenting a distinct scenario for Poiseuille-like, that is, negative shear ratio, or Couette-like, that is, positive shear ratio, configurations. This observation agrees with the fact that very large streamwise rolls, which remain fixed in the spanwise direction, are present in Couette-like configurations. Thus, a point remains open regarding the causality between the presence of streamwise rolls and the changes in coherent structure characteristics. Despite the comprehensive comparison carried out in this study, several limitations should be acknowledged. The simulations were performed at a relatively low-friction Reynolds number, which restricts the scale separation and may affect the generality of the results at higher Reynolds numbers where outer and inner interactions are stronger. Moreover, the identification of streaks relies on a global threshold that does not account for the wall-normal variation of turbulent intensity, and the detection of Q structures uses a percolation methodology that introduces an arbitrary parameter. These limitations suggest several promising directions for further research, including extending the analysis to higher Reynolds numbers, refining the detection criteria for coherent structures, and

https://doi.org/10.3390/fluids11010018

using complementary tools such as spectral energy budget analysis or causal inference to better resolve the link between the emergence of large streamwise rolls and the spatial organisation of momentum-carrying structures.

**Author Contributions:** Methodology, S.H.; Software, S.G.-B.; Validation, S.G.-B.; Formal analysis, S.G.-B. and S.H.; Investigation, S.G.-B.; Resources, S.H.; Writing—original draft, S.G.-B.; Writing—review & editing, S.H.; Supervision, S.H.; Project administration, S.H. All authors have read and agreed to the published version of the manuscript.

**Funding:** The authors acknowledge support by CIN/AEI/10.13039/501100011033 and by ERDF, "A way of making Europe", under project PID2024-162480OB-I00.

**Data Availability Statement:** Data of this study is available by contacting the corresponding author.

**Acknowledgments:** The computations of new simulations were made possible by a generous grant of computing time from the supercomputing centre of the Universitat Politècnica de València.

**Conflicts of Interest:** The authors declare no conflicts of interest.

## Abbreviations

The following abbreviations are used in this manuscript:

| | |
|---|---|
| DNS | Direct Numerical Simulation |
| C-P | Couette–Poiseuille |
| PDF | Probability Density Function |
| ETT | Eddy Turnover Time |
| TKE | Turbulent Kinetic Energy |
| SHAP | SHapley Additive exPlanations |
| Re | Reynolds number |
| CFD | Computational Fluid Dynamics |

## References


1. Jiménez, J. Coherent structures in wall-bounded turbulence. *J. Fluid Mech.* **2018**, *842*, P1. [CrossRef]
2. Kline, S.J.; Reynolds, W.C.; Schraub, F.A.; Runstadler, P.W. The structure of turbulent boundary layers. *J. Fluid Mech.* **1967**, *30*, 741–773. [CrossRef]
3. Lu, S.S.; Willmarth, W.W. Measurements of the structure of the Reynolds stress in a turbulent boundary layer. *J. Fluid Mech.* **1973**, *60*, 481–511. [CrossRef]
4. del Álamo, J.C.; Jiménez, J.; Zandonade, P.; Moser, R. Self-similar vortex clusters in the turbulent logarythmic region. *J. Fluid Mech.* **2006**, *561*, 329–358. [CrossRef]
5. Lozano-Durán, A.; Flores, O.; Jiménez, J. The three-dimensional structure of momentum transfer in turbulent channels. *J. Fluid Mech.* **2012**, *694*, 100–130. [CrossRef]
6. Osawa, K.; Jiménez, J. Intense structures of different momentum fluxes in turbulent channels. *Phys. Rev. Fluids* **2018**, *3*, 084603. [CrossRef]
7. Hwang, J.; Sung, H.J. Wall-attached structures of velocity fluctuations in a turbulent boundary layer. *J. Fluid Mech.* **2018**, *856*, 958–983. [CrossRef]
8. Cheng, C.; Li, W.; Lozano-Durán, A.; Liu, H. Uncovering Townsend's wall-attached eddies in low-Reynolds-number wall turbulence. *J. Fluid Mech.* **2020**, *889*, A29. [CrossRef]
9. Yoon, M.; Hwang, J.; Yang, J.; Sung, H.J. Wall-attached structures of streamwise velocity fluctuations in an adverse-pressure-gradient turbulent boundary layer. *J. Fluid Mech.* **2020**, *885*, A12. [CrossRef]
10. Bae, H.J.; Lee, M. Life cycle of streaks in the buffer layer of wall-bounded turbulence. *Phys. Rev. Fluids* **2021**, *6*, 064603. [CrossRef]
11. Gandía-Barberá, S.; Alcántara-Ávila, F.; Hoyas, S.; Avsarkisov, V. Stratification effect on extreme-scale rolls in plane Couette flows. *Phys. Rev. E* **2021**, *6*, 034605. [CrossRef]
12. Pope, S.B. *Turbulent Flows*; Cambridge University Press: Cambridge, UK, 2000.
13. Hoyas, S.; Jiménez, J. Scaling of the velocity fluctuations in turbulent channels up to $Re_\tau = 2003$. *Phys. Fluids* **2006**, *18*, 011702. [CrossRef]


https://doi.org/10.3390/fluids11010018

14. Lee, M.; Moser, R. Direct numerical simulation of turbulent channel flow up to $Re_\tau \approx 5200$. *J. Fluid Mech.* **2015**, *774*, 395–415. [CrossRef]
15. Hoyas, S.; Oberlack, M.; Alcántara-Ávila, F.; Kraheberger, S.V.; Laux, J. Wall turbulence at high friction Reynolds numbers. *Phys. Rev. Fluids* **2022**, *7*, 014602. [CrossRef]
16. Avsarkisov, V.; Hoyas, S.; Oberlack, M.; García-Galache, J. Turbulent plane Couette flow at moderately high Reynolds number. *J. Fluid Mech.* **2014**, *751*, R1. [CrossRef]
17. Lee, M.; Moser, R. Extreme-scale motions in turbulent plane Couette flows. *J. Fluid Mech.* **2018**, *842*, 128–145. [CrossRef]
18. Hoyas, S.; Oberlack, M. Turbulent Couette flow up to $Re_\tau = 2000$. *J. Fluid Mech.* **2024**, *987*, R9. [CrossRef]
19. Pirozzoli, S.; Bernardini, M.; Orlandi, P. Large-scale motions and inner/outer layer interactions in turbulent Couette—Poiseuille flows. *J. Fluid Mech.* **2011**, *680*, 534–563. [CrossRef]
20. Gandía-Barberá, S.; Hoyas, S.; Oberlack, M.; Kraheberger, S. The link between the Reynolds shear stress and the large structures of turbulent Couette-Poiseuille flow. *Phys. Fluids* **2018**, *30*, 041702. [CrossRef]
21. Cheng, W.; Pullin, D.; Samtaney, R.; Luo, X. Numerical simulation of turbulent, plane parallel Couette–Poiseuille flow. *J. Fluid Mech.* **2023**, *955*, A4. [CrossRef]
22. Lluesma-Rodríguez, F.; Hoyas, S. Approximation of the Couette-Poiseuille flow on an annular domain in compressible regime for a hyperloop pod. *Results Eng.* **2024**, *21*, 101680. [CrossRef]
23. Tillmark, N. Experiments on Transition and Turbulence in Plane Couette Flow. Ph.D. Thesis, KTH, Royal Institute of Technology, Stockholm, Sweden, 1995.
24. Kitoh, O.; Nakabyashi, K.; Nishimura, F. Experimental study on mean velocity and turbulence characteristics of plane Couette flow: Low-Reynolds-number effects and large longitudinal vortical structure. *J. Fluid Mech.* **2005**, *539*, 199–227. [CrossRef]
25. Bech, K.; Tillmark, N.; Alfredsson, P.; Andersson, H. An investigation of turbulent plane Couette flow at low Reynolds numbers. *J. Fluid Mech.* **1995**, *286*, 291–325. [CrossRef]
26. Komminaho, J.; Lundbladh, A.; Johansson, A. Very large structures in plane turbulent Couette flow. *J. Fluid Mech.* **1996**, *320*, 259–285. [CrossRef]
27. Tsukahara, T.; Kawamura, H.; Shingai, K. DNS of turbulent Couette flow with emphasis on the large-scale structure in the core region. *J. Turbul.* **2006**, *7*, N19. [CrossRef]
28. Bernardini, M.; Pirozzoli, S.; Orlandi, P. The effect of large-scale turbulent structures on particle dispersion in wall-bounded flows. *Int. J. Multiph. Flow* **2013**, *51*, 55–64. [CrossRef]
29. Pirozzoli, S.; Bernardini, M.; Orlandi, P. Turbulence statistics in Couette flow at high Reynolds number. *J. Fluid Mech.* **2014**, *758*, 323–343. [CrossRef]
30. Kraheberger, S.; Hoyas, S.; Oberlack, M. DNS of a turbulent Couette flow at constant wall transpiration up to $Re_\tau = 1000$. *J. Fluid Mech.* **2018**, *835*, 421–443. [CrossRef]
31. Hamilton, J.; Kim, J.; Waleffe, F. Regeneration mechanisms of near-wall turbulence structures. *J. Fluid Mech.* **1995**, *287*, 317–348. [CrossRef]
32. Alcántara-Ávila, F.; Gandía-Barberá, S.; Hoyas, S. Evidences of persisting thermal structures in Couette flows. *Int. J. Heat Fluid Flow* **2019**, *76*, 287–295. [CrossRef]
33. Chong, M.; Perry, A.; Cantwell, B. A general classification of three-dimensional flow fields. *J. Phys. A* **1990**, *2*, 765–777. [CrossRef]
34. Cremades, A.; Hoyas, S.; Deshpande, R.; Quintero, P.; Lellep, M.; Lee, W.J.; Monty, J.P.; Hutchins, N.; Linkmann, M.; Marusic, I.; et al. Identifying regions of importance in wall-bounded turbulence through explainable deep learning. *Nat. Commun.* **2024**, *15*, 3864. [CrossRef] [PubMed]
35. Schmekel, D.; Alcántara-Ávila, F.; Hoyas, S.; Vinuesa, R. Predicting coherent turbulent structures via deep learning. *Front. Phys.* **2022**, *10*, 888832. [CrossRef]
36. Lellep, M.; Prexl, J.; Eckhardt, B.; Linkmann, M. Interpreted machine learning in fluid dynamics: Explaining relaminarisation events in wall-bounded shear flows. *J. Fluid Mech.* **2022**, *942*, A2. [CrossRef]
37. Cremades, A.; Hoyas, S.; Vinuesa, R. Classically studied coherent structures only paint a partial picture of wall-bounded turbulence. *Nat. Commun.* **2025**, *16*, 10189. [CrossRef]
38. Orr, W.M. The Stability or Instability of the Steady Motions of a Perfect Liquid and of a Viscous Liquid. Part II: A Viscous Liquid. *Proc. R. Ir. Acad. Sect. A Math. Phys. Sci.* **1907**, *27*, 69–138.
39. Motoori, Y.; Goto, S. Hierarchy of coherent structures and real-space energy transfer in turbulent channel flow. *J. Fluid Mech.* **2021**, *911*, A27. [CrossRef]
40. Kawata, T.; Alfredsson, P.H. Inverse interscale transport of the Reynolds shear stress in plane Couette turbulence. *Phys. Rev. Lett.* **2018**, *120*, 244501. [CrossRef]
41. Lee, M.; Moser, R.D. Spectral analysis of the budget equation in turbulent channel flows at high Reynolds number. *J. Fluid Mech.* **2019**, *860*, 886–938. [CrossRef]


https://doi.org/10.3390/fluids11010018

42. Bauer, C.; von Kameke, A.; Wagner, C. Kinetic energy budget of the largest scales in turbulent pipe flow. *Phys. Rev. Fluids* **2019**, *4*, 064607. [CrossRef]
43. Kawata, T.; Tsukahara, T. Scale interactions in turbulent plane Couette flows in minimal domains. *J. Fluid Mech.* **2021**, *911*, A55. [CrossRef]
44. Wang, H.; Yang, Z.; Wu, T.; Wang, S. Coherent structures associated with interscale energy transfer in turbulent channel flows. *Phys. Rev. Fluids* **2021**, *6*, 104601. [CrossRef]
45. Kawata, T.; Tsukahara, T. Spectral analysis on dissimilarity between turbulent momentum and heat transfers in plane Couette turbulence. *Phys. Fluids* **2022**, *34*, 075135. [CrossRef]
46. Oberlack, M.; Hoyas, S.; Kraheberger, S.V.; Alcántara-Ávila, F.; Laux, J. Turbulence Statistics of Arbitrary Moments of Wall-Bounded Shear Flows: A Symmetry Approach. *Phys. Rev. Lett.* **2022**, *128*, 024502. [CrossRef]
47. Lozano-Durán, A.; Jiménez, J. Effect of the computational domain on direct simulations of turbulent channels up to $Re_\tau = 4200$. *Phys. Fluids* **2014**, *26*, 011702. [CrossRef]
48. Sánchez-Roncero, A.; Garibo-i Orts, Ò.; Conejero, J.A.; Eivazi, H.; Mallor, F.; Rosenberg, E.; Fuso-Nerini, F.; García-Martínez, J.; Vinuesa, R.; Hoyas, S. The sustainable development goals and aerospace engineering: A critical note through artificial intelligence. *Results Eng.* **2023**, *17*, 100940. [CrossRef]
49. Lele, S.K. Compact finite difference schemes with spectral-like resolution. *J. Comput. Phys.* **1992**, *103*, 16–42. [CrossRef]
50. Spalart, P.R.; Moser, R.D.; Rogers, M.M. Spectral methods for the Navier-Stokes equations with one infinite and two periodic directions. *J. Comput. Phys.* **1991**, *96*, 297–324. [CrossRef]
51. Lluesma-Rodríguez, F.; Álcantara Ávila, F.; Pérez-Quiles, M.; Hoyas, S. A code for simulating heat transfer in turbulent channel flow. *Mathematics* **2021**, *9*, 756. [CrossRef]
52. Tang, S.; Antonia, R. Similarity for dissipation-scaled wall turbulence. *J. Fluid Mech.* **2023**, *960*, A18. [CrossRef]
53. Spalart, P. The Conjecture of a General Law of the Wall for Classical Turbulence Models, Implying a Structural Limitation. *Flow Turbul. Combust.* **2023**, *112*, 443–457. [CrossRef]
54. Yu, L.; Chen, Y.; Yousif, M.Z.; Lim, H.C. Three-dimensional super-resolution reconstruction of turbulent flow using 3D-ESRGAN with random sampling strategy. *Comput. Fluids* **2025**, *305*, 106890. [CrossRef]
55. Palasis, A.; Liakopoulos, A.; Sofiadis, G. From Direct Numerical Simulations to Data-Driven Models: Insights into Mean Velocity Profiles and Turbulent Stresses in Channel Flows. *Modelling* **2025**, *6*, 18. [CrossRef]
56. Hoyas, S.; Vinuesa, R.; Schmid, P.; Nagib, H. Sensitivity study of resolution and convergence requirements for the extended overlap region in wall-bounded turbulence. *Phys. Rev. Fluids* **2024**, *9*, L082601. [CrossRef]
57. Gandía-Barberá, S.; Cremades, A.; Vinuesa, R.; Hoyas, S.; Pérez-Quiles, M.J. Sequential and Parallel Algorithms to Compute Turbulent Coherent Structures. *Mathematics* **2024**, *12*, 3325. [CrossRef]
58. Lozano-Durán, A.; Borrel, G. Algorithm 964: An Efficient Algorithm to Compute the Genus of Discrete Surfaces and Applications to Turbulent Flows. *Assoc. Comput. Mach.* **2016**, *42*, 34. [CrossRef]
59. Kevin, K.; Monty, J.; Hutchins, N. The meandering behaviour of large-scale structures in turbulent boundary layers. *J. Fluid Mech.* **2019**, *865*, R1. [CrossRef]
60. Rawat, S.; Cossu, C.; Hwang, Y.; Rincon, F. On the self-sustained nature of large-scale motions in turbulent Couette flow. *J. Fluid Mech.* **2015**, *782*, 515–540. [CrossRef]
61. Jiménez, J. The streaks of wall-bounded turbulence need not be long. *J. Fluid Mech.* **2022**, *945*, R3. [CrossRef]
62. Dong, S.; Lozano-Durán, A.; Sekimoto, A.; Jiménez, J. Coherent structures in statistically stationary homogeneous shear turbulence. *J. Fluid Mech.* **2017**, *816*, 167–208. [CrossRef]
63. Cheng, C.; Li, W.; Lozano-Durán, A.; Liu, H. On the structure of streamwise wall-shear stress fluctuations in turbulent channel flows. *J. Fluid Mech.* **2020**, *903*, A29. [CrossRef]
64. Bae, H.; Encinar, M.; Lozano-Durán, A. Causal analysis of self-sustaining processes in the logarithmic layer of wall-bounded turbulence. *J. Phys. Conf. Ser.* **2018**, *1001*, 012013. [CrossRef]
65. Moisy, F.; Jiménez, J. Geometry and clustering of intense structures in isotropic turbulence. *J. Fluid Mech.* **2004**, *513*, 111–133. [CrossRef]
66. Leung, T.; Swaminathan, N.; Davidson, P.A. Geometry and interaction of structures in homogeneous isotropic turbulence. *J. Fluid Mech.* **2012**, *710*, 453–481. [CrossRef]
67. Corrsin, S. Progress report on some turbulent diffusion research. *Adv. Geophys.* **1959**, *6*, 441–451. [CrossRef]




https://doi.org/10.3390/fluids11010018

fluids

MDPI

*Review*

# Modelling Water Waves on Graphs

**André Nachbin**

Department of Mathematical Sciences, Worcester Polytechnic Institute, 100 Institute Rd, Worcester, MA 01609, USA; anachbin@wpi.edu

**Abstract:** Waves on graphs are a current subject of research interest. As opposed to flows on graphs, the reflection–transmission of waves at the graph's vertex is a problem that needs to be further modelled mathematically. The literature on the reflection and transmission of waves at a vertex is scarce. Some articles are reviewed and discussed. Water waves are a good topic for comparing different mathematical models, from hyperbolic conservation laws to weakly nonlinear, weakly dispersive systems of partial differential equations on a two-dimensional fattened (thick) graph and the respective one-dimensional graph-model reduction. In this study, we present a particular water wave model in which junction angles are systematically included in the mathematical model. Comparing the solutions with the fattened-graph model gave rise to a more general compatibility condition at the vertex. Current research topics of interest are outlined at the end.

**Keywords:** water waves; channel network; quantum graphs; mathematical modelling

## 1. Introduction

Water waves are a topic of great importance in environmental modelling, as well as in the mathematical modelling and theory of partial differential equations (PDEs). Applications range from surface and internal waves in the ocean to waves in rivers or canal networks. Water waves are also of great importance in Mathematics, where they can represent many challenging problems in a range of areas, from Analysis to mathematical modelling and computing. As an area of PDEs, they can range from hyperbolic conservation laws to nonlinear, dispersive wave equations. From linear to nonlinear equations, constant or variable coefficients, containing local or nonlocal (differential or pseudo-differential) operators, may be present. Namely, a rich and broad area of PDEs, in which water waves on graphs is quite unexplored, will be reported here.

Irrespective of the fluids' application, flow on graphs is a topic of current mathematical interest. It falls in a category of problems which has attracted a great deal of attention recently: quantum graphs. Quantum graphs are defined as metric graphs equipped with a differential equation along it's edges [1–3]. Very recently, Goodman et al. [4] published an article pointing to a MATLAB (version R2023b) package for computations on quantum graphs, mostly specialized for elliptic problems. Goodman et al. present a brief introduction to quantum graphs, containing many references to recent work. The mathematical literature on wave propagation on graphs is not as abundant as for flows on graphs [5,6] or steady waves on graphs [1]. It is almost inexistent for wave reflection and transmission at the forked region.

The present article calls attention to wave models on graphs, discussing two modelling levels, where solutions of 2D thick graphs (fattened graphs) and 1D graphs should be compared. As displayed in Figure 1, the thick-graph domain has a small lateral width $B$

when compared to the wavelength (or pulse-width) $\lambda > B$. The 1D graph, depicted in red, has a vertex (a node) where compatibility conditions need to be imposed.

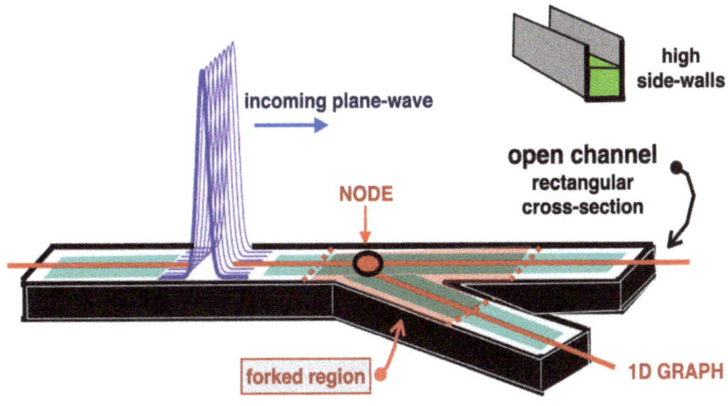


**Figure 1.** A schematic picture for the geometrical setting of this water-waves problem: a 2D open channel has a water-wave pulse propagating towards the forked region from the left. The black side-walls of the channel are provided for better visualization. The reduced 1D model is defined on the graph (in red) with a dot at its vertex (node), which will be denoted as $P$.

Another important point calls attention to diverse applications of waves in graphs. To the best of our knowledge, there are no articles in the literature which present a systematic modelling of branching angles into the 1D-graph's PDEs. Actually, the majority of the models encountered in the literature do not take angles into account, which for water waves, among other applications, is very important to consider. Moreover, we are not aware of any article that performs a comprehensive comparison between solutions of the parent 2D system of PDEs with those of the reduced 1D system. As reported below, some authors comment on the importance of such a comparative study. The solutions for the reflection–transmission of waves at the forked region are not always in agreement. By comparing solutions on different domains (2D×1D) Nachbin and Simões [7] deduced a new compatibility condition that generalizes the (well-established) Neumann–Kirchhoff conditions [1,3].

## 2. The Mathematical Background of Water Waves on Graphs

In this section, we present the basic equations for water waves on graphs and review some of the existing articles relating to fluid dynamics on graph-like domains. We also discuss some articles of interest from waves on graphs that are not fluid dynamics-related.

In 1981, Wu [8] worked with Euler equations using the potential theory framework by considering an inviscid fluid flow to be incompressible and irrotational in an open channel. The 3D fluid domain is displayed in detail in the upper-right of Figure 1. The open channel has a rectangular cross-section with high side-walls so that there is no spillage from the channel. The velocity potential $\phi(x, y, z, t)$ satisfies the 3D Laplace equation in $\Omega(t)$, the time-dependent fluid body (in green), together with two nonlinear free surface conditions along the wave elevation $z = \eta(x, y, t)$. At the impermeable side walls and flat bottom, a homogeneous Neumann condition has to be satisfied. The velocity field in the bulk of the fluid is given by $(u, v, w) = \nabla\phi(x, y, z, t)$. The 3D potential theory equations are [8,9]:

$$\phi_{xx} + \phi_{yy} + \phi_{zz} = 0, \text{ in } \Omega(t), \tag{1}$$

$$\phi_t + \frac{1}{2}\|\nabla\phi\|^2 + g\eta = 0, \text{ at } z = \eta(x,y,t), \tag{2}$$

$$\eta_t + \nabla\phi \cdot \nabla\eta - \phi_z = 0, \text{ at } z = \eta(x,y,t), \tag{3}$$

$$\frac{d\phi}{dn} = 0, \text{ at the bottom } z = -h_0 \text{ and along the side-walls.} \tag{4}$$

The (outward) normal derivative in (4) indicates a lack of flow through the impermeable boundaries. Initial conditions are provided along the free surface where the evolution operators are imposed: $\eta(x,y,0) = \eta_0(x,y)$ is the initial free surface disturbance, defining the initial domain configuration $\Omega(0)$, while $\phi(x,y,\eta_0(x,y),0) = \phi_0(x,y)$ is the Dirichlet data for the harmonic function to be defined in the bulk.

Through a depth-averaged asymptotic model reduction of the Euler equations, Wu [8] formulated the following weakly nonlinear, weakly dispersive Boussinesq system:

$$\eta_t + (1+\eta)(\phi_{xx} + \phi_{yy}) + \eta_x\phi_x + \eta_y\phi_y = 0, \tag{5}$$

$$\phi_t + \eta + \frac{\phi_x^2 + \phi_y^2}{2} - \frac{\phi_{xxt} + \phi_{yyt}}{3} = 0. \tag{6}$$

For simplicity, we have used the notation and denoted the reduced potential as $\phi(x,y,t)$. The reduced potential is not a harmonic function, or else the term $(\phi_{xx} + \phi_{yy})$ in (5) would vanish. Therefore it is different from full velocity potential $\phi(x,y,z,t)$ given earlier in system (1)–(4). It is a reduced velocity potential in the sense that the 2D horizontal velocity field is given by $(u,v) = \nabla\phi$. The system has been normalized to have a unit shallow water speed.

This Boussinesq system (5) and (6) was used in the only paper on forked channel regions we are aware of, for nonlinear, dispersive water waves. Shi et al. [10] considered a Y-shaped thick-graph, as depicted in Figure 1. No systematic reduction to a 1D graph-model was attempted in [10], or elsewhere. It is also worth mentioning an earlier work by the group [11] which focused on solitary waves on significantly curved channels. Using Cartesian coordinates, the authors were only able to perform numerical simulations for branching angles that were multiples of $\pi/4$. In these cases, the finite difference nodes fell properly along the walls of the secondary channel reaches, as depicted in the schematic Figure 2. Otherwise, one needed to scale the $\Delta x$-$\Delta y$ grid accordingly with the branching angle of the reach. But in the case of two branching reaches at different angles, this strategy would fail.

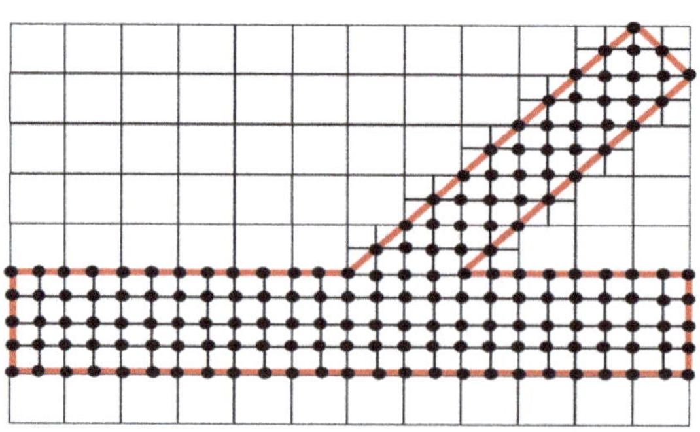

**Figure 2.** A schematic picture for a finite difference grid in the physical domain. The red lines depict the channel boundaries while the black circles the nodes for the finite difference scheme.

At the corners of the branching region, one has singularities. In their appendix, Shi et al. [10] report an averaging strategy at corner nodes of the branching region, in order to deal with the lack of a well-defined normal state for the Neumann condition at the side walls. The averaging strategy, using two other neighboring nodes, worked well for channels of moderate width. As the channel widths became smaller, numerical spikes were observed in the simulations. These spikes remained localized near the corner, and would not propagate away.

Next, we present a series of model reductions. We restrict Equations (5) and (6) to a one-dimensional domain:

$$\eta_t + (1+\eta)\phi_{xx} + \eta_x\phi_x = 0, \tag{7}$$

$$\phi_t + \eta + \frac{\phi_x^2}{2} - \frac{\phi_{xxt}}{3} = 0. \tag{8}$$

Then, we differentiate (8) with respect to $x$. By the definition of the velocity potential, we substitute $\phi_x = u$, and the system can be rewritten as follows:

$$\eta_t + (1+\eta)u_x + u\eta_x = 0, \tag{9}$$

$$u_t + \eta_x + uu_x - \frac{1}{3}u_{xxt} = 0. \tag{10}$$

This is a nonlinear, dispersive Boussinesq system. This system can be further specialized to unidirectional waves, propagating to the right [9]. To obtain the leading order in nonlinearity and dispersion, one can formulate the Korteweg–de Vries (KdV) equation

$$\eta_t + \eta_x + \eta\eta_x + \frac{1}{3}\eta_{xxx} = 0, \tag{11}$$

or the Benjamin–Bona–Mahony (BBM) equation

$$\eta_t + \eta_x + \eta\eta_x - \frac{1}{3}\eta_{xxt} = 0. \tag{12}$$

Equations (11) and (12) allow for solitary (travelling) waves solutions. System (9) and (10) allows for solitary waves in an approximate fashion [11]. Let the total water depth be $h(x,t) = H + \eta(x,t)$, where we can set $H = 1$ as the normalized undisturbed depth. Drop the (third-order) dispersive term in (10) to obtain the hyperbolic system

$$h_t + hu_x + uh_x = 0, \tag{13}$$

$$u_t + h_x + uu_x = 0. \tag{14}$$

We have therefore briefly outlined a modelling hierarchy which started with a 3D, fully dispersive, nonlinear potential theory formulation, and ended with a 1D hyperbolic system. This hyperbolic system is the starting point for our review on a few references for water waves on graphs.

### 2.1. Fluid Flow Problems on a Graph

We begin by reviewing a few papers we have encountered which relate to fluid flow on graphs. The shallow water (long wave) model presented by Jacovkis [6] was the starting point for our interest in this topic. Two typical junction configurations are considered as shown in Figure 3.

Jacovkis [6] considers the following PDE-system of conservation laws:

$$\frac{\partial u}{\partial t} + u\frac{\partial u}{\partial x} + g\frac{\partial h}{\partial x} = 0, \tag{15}$$

$$\frac{\partial h}{\partial t} + h\frac{\partial u}{\partial x} + u\frac{\partial h}{\partial x} = 0. \tag{16}$$

As opposed to the process described above for the hyperbolic system, now, the wave speed is not normalized and will depend on $g$, which is the acceleration due to gravity. This is a dimensional version and the shallow water speed is $C_0 = \sqrt{gH}$, which is not necessarily equal to 1. This is a strictly hyperbolic system with two real eigenvalues for the system's matrix

$$A = \begin{bmatrix} u & g \\ h & u \end{bmatrix}, \text{where } \frac{\partial \vec{w}}{\partial t} = A\frac{\partial \vec{w}}{\partial x}, \text{ and } \vec{w}(x,t) = [u,h]^T. \tag{17}$$

Consider junction configurations such as those displayed in Figure 3. Jacovkis presents a theorem for the linear model in the subcritical regime, where one characteristic points upstream while the other points downstream. The slope $dx/dt$ of the characteristics are given by the eigenvalues $\lambda_1 = U - C_0$ and $\lambda_2 = U + C_0$, where $U$ and $H$ are (constant) reference values for the velocity and depth. The linear shallow water system has the form

$$A = \begin{bmatrix} U & g \\ H & U \end{bmatrix}, \text{where } \frac{\partial \vec{w}}{\partial t} = A\frac{\partial \vec{w}}{\partial x}, \text{ and } \vec{w}(x,t) = [\tilde{u},\tilde{h}]^T. \tag{18}$$

Along each reach, we consider the initial conditions

$$\vec{w}_0 = \vec{w}(x,0) = [\tilde{u}(x,0),\tilde{h}(x,0)]^T, \tag{19}$$

where $x$ is a local coordinate within the respective reach. The Neumann–Kirkhhoff compatibility conditions are as follows:

$$B_i\tilde{u}(x_i,t) \pm B_j\tilde{u}(x_j,t) = B_k\tilde{u}(x_k,t), \tag{20}$$

$$\tilde{h}(x_i,t) = \tilde{h}(x_j,t) = \tilde{h}(x_k,t). \tag{21}$$

At the vertex $P$, the respective reach coordinate is $x_i$, $x_j$ and $x_k$. The widths of the respective reaches are displayed in Figure 3 and denoted by $B_{\text{index}}$, index $\in \{i,j,k\}$. Conservation of mass is given by (20), while a first approximation for conservation of momentum is expressed by the continuity of water elevation (21). For simplicity, we now drop the tilde from the velocity and water height perturbations $\tilde{u}$ and $\tilde{h}$. Jacovkis considers a type-I junction, as depicted in Figure 3, in order to present the following theorem.

**Theorem 1** ([6]). *Consider the linear shallow water system (18), with initial conditions (19) along each reach, satisfying the Neumann–Kirchhoff compatibility conditions (20) and (21) at the junction point P, together with boundary conditions*

$$\vec{b}_B \cdot \vec{w}(t,x_P) = [b_{1B},b_{2B}] \cdot [u(x_B,t),h(x_B,t)] = f_B(t),$$

*at each open extreme point $x_B$ of the Y-shaped network. This problem has a unique solution $\vec{w} \in C^1$, at each reach, provided that the following condition apply:*

*(i)* $\vec{w}_0$ *and all the* $f_B$ *are* $C^1$ *and consistent with the initial conditions at* $t = 0$*;*

*(ii)* *the matrix A has real eigenvalues* $\lambda_1 < 0 < \lambda_2$*;*

*(iii)* $b_{1B}$ *and* $b_{2B}$ *are 0/1 binary variables, such that* $b_{1B} + b_{2B} = 1$.

*The solution exists for all times $t > 0$.*

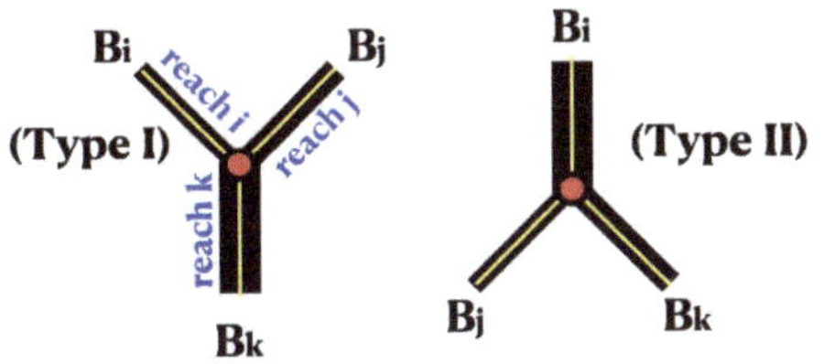


**Figure 3.** Two typical junction configurations are considered. A converging junction is indicated as type I and a diverging junction as type II. The reaches' widths are respectively denoted by $(B_i; B_j; B_k)$, which can be different from each other. The thinner yellow lines represent the graph's edges. The thicker black rectangles highlight the different widths associated with each reach of the 1D model. The (red) circle indicates the vertex $P$ where compatibility conditions must be applied to the graph solutions along the one-dimensional segments.

We outline some points from the proof, which are relevant for our discussion. The matrix $A$ is diagonalizable, with

$$\Lambda = S^{-1}AS; \ \Lambda = \begin{bmatrix} U - C_0 & 0 \\ 0 & U + C_0 \end{bmatrix}, \ S = \begin{bmatrix} -C_0 & H \\ C_0 & H \end{bmatrix}. \tag{22}$$

System (18) can be rewritten as two decoupled wave equations for each component of $\vec{y} = [v, z]^T$, where $\vec{y} = S^{-1}\vec{w}$:

$$v_t + \lambda_1 v_x = 0, \ \lambda_1 < 0, \tag{23}$$

$$z_t + \lambda_2 z_x = 0, \ \lambda_2 > 0. \tag{24}$$

Consider the Neumann–Kirchhoff compatibility conditions (20) and (21) at the vertex $P$, with reaches $i$ and $j$ merging onto reach $k$, in the Y-configuration, type-I. We state that $z_i = z(x_i, t)$, $z_j = z(x_j, t)$ and $v_k = v(x_k, t)$ are **incoming modes** at the vertex, while $v_i = z(x_i, t)$, $v_j = z(x_j, t)$ and $z_k = z(x_k, t)$ are the respective **outgoing modes**. The compatibility conditions yield a functional relation between input and output from the vertex $P$:

$$\begin{bmatrix} B_iC_0 & B_jC_0 & -B_k C_0 \\ H & -H & 0 \\ H & 0 & -H \end{bmatrix} \begin{bmatrix} v_i \\ v_j \\ z_k \end{bmatrix} = \begin{bmatrix} -B_iC_0 & -B_jC_0 & B_k C_0 \\ -H & H & 0 \\ -H & 0 & H \end{bmatrix} \begin{bmatrix} z_i \\ z_j \\ v_k \end{bmatrix}. \tag{25}$$

The input (incoming modes) appear on the right. Recall that $B_i$, $B_j$ and $B_k$ are the widths of the respective reaches $(i, j, k)$. It is straightforward to show that theses matrices are nonsingular, so that one can write

$$\begin{bmatrix} v_i \\ v_j \\ z_k \\ \textbf{(outgoing)} \end{bmatrix} = R_{3\times 3} \begin{bmatrix} z_i \\ z_j \\ v_k \\ \textbf{(incoming)} \end{bmatrix}. \tag{26}$$

$R_{3\times 3}$ is a scattering matrix for the vertex, mapping incoming information to outgoing information. The linear system of Equation (25) indicates that the linear hyperbolic system, defined on this Y-shaped channel region, has a solution. Mathematically, the compatibility conditions allow us to connect the solutions from three different half-lines (semi-infinite reaches).

But this model does not have any information regarding the angles between reaches at the junction. This is an interesting wave model on a graph, which is well defined mathematically, but incomplete with regard to the effect of angles. For example, the model cannot properly account for the back-flow promoted by wave reflection at the junction. The following simple discussion, to the best of our knowledge, is not considered in the work of Jacovkis [6], nor elsewhere in the literature.

Consider three type-I junctions, as depicted in Figure 4. The configuration of the Y-shaped domain, whether it is symmetric or not, should play a role in the flow distribution along the reaches. Also, this should be applicable for wave reflection, as opposed to a pure streaming flow. As an extreme, consider case (A), which is depicted in Figure 4, and case (C). The incoming reaches (both $i$ and $j$) have half the width of the outgoing reach $k$. Imagine case (A) with a very small angle upstream (between reaches $i$ and $j$). For this case, we should not expect back-flow (wave reflection), as opposed to a T-shaped configuration, where there should be a nontrivial flow upstream. Consider expression (26) with the incoming vector given by $[a(t), a(t), 0]^T$. Namely, the flow arriving from reaches $i$ and $j$ is the same (at all times), while there is no upstream flow arriving from reach $k$. For the case $B_i = B_j = B_k/2$ (say, with $H = 1$ and $C_0 = 1$, for simplicity) it is straightforward to show that the outgoing vector given by (26) is $[v_i = v(x_i, t) = 0,\ v_j = v(x_j, t) = 0, z_k = z(x_k, t) = a(t)]^T$.

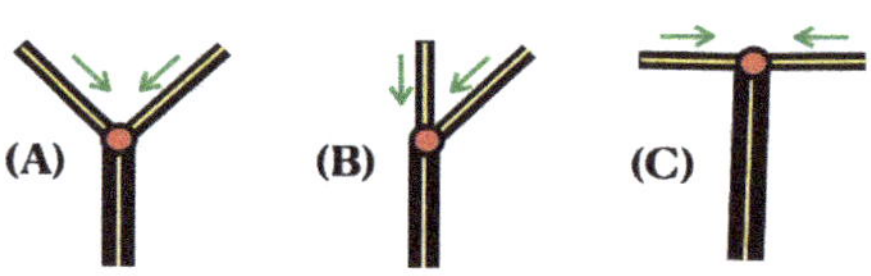


**Figure 4.** Three converging-junction configurations: (**A**) symmetric Y-shaped; (**B**) asymmetric Y-shaped; (**C**) T-shaped. The arrows indicate data which enter the vertex $P$ from upstream, without any flow coming from downstream. In all cases, let $B_i = B_j = B_k/2$.

Waves merging onto reach $k$ should generate quite different reflections (and back-flow) in these two extreme cases, as well as the asymmetric case (B) of Figure 4. Comparisons between the 2D fattened-graph model and the 1D simplification help to resolve modelling issues. This was taken into account by Nachbin and Simões [7] for the weakly nonlinear, weakly dispersive PDEs. In Jacovkis [6], as well as many other references, no information relating to angles is built into the models, nor are comparisons drawn between 1D graph-solutions with those of the parent 2D model. It is worth mentioning the paper by Caputo et al. [12], where the authors compared the results of the 2D shallow water equations with its restriction on a graph for a T-junction. The authors report that for the linear case, the Stoker conditions hold, indicating continuity of the wave elevation at the junction point of three reaches with the same width. A wave propagates up the vertical part of the T branch into the two horizontal reaches of the T. The effective width of the fluid domain doubles. Therefore, it is unclear how the wave heights remain the same as they propagate away from each other, while conserving the total excess mass.

The asymmetric configuration of Figure 4B has been addressed through a complex-variables formulation by Milne-Thomson [13]. It is revealing regarding the effect of the branching angle. The 2D shallow water system is of the form

$$\frac{\partial u}{\partial t} + u\frac{\partial u}{\partial x} + v\frac{\partial u}{\partial y} + g\frac{\partial h}{\partial x} = 0,$$
$$\frac{\partial v}{\partial t} + u\frac{\partial v}{\partial x} + v\frac{\partial v}{\partial y} + g\frac{\partial h}{\partial y} = 0,$$
$$\frac{\partial h}{\partial t} + \frac{\partial (hu)}{\partial x} + \frac{\partial (hv)}{\partial y} = 0.$$

The corresponding linear system is given by

$$\frac{\partial u}{\partial t} + g\frac{\partial h}{\partial x} = 0, \tag{27}$$

$$\frac{\partial v}{\partial t} + g\frac{\partial h}{\partial y} = 0, \tag{28}$$

$$\frac{\partial h}{\partial t} + H\frac{\partial u}{\partial x} + H\frac{\partial v}{\partial y} = 0. \tag{29}$$

For the fattened (2D) graph-like domain, impose the very simple regime of a steady flow $(u(x,y), v(x,y))$, with no wave elevation, so that $h \equiv H$. For this regime, system (27)–(29) reduces to the incompressibility condition $u_x + v_y = 0$. Adding the hypothesis of an irrotational flow, $v_x - u_y = 0$, allows for the existence of a velocity potential $\phi(x,y)$, such that $(u,v) = \nabla\phi$. Incompressibility implies that the velocity potential is a harmonic function, which has the stream function $\psi(x,y)$ as its harmonic conjugate, where $(u,v) = \nabla^{\perp}\psi$. Using a complex potential $\Phi(z) = \phi(x,y) + i\psi(x,y)$, with $z = x + iy$, leads to a complex velocity $d\Phi/dz = u(x,y) - iv(x,y)$ which automatically satisfies the constraints for an incompressible, irrotational flow, due to the Cauchy–Riemann equations for $\phi$ and $\psi$. Milne-Thomson [13] then uses a sequence of conformal-mapping compositions to express the solution for the streaming flow in an asymmetric junction with angle $\theta$, as shown in Figure 5. Very quickly, at a small distance away from the branching region, the flow aligns uniformly, parallel to the side walls which have no lateral variations. Milne-Thomson [13] provides the following equation that relates the three limiting uniform speeds along each reach:

$$\left(\frac{U_2}{U_1}\right)^{1-\pi/\theta} B_2 - \left(\frac{U_3}{U_1}\right)^{1-\pi/\theta} B_3 = B_1. \tag{30}$$

Define $\chi \equiv (U_2/U_1)$ and consider the case $B_1 = B_2 = B_3 =$L, which yields $U_1 = U_2 + U_3$ from the conservation of mass. Equation (30) becomes

$$(\chi)^{1-\pi/\theta} + (1-\chi)^{1-\pi/\theta} = 1. \tag{31}$$

This nonlinear equation for the speed ratio in the main reach depends nontrivially on the branching angle $\theta$. The respective values of $\chi$ are displayed in the top part of Table 1, given a few values of the branching angle. We may also consider the case $B_1/2 = B_2 = B_3 =$ L$/2$, which yields $U_3 = U_1(2-\chi)$. Equation (30) becomes

$$(\chi)^{1-\pi/\theta} + (2-\chi)^{1-\pi/\theta} = 2. \tag{32}$$

The respective values of $\chi$ are displayed in the bottom part of Table 1. For simplicity, let $U_1 = 1$, so that columns 5 and 6 in Table 1 give the speeds $U_2$ along the main reach, after the branching point, as well as $U_3$ in the secondary reach. When all reaches have the same width, L, and the branching angle $\theta$ is small, the speed after the forked region falls to half of the incoming value, as shown in the first row of Table 1. In the schematic, Figure 6 the total channel width doubles at the forked region, and therefore the speed is halved so that the flux is maintained. When the widths of the secondary reaches reduce by half (right of Figure 6), the total width remains unchanged, and therefore all speeds are (effectively) the same as shown in the first row of the bottom part of Table 1. In both cases, the speeds $U_2$ and $U_3$ change as the branching angle increases. This shows that the streaming flow, through a (harmonic) velocity potential, captures the effect of a branching angle. Nevertheless, as the angle increases further, the speed $U_3$ in the secondary reach becomes larger than $U_2$ in the main reach. As the angle increases, the flow separatrix, depicted in Figure 5, moves downwards, promoting a smaller flux along the lower part of the channel. We have an ideal fluid that slides along boundaries. No vorticity generation mechanism is present, which likely would occur near the outer corner of the inclined secondary reach. This model also has its limitations, but captures angle variations at the forked region. In our wave model [7], in contrast, we observe that as the angle increases, the preferred wave direction is through the main reach. This will be shown in Section 3.

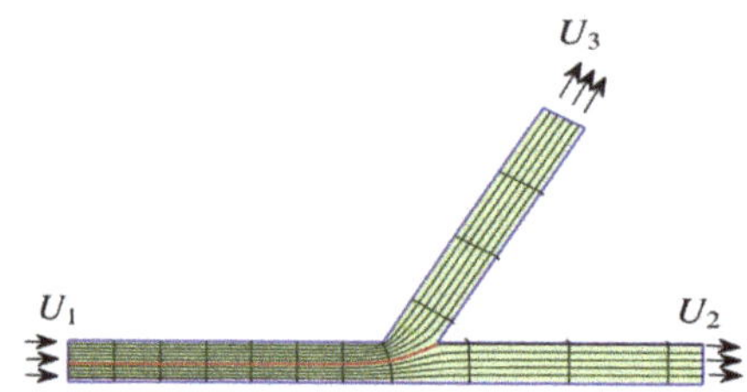


**Figure 5.** Streaming flow past an asymmetric forked region at angle $\theta$. The incoming reach has a width of $B_1$, while the other two reaches have, respectively, widths of $B_2$ and $B_3$. The separatrix (red curve), emanating from the forking corner, divides the laminar flow region into two parts: the bottom part, flowing towards the main reach, and the top part flowing towards the secondary reach.

**Table 1.** Parameters: $C_0 = 1.0$.

| $B_1$ | $B_2$ | $B_3$ | $\theta$ | $\chi = U_2/U_1$ | $U_3/U_1$ |
|---|---|---|---|---|---|
| L | L | L | $\pi/10$ | 0.4999 | 0.5001 |
| L | L | L | $\pi/4$ | 0.4896 | 0.5104 |
| L | L | L | $\pi/3$ | 0.469 | 0.5310 |
| L | L | L | $\pi/2$ | 0.382 | 0.6180 |
| L | L/2 | L/2 | $\pi/100$ | 0.991 | 1.009 |
| L | L/2 | L/2 | $\pi/10$ | 0.905 | 1.095 |
| L | L/2 | L/2 | $\pi/3$ | 0.625 | 1.375 |

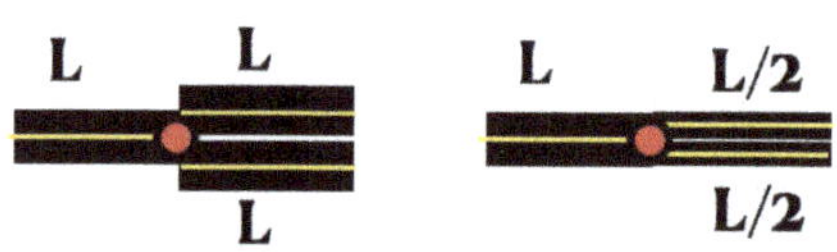


**Figure 6.** Two test cases of Table 1, in the small angle limit. Left: when $B_1 = B_2 = B_3 = \text{L}$, the total width of the channel doubles. Right: when $B_1 = B_2 = B_3 = \text{L}/2$, the total width of the channel remains the same.

In 2008, Bona and Cascaval [14] presented an analysis of weakly nonlinear, weakly dispersive water waves on 1D graphs. The main application mentioned was blood flow in a branching blood system. The model chosen was the Benjamin–Bona–Mahony (BBM) equation. This is a unidirectional long-wave model which allows for solitary wave propagation. It is a regularized version of the Korteweg–de Vries (KdV) equation. The BBM's dispersion relation is a Padé approximation of the potential theory's full dispersion relation [9], which is better equipped to handle the short waves of the solution's Fourier spectrum.

Let $\eta(x,t)$ represent the free surface displacement at location $x$ and time $t$. The BBM equation is defined as [14]

$$\eta_t + \eta_x + \eta\eta_x - \eta_{xxt} = 0, \ x \in \text{reach}(i,j,k), \ t \geq 0, \tag{33}$$

with the Neumann–Kirchhoff conditions at the vertex $P$, of the flipped Y-domain (type-II in Figure 3):

$$\eta(x_i, t) = \eta(x_j, t) = \eta(x_k, t), \tag{34}$$

$$\eta_x(x_i, t) = \eta_x(x_j, t) + \eta_x(x_k, t). \tag{35}$$

As mentioned by the authors, in comparison with the KdV equation, the BBM equation seems more amenable to the functional analysis performed, since the number of boundary conditions needed for well-posedness over finite intervals is two, whereas the KdV requires three boundary conditions. The solitary wave profiles for both equations are similar, even though the KdV is an integrable system and the BBM is not.

The above BBM model has no information regarding the angle at a vertex. The authors mention that angles should play a role in the solution, and this is a topic of future interest. A preliminary 1D numerical study is presented at the end of their paper, with the solution displaying reflected waves at the vertex. Wave reflection should be better modeled by a bidirectional PDE system, such as a Boussinesq system. No comparisons are mentioned with 2D solutions in a fattened-graph domain. The functional-analytic study presented [14] is nontrivial and important, and it is necessary to consider the function-space analysis on the half-line for each reach. The authors draw conclusions about the well-posedness of the solutions in its current configuration. As discussed above, the model has limitations but establishes an important step in understanding mathematical aspects of this nontrivial problem of nonlinear evolution equations on a nonstandard domain.

In 2014, Bressan et al. [5] reviewed flows on networks with examples such as traffic flow and blood flow. Their mathematical framework is achieved through conservation laws, in the form of a balance equation such as

$$\partial_t u + \partial_x f(u) = g(t, x, u), \tag{36}$$

where $t > 0$, $x$ is the one-dimensional spatial variable, $f$ is the flow, and $g$ is the forcing term. Results on the existence for solutions are presented for scalar conservation laws as well as systems of conservation laws, where $u$, $f$ and $g$ are vector quantities. In most cases, the conservation law is equipped with algebraic conditions at the vertices of a directed graph. The authors mention that in most applications considered, there is a preferred direction for the network flow. Section 3.1 in [5], which relates to traffic flow on road networks, reviews different unidirectional traffic flow models and different nodal conditions at the single vertex of the model. The authors comment that the existence and well-posedness properties for the Cauchy problem of $m_1$ incoming roads on a vertex, together with $m_2$ outgoing roads, depend on the choice of the nodal (compatibility) conditions at $x = 0$. The Cauchy problem considers $m_1$ Equation (36) defined on the half-line $(-\infty, 0)$, together with $m_2$ Equation (36) defined on $(0, +\infty)$.

The review in [5] does not discuss information on angles between edges or comparisons with their parent 2D thick-graph model. Bressan et al. present an interesting and extensive analytical review of 1D hyperbolic graph-models. For example, in Section 8 [5] on *Future perspectives* (page 101), the authors call attention to the importance of imposing proper coupling conditions between different branches of the graph. The consistency of the coupling conditions between edges is very important, where consistency infers comparing with solutions of the full multidimensional problem. Proving that a reduced model is consistent with the higher dimensional one is, in general, an open problem from the rigorous analysis point of view [5]. The rigorous analysis presented in [5] regards 1D solution properties, which are important and nontrivial. Our approach has been to investigate the consistency between the two-level models (1D $\times$ 2D) using PDE modelling and theory together with scientific computing. This has led to a novel nonlinear coupling condition at the vertex of the graph. This was performed in Nachbin and Simões [7], where comparisons between the reduced 1D model and the parent model led to a novel and more accurate compatibility condition between edges. This will be discussed in the Results Section 3.

### 2.2. Non-Fluid Mechanics-Related Problems

Below, we comment on some recent review articles for waves on graphs. We highlight points that call attention to current mathematical problems of interest.

Kuchment [2,3] has written a few review articles (in 2002 and 2008) on quantum graph models for waves in thin structures. We recall the general definition for a quantum graph as a metric graph equipped with a differential equation along its edges, as the present case of water waves. One of the author's main interests relates to the spectral behavior of some self-adjoint operators on a graph-like domain. It is challenging to understand the convergence of the spectral features in the limiting case when the thin graph-like structure collapses onto a 1D graph, which Kuchment calls a one-dimensional singular variety, also known as the singular nature residing on the presence of a vertex. Mathematically speaking, Kuchment analyses differential operators, such as a Laplace operator, or a Schrödinger operator, with Dirichlet or Neumann boundary conditions, defined on a narrow branching domain. This leads to highly nontrivial problems depending on the graph topology. These questions have attracted the attention of many researchers. Kuchment remarks that vertex conditions can be imposed in different forms, mostly through the Neumann–Kirchhoff condition and some of its variations. The choice of a vertex (compatibility) condition involves prescribing how waves scatter at each vertex. Wave scattering at a vertex sets a difference between flow problems through a vertex, where reflection or a backward flow are less likely to be established by a compatibility condition. This is an important point that we were able to address in the water wave problem by providing vertex conditions together with information on the angle between the respective edges [7].

A recent (2022) review article by Kairzhan, Noja and Pelinovsky [1] considers standing waves of the nonlinear Schrödinger equation on star-shaped quantum graphs. The authors review the existence and stability results for standing waves trapped on quantum graphs. They consider some variations in the Neumann–Kirchhoff condition and mention that angles do not play a role in their applications. Restricting a differential operator to the intervals defined by each edge is trivial. Kairzhan et al. [1] call attention to the ambiguity that can arise through the choice of the compatibility conditions and also that the modelling choices made should be based on the underlying physical problem. This is along the lines of our modelling investigation for the water waves problem. On the other hand, the authors admit that it is unclear which conditions at nodes are appropriate for a good agreement with solutions of the concrete underlying problem. There are different choices that can render a mathematically well-posed problem. But a realistic reduced modelling from the higher-dimensional problem to the one-dimensional metric graph is a fundamental step [1].

The work by Kevrekidis et al. [15] considers matter waves on Y-junctions. The authors also comment on several points which are currently of interest, with an effort to model junction angles. A potential-like term is introduced in a discrete (lattice) nonlinear Schrödinger equation in order to numerically study the effects of graph-asymmetry and of branching angles. This term at the junction is responsible for guiding localized solitary-wave excitations that are studied in the context of optical lattices in Bose–Einstein condensates. They find that the fraction of transmitted condensate may strongly depend on the angle. The authors also comment on the subtleties of one-dimensionality versus small but finite edge-width. And they note that dimensionality comparisons are important and a topic of future investigation.

Many comments from the related works mentioned above corroborate with the direction we have been pursuing: a better understanding of angle modelling for waves on graphs and the effects of dimensionality through a thin fattened-graph and its companion 1D graph. Is the vanishing-edge-width regime a singular limit? In which cases do solutions of the 2D system converge to solutions of the 1D graph? Angles and different compatibility conditions should play a role and need to be investigated more carefully. These are some of the fundamental questions we want to address in ongoing and future work. Water waves represent a proper mathematical setup for this study, where angle modelling, nonlinearity and other features can be added to the PDE study in a natural, physically motivated fashion.

## 3. Results

In this section, we review our results on weakly nonlinear, weakly dispersive water waves on graphs [7,16]. In particular, we describe how we introduce information about the angles of a forked region in a systematic fashion. The mathematical modelling which improved the compatibility conditions on the vertex of a 1D graph grew out of comparisons with the parent 2D fattened-graph simulations.

In a more general framework, Nachbin and Simões [7,16] used Wu's Boussinesq system (5) and (6) together with a Schwarz–Christoffel mapping [17], in order to conformally map the physical fluid domain onto a uniform strip. As depicted schematically in Figure 7, the (MATLAB; version R2023b) Schwarz–Christoffel Toolbox [17] maps the canonical domain (on the right) onto the physical domain (on the left). The canonical domain is defined in the complex $w$-plane ($w = \xi + i\zeta$), while the physical domain is defined in the complex $z$-plane ($z = x + iy$). The Schwarz–Christoffel mapping $z = F(w)$ is tailored for mapping polygonal-shaped domains; in this case, the forked region shown in Figure 7. The geometrical singularities at the corners of the physical domain are changed in the canonical domain. The outer corner is straightened out and the Jacobian of the coordinate transformation is zero at that point [18]. Clearly, the mapping is not conformal at that point, since it does not preserve the angle. At the inner corner, the Jacobian blows up [18].

**Figure 7.** This is a schematic picture of the Schwarz–Christoffel (SCh) conformal mapping between the physical domain (left) and the canonical domain (right). The physical forked domain is defined in the complex $z$-plane while the canonical (computational) domain is defined in the complex $w$-plane. The SCh mapping is defined as $z = F(w)$. The inverse function $F^{-1}(z)$ maps the forked region onto a uniform strip with a slit. The slit (defined in blue) arises from the conformal mapping having fractional exponents, and therefore a complex function with a branch cut [18]. The slit is connected to the domain's separatrix, depicted in light green.

In order to avoid the presence of these singularities in our computational domain, we map a ($z$-plane) domain with a slightly larger width than the desired physical domain. This wider region is shown in Figure 8. We then evaluate the Jacobian on $(\xi_i, \zeta_j)$ grid points away from the boundary and the singularity. These grid points are those within the forked region bounded by the thick black lines, as depicted in Figure 8. This strategy was tested numerically for both a sharp turn [16] as well as the forked region [7].

Using the Schwarz–Christoffel Toolbox [17] we evaluate and plot the Jacobian of an asymmetric forked region, as shown in Figure 9. Note that the singularity is felt near the branching region, shown by a spike. But the spike is finite, since the singularity is outside the domain. Away from the forked region, the Jacobian very quickly reaches plateaus of constant value. This is related to the discussion presented together with Figure 5.

The roles played by the coordinates $\xi(x,y)$ and $\zeta(x,y)$, as functions of the inverse mapping $w = F^{-1}(z)$, are analogous to the roles played by $\phi(x,y)$ and $\psi(x,y)$ in the steady streaming flow presented by Milne-Thomson [13]. Note that in Figure 5, a short distance away from the forked region, the streamlines (aligned with the walls), together with potential level-curves, form a Cartesian grid. This means that the Jacobian matrix away from the forked region is a constant multiple of the identity. Namely, it is a rotation and/or rescaling of the standard Cartesian coordinates.

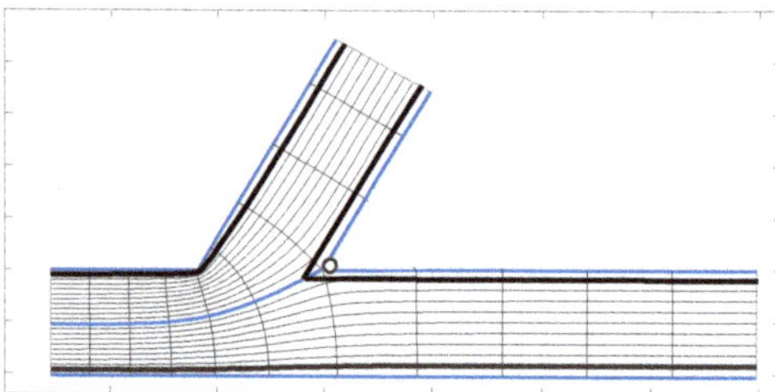

**Figure 8.** In order to deal with the conformal mapping's singularity, we consider a slightly wider forked channel to be mapped by $F^{-1}$ onto a uniform strip in the canonical $w$-plane, as shown in Figure 7. The Jacobian of this mapping blows up at the corner highlighted with a circle. To avoid the singularity, the physical domain is defined by the slightly smaller region within the region bounded by the thicker inner curves. The mapping's separatrix is depicted emanating from the singular point. For the numerical simulations, the values of the Jacobian are finite and taken from this inner region, which has the singularity nearby, but outside the domain of interest. This is depicted in Figure 9.

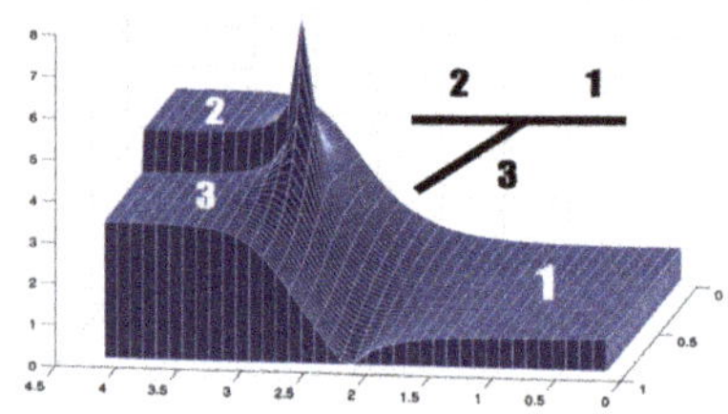


**Figure 9.** The Jacobian produced by the Schwarz–Christoffel Toolbox [17]: $|J|(\xi,\zeta) = |dF/dw|^2 = x_\xi^2 + y_\xi^2$, by the Cauchy–Riemann equations. At the corner between reaches 1 and 3, the Jacobian is zero. At the corner between reaches 2 and 3, the Jacobian blows up [18].

Bearing the above conformal mapping in mind, we apply this curvilinear, orthogonal change of coordinates to the Boussinesq system formulated by Wu, which we repeat here for convenience:

$$\eta_t + (1+\eta)(\phi_{xx} + \phi_{yy}) + \eta_x\phi_x + \eta_y\phi_y = 0,$$
$$\phi_t + \eta + \frac{\phi_x^2 + \phi_y^2}{2} - \frac{\phi_{xxt} + \phi_{yyt}}{3} = 0.$$

Recall that the $\phi(x,y,t)$ is not a harmonic function, or else the term $(\phi_{xx} + \phi_{yy})$ in the first equation would be zero. This Boussinesq system is not invariant under the Schwarz–Christoffel change of variables. But the $\xi\zeta$-coordinate system is boundary-fitted and the respective Jacobian encodes geometrical angle-information in an algebraic form, through variable coefficients in the system of PDEs. Under the conformal change of coordinates, the variable-coefficient Boussinesq system takes the form

$$|J|\eta_t + (1+\eta)(\phi_{\xi\xi} + \phi_{\zeta\zeta}) + \eta_\xi\phi_\xi + \eta_\zeta\phi_\zeta = 0, \tag{37}$$

$$\phi_t + \eta + \frac{\phi_\xi^2 + \phi_\zeta^2}{2|J|} - \frac{\phi_{\xi\xi t} + \phi_{\zeta\zeta t}}{3|J|} = 0. \tag{38}$$

The Jacobian $|J|$ of the Schwarz–Christoffel mapping is given by $|J|(\xi,\zeta)| = |dF/dw|^2$. As a boundary-fitted coordinate system, it is easy to impose the Neumann conditions along the walls of the secondary reaches. Therefore, forked regions with reaches of different angles are easily treated in our numerical computations, with the discretization being in the (Cartesian) canonical variables. An outline of the numerical scheme used is provided in Section 4.

For the regime of interest (long waves in narrow channels), the 2D fattened-graph simulations displayed solutions which are essentially one-dimensional [7] when they are away from the forked region. The narrow channel is a wave-guide and the water waves propagate as a plane-waves along each reach. The Neumann conditions along the side walls act in this regard, aligning the wavefront with the propagation direction of the reach. This means that away from the forked region, we can disregard lateral variations of the wave. In other words, the $\zeta$-dependence of solutions of the Boussinesq system (37) and (38) can be dropped. With this in mind, a 1D water-wave graph-model was deduced by Nachbin and Simões [7] along the axial direction of the reaches:

$$\overline{|J|}\bar{\eta}_t + (1+\bar{\eta})\bar{u}_\xi + \bar{\eta}_\xi\bar{u} = 0, \tag{39}$$

$$\bar{u}_t + \bar{\eta}_\xi + \left(\frac{\bar{u}^2}{2\overline{|J|}}\right)_\xi - \left(\frac{\bar{u}_{\xi t}}{3\overline{|J|}}\right)_\xi = 0. \tag{40}$$

The bars denote that these functions do not depend on $\zeta$; only on $\xi$. The 1D horizontal velocity along the channel axis is denoted by $\bar{u}(\xi,t) = \bar{\phi}_\xi(\xi,t)$, while the wave elevation by $\bar{\eta}(\xi,t)$. The lateral averaging of the Jacobian is denoted by $\overline{|J|}$. In this 1D equation, the Jacobian is taken as a piecewise-constant coefficient which encodes angle-information.

By observing the 2D fattened-graph solution, we see that continuity of the wave elevation does not hold, as clearly depicted in Figure 10. There is a sudden change in wave height at the forking point, and therefore a Rankine–Hugoniot jump condition was imposed in [7]. This jump condition gave rise to our compatibility condition, generalizing the established Neumann–Kirchhoff condition. Writing Equation (40) in divergence form leads to a Rankine–Hugoniot jump condition [7] which is deduced from a stationary shock at the vertex. Rankine–Hugoniot conditions along each reach connection (1-2; 1-3) yield the nonlinear compatibility conditions (41) and (42), which generalize the Neumann–Kirchhoff condition:

$$\bar{\eta}(P_1,t) - \bar{\eta}(P_2,t) = \left[\frac{\bar{u}(P_2,t)^2}{2\overline{|J|}_{P_2}} - \frac{\bar{u}(P_1,t)^2}{2\overline{|J|}_{P_1}}\right], \tag{41}$$

$$\bar{\eta}(P_1,t) - \bar{\eta}(P_3,t) = \left[\frac{\bar{u}(P_3,t)^2}{2\overline{|J|}_{P_3}} - \frac{\bar{u}(P_1,t)^2}{2\overline{|J|}_{P_1}}\right], \tag{42}$$

$$Q(P_1,t) = Q(P_2,t) + Q(P_3,t). \tag{43}$$

$Q(\xi,t)$ is the fluid flux at a given point of the graph, in canonical coordinates. At the vertex $P$, condition (43) implies conservation of mass, where the $P_{\text{index}}$ indicates the respective endpoint of reach $i = 1,2,3$. When the nonlinear terms are discarded from (41) and (42), the Neumann–Kirchhoff condition [1,3] is recovered, which is also called the Stoker condition mentioned by Jacovkis [6]. The Neumann–Kirchhoff condition imposes the continuity of the solution at the graph's vertex: $\bar{\eta}(P_1, t) = \bar{\eta}(P_2, t) = \bar{\eta}(P_3, t)$. Nachbin and Simões [7] (Figure 7) present a numerical simulation with the Neumann–Kirkhhoff (Stoker) condition for a moderate angle of $\pi/6$. We compare the 2D solution with the 1D graph-solution. The agreement for the reflected and transmitted waves is not very good. This is particularly true for the reflected wave. Then, in Figure 8 [7], using the new compatibility conditions with the same data as before, the 1D-2D wave reflection-and-transmission agreement is excellent. The new compatibility conditions are further numerically validated by comparing 1D with 2D solutions for different sets of angles and widths [7], in symmetric and asymmetric junctions. A mean-square error analysis for the wave elevation is presented in Figure 11 [7]. It considers different channel widths (all three reaches with $B = 1$ or all with $B = 3$), as well as different forking angles ($\pi/6$ and $\pi/3$). This comprehensive 1D-2D comparison of solutions in both directions, for several different junction configurations, is not found in the literature of waves on graphs, including water waves.

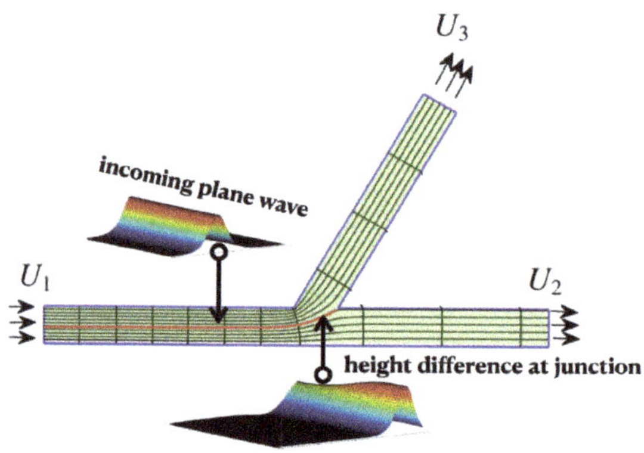


**Figure 10.** Snapshots of the abrupt wave-height change while passing through the forked region. The background displays the streaming flow solution [13], which allows less flow ($U_2 < U_3$) through the horizontal reach. A plane waves approaches the forked region from the left, as shown in the left snapshot. At the forked region, the other snapshot displays a larger wave propagating through the main reach, while a smaller wave is about to propagate through the secondary reach. The two wave profiles, depicted in the snapshots, are from 2D Boussinesq computations by Simões [19]. These wave snapshots clearly illustrate that the Neumann–Kirchhoff condition, which requires continuity of the wave elevation at the forking point, is not a proper condition in this case.

## 4. Numerical Method for the 2D Boussinesq System

We briefly review the numerical method developed by Nachbin and Simões [16] and used for forked regions in [7]. It is convenient to write system (37) and (38) as follows:

$$\eta_t = -\frac{1}{|J|}\left[(1+\eta)(\phi_{\xi\xi} + \phi_{\zeta\zeta}) + (\eta_\xi \phi_\xi + \eta_\zeta \phi_\zeta)\right], \tag{44}$$

$$\left(\phi - \frac{\phi_{\xi\xi} + \phi_{\zeta\zeta}}{3|J|}\right)_t = -\eta - \frac{(\phi_\xi)^2 + (\phi_\zeta)^2}{2|J|}. \tag{45}$$

We use centered finite differences for all spatial derivatives, where the grid functions are denoted accordingly, as for example, $\phi^n(i,j) \equiv \phi(\xi_i, \zeta_j, t^n)$, at mesh points $(\xi_i, \zeta_j)$. Then,

$$d_\xi \phi^n := \frac{\phi^n(i+1,j) - \phi^n(i-1,j)}{2\Delta\xi} = \phi^n_\xi(i,j) + O(\Delta\xi^2), \tag{46}$$

$$d_\zeta \phi^n := \frac{\phi^n(i,j+1) - \phi^n(i,j-1)}{2\Delta\eta} = \phi^n_\eta(i,j) + O(\Delta\zeta^2), \tag{47}$$

$$d^2_\xi \phi^n := \frac{\phi^n(i+1,j) - 2\phi^n(i,j) + \phi^n(i-1,j)}{\Delta\xi^2} = \phi^n_{\xi\xi} + O(\Delta\xi^2), \tag{48}$$

$$d^2_\zeta \phi^n := \frac{\phi^n(i,j+1) - 2\phi^n(i,j) + \phi^n(i,j-1)}{\Delta\zeta^2} = \phi^n_{\zeta\zeta} + O(\Delta\zeta^2). \tag{49}$$

Let the right-hand side of (44) be denoted by $E^n(i,j)$ and the right hand side of (45) be denoted by $G^n(i,j)$:

$$E^n(i,j) \quad = \quad -\tfrac{1}{|J|_{ij}}((1+\eta^n_{ij}) * (d^2_\xi \phi^n + d^2_\zeta \phi^n) + d_\xi \eta^n d_\xi \phi^n + d_\zeta \eta^n d_\zeta \phi^n),$$

$$G^n(i,j) \quad = \quad -\eta^n_{ij} - [(d_\xi \phi^n)^2 + (d_\zeta \phi^n)^2]/(2|J|_{ij}).$$

Define the auxiliary function $F \equiv \phi - (\phi_{\xi\xi} + \phi_{\zeta\zeta})/(3|J|)$, which we will use to evolve our ODE system. This function arises from the the time-derivative term in Equation (45). After the ODE solver's time step, the potential is recovered by solving an elliptic equation at each time step. We start with a predictor scheme for the wave elevation and the auxiliary function:

$$\eta^{n+1}_0(i,j) = \eta^n(i,j) + \Delta t\, E^n(i,j), \tag{50}$$

$$F^{n+1}_0(i,j) = F^n(i,j) + \Delta t\, G^n(i,j). \tag{51}$$

These values provide an initial guess for the (implicit) corrector scheme. The corrector scheme, at the iteration step $\mu \geq 1$, is given by

$$\eta^{n+1}_\mu(i,j) = \eta^n(i,j) + \frac{\Delta t}{2}[E^{n+1}_{\mu-1}(i,j) + E^n(i,j)], \tag{52}$$

$$F^{n+1}_\mu(i,j) = F^n(i,j) + \frac{\Delta t}{2}[G^{n+1}_{\mu-1}(i,j) + G^n(i,j)]. \tag{53}$$

The iterations are performed until

$$\frac{\max(\eta^{n+1}_\mu - \eta^{n+1}_{\mu-1})}{\max \eta^0} < 10^{-5}.$$

Once $F^{n+1}$ is obtained, we recover the potential $\phi^{n+1}$ from the discrete elliptic problem

$$\phi^{n+1} - (d^2_\xi \phi^{n+1} + d^2_\zeta \phi^{n+1}) = 3|J|\; F^{+1}n, \tag{54}$$

with trivial Neumann boundary conditions $d\phi^{n+1}/d\zeta = 0$ along the walls. Further details can be found in the work of Nachbin and Simões [16].

The 1D Boussinesq graph-model uses a similar predictor–corrector scheme, also defining an auxiliary function to deal with the dispersion term. Details of the implementation of the compatibility condition are given in the appendix of Nachbin and Simões [7].

## 5. Discussion and Future Work

Flows on graphs are an interesting research problem discussed in a diverse range of mathematical literature, including some review articles that have been published recently. Here, we commented on two of these articles [1,3]. In contrast with flow on graphs, we called attention to the problem of waves on graphs, where one needs to deal with the reflection and transmission of waves through the vertex of the graph. Existing wave models lack information about the junction's angles as well as comparisons with the (parent) 2D fattened-graph model. This has an impact on the reflection–transmission dynamics at the vertex of the reduced 1D graph model, and how it compares to the 2D mesoscale model.

As described in Section 2, there are many mathematical problems of interest in the literature which are currently being investigated. These range from PDE modelling, numerical methods and simulations to theoretical studies. In this manuscript, we addressed some issues that are not fully resolved. In order to address some of these interesting questions, we plan to step back from the weakly nonlinear, weakly dispersive Boussinesq system and study (nondispersive) hyperbolic systems. As mentioned in Section 2, this means working with the 2D shallow water equations. To the best of our knowledge, there has not been a comprehensive comparison between thick-Y-graph solutions with that of the reduced (variable angle) 1D tree-like graph, exploring a variety of junction configurations. Also, angles have never been considered systematically in 1D hyperbolic fluid-flow models, and certainly not for shallow water waves on graphs. We believe that linearized systems should take into account junction configurations within the equations. This is currently being addressed in a research project.

Another modelling ingredient of interest is channels with different cross-sections. The Boussinesq model considered in this paper is for a rectangular cross-section, which does not vary along the canal. These are two modelling variations to be considered: (i) non-rectangular cross sections; (ii) spatially varying cross sections.

By no means do we intend to exhaust the list of possibilities, but another topic of interest is the consideration of different types of solutions. Up to now, we have only studied solitary waves. With solitary waves, we observed [7] that the accuracy of the compatibility conditions (41) and (42) deteriorated for an asymmetric junction, as the forking angle was approximately $\pi/3$. We want to take a closer look to understand why this loss of accuracy takes place between the 1D and the 2D models.

**Funding:** The author is on leave from the Instituto de Matemática Pura e Aplicada (IMPA), Brazil. This research was partially funded by the Conselho Nacional de Desenvolvimento Científico e Tecnológico (PQ1D 307078/2021-3) and by Fundação Carlos Chagas Filho de Amparo à Pesquisa do Estado do Rio de Janeiro (CNE E-26/201.156/2021), Brazil.

**Conflicts of Interest:** The author declares no conflicts of interest.

## References

1. Kairzhan, A.; Noja, D.; Pelinovsky, D.E. Standing waves on quantum graphs. *J. Phys. A Math. Theor.* **2022**, *55*, 2430001. [CrossRef]
2. Kuchment, P. Graph models for waves in thin structures. *Waves Random Media* **2002**, *12*, R1–R24. [CrossRef]
3. Kuchment, P. Quantum graphs: An introduction and a brief survey. *Proc. Symp. Pure Math.* **2008**, *77*, 291–314.
4. Goodman, R.H.; Conte, G.; Marzuola, J.L. QGLAB: A MATLAB package for computations on quantum graphs. *SIAM J. Sci. Comput.* **2025**, *47*, B428–B453. [CrossRef]
5. Bressan, A.; Canić, S.; Garavello, M.; Herty, M.; Piccoli, B. Flows on networks: Recent results and perspectives. *EMS Surv. Math. Sci.* **2014**, *1*, 47–111. [CrossRef]
6. Jacovkis, P.M. One-dimensional hydrodynamic flow in complex networks and some generalizations. *SIAM J. Appl. Math.* **1991**, *51*, 948–966. [CrossRef]
7. Nachbin, A.; Simões, V.S. Solitary waves in forked channel regions. *J. Fluid Mech.* **2015**, *51*, 544–568. [CrossRef]
8. Wu, T.Y. Long waves in ocean and coastal waters. *J. Engng Mech. ASCE* **1981**, *107*, 501–522. [CrossRef]
9. Whitham, G.B. *Linear and Nonlinear Waves*; John Wiley & Sons: Hoboken, NJ, USA, 1974.
10. Shi, A.; Teng, M.H.; Sou, I.M. Propagation of solitary waves in branching channels. *J. Engng Mech. ASCE* **2005**, *131*, 859–871. [CrossRef]
11. Shi, A.; Teng, M.H.; Wu, T.Y. Propagation of solitary waves through significantly curved shallow water channels. *J. Fluid Mech.* **1998**, *362*, 157–176. [CrossRef]
12. Caputo, J.-G.; Dutykh, D.; Gleyse, B. Coupling Conditions for Water Waves at Forks. *Symmetry* **2019**, *11*, 434. [CrossRef]

13. Milne-Thomson, L.M. *Theoretical Hydrodynamics*, 5th ed.; Dover Publications: New York, NY, USA, 1968.
14. Bona, J.L.; Cascaval, R.C. Nonlinear dispersive waves in trees. *Canad. Appl. Math. Quart.* **2008**, *16*, 1–18 .
15. Kevrekidis, P.G.; Frantzeskakis, D.J.; Theocharis, G.; Kevrekidis, I.G. Guidance of matter waves through Y-junctions. *Phys. Lett. A* **2003**, *317*, 513–522. [CrossRef]
16. Nachbin, A.; Simões, V.S. Solitary waves in open channels with abrupt turns and branching points. *J. Nonlinear Math. Phys.* **2012**, *19*, 1240011. [CrossRef]
17. Driscoll, T.A.; Trefethen, L. *Schwarz–Christoffel Mapping*; Cambridge University Press: Cambridge, UK, 2002.
18. Howell, L.H. *Schwarz-Christoffel Methods for Multiply-Elongated Regions*; Lawrence Livermore National Laboratory (LLNL): Livermore, CA, USA, 1994.
19. Simões, V.S. Evolution of Solitary Waves in Channels with Abrupt Turns and Branching Points. Ph.D. Thesis, Instituto de Matemática Pura e Aplicada (IMPA), Rio de Janeiro, Brazil, 2013.

MDPI AG
Grosspeteranlage 5
4052 Basel
Switzerland
Tel.: +41 61 683 77 34

*Fluids* Editorial Office
E-mail: fluids@mdpi.com
www.mdpi.com/journal/fluids

www.ingramcontent.com/pod-product-compliance
Lightning Source LLC
Chambersburg PA
CBHW061835210926
51850CB00006B/894